Molecular Pathology

Molecular Pathology

A Primer for Laboratory Scientists

Andrew Blann

School of Applied Sciences
Huddersfield University
Huddersfield, UK

WILEY

This edition first published 2025

Registered Offices
John Wiley & Sons, Inc., 111 River Street, Hoboken, NJ 07030, USA
John Wiley & Sons Ltd, New Era House, 8 Oldlands Way, Bognor Regis, West Sussex, PO22 9NQ, UK
John Wiley & Sons Singapore Pte. Ltd, 134 Jurong Gateway Road, #04-307H, Singapore 600134

For details of our global editorial offices, customer services, and more information about Wiley products visit us at www.wiley.com.

The manufacturer's authorized representative according to the EU General Product Safety Regulation is Wiley-VCH GmbH, Boschstr. 12, 69469

Weinheim, Germany, e-mail: Product_Safety@wiley.com.

Wiley also publishes its books in a variety of electronic formats and by print-on-demand. Some content that appears in standard print versions of this book may not be available in other formats.

Library of Congress Cataloging-in-Publication Data

Names: Blann, Andrew D. author
Title: Molecular pathology : a primer for the laboratory scientist / Dr. Andrew Blann, University Department of Medicine, City Hospital NHS Trust, Birmingham, UK.
Description: Hoboken, NJ, USA : Wiley, 2025. | Includes bibliographical references and index.
Identifiers: LCCN 2024055147 | ISBN 9781394254637 paperback | ISBN 9781394254651 adobe pdf | ISBN 9781394254644 epub
Subjects: LCSH: Pathology, Molecular–Textbooks | Diseases–Molecular aspects–Textbooks
Classification: LCC RB113 .B553 2025 | DDC 616.07–dc23/eng/20250226
LC record available at https://lccn.loc.gov/2024055147

Cover Design: Wiley
Cover Images: © Anna/stock.adobe.com, © Love Employee/Shutterstock

Set in 9.5pt STIX Two by Lumina Datamatics, Inc.

Printed and bound by CPI Group (UK) Ltd, Croydon, CR0 4YY

C9781394254637_110425

Contents

Preface

The understandable desire to fully come to grips with the functions of our bodies has led to the development of molecular genetics. This discipline of medical science explains the form and function of genes at the level, not of the individual (although this naturally follows), but at the level of the molecule. The principal molecule in this investigation is the macromolecular deoxyribonucleic acid (DNA), followed very closely by its near relative, ribonucleic acid (RNA, of which there are several forms).

However, scientific knowledge is nothing without application. Molecular pathology may be seen as a point at which decades of biomedical research into molecular genetics has a direct impact on the practice of medicine. This is pertinent as perhaps 1–2% of the population carries a genetic variant that brings a high lifetime risk of cardiovascular disease or cancer, while many more have a variant that predisposes them to many other diseases and conditions. As a developing discipline, molecular pathology has a place in each of the other major pathology disciplines (Figure 1).

The purpose of this book is to provide a sound and broad outline of the subject for the laboratory scientist, which will be laid out in a series of chapters.

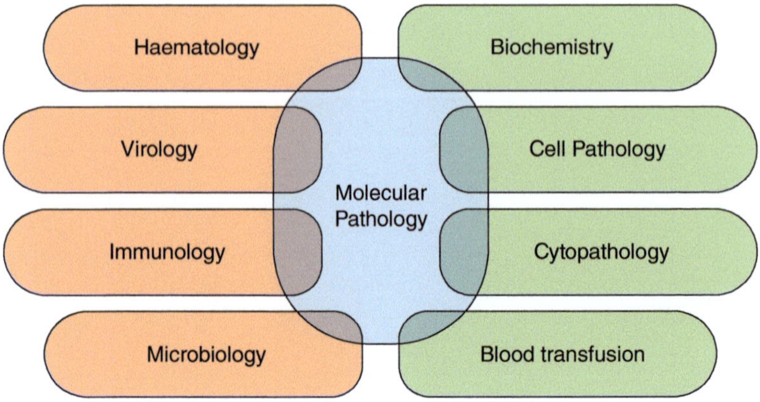

Figure 1 Impact of molecular pathology on other pathology disciplines.

What This Book Is...

This book will provide awareness among health-care professionals, both in training and in practice, with a solid understanding of the role of molecular pathology in modern medicine.

What This Book Is Not...

Molecular genetics and pathology are complex branches of biomedical science and require a basic knowledge of cell biology, physiology and pathology. A basic knowledge of these subdivisions is expected of the reader, and numerous texts on the basics of these three aspects of biology are commonly available, the contents of which are essential prerequisites.

How This Book Is Organised...

Many readers find molecular genetics and molecular pathology particularly taxing. Accordingly, the opening chapter of the book provides a general introduction to molecular pathology, with a historical perspective of the key findings leading to our current view of molecular pathology in 2024. This will be followed by a look at the scope of chromosomal conditions (such as Down's syndrome) and genetic diseases (such as sickle cell disease), which will provide information about the material in chapters to come, and an introduction to the modern practice of the discipline.

Having refreshed ourselves on the basics of nucleic acid biology, we move to the laboratory. The complex nature of many analyses demands a full explanation of each, and as there are many such platforms, it is convenient to describe them in the broad order in which they were developed. Thus, Chapter 2 will describe the numerous analytical tools in the early stages of molecular pathology. Again, a historical perspective will serve to explain how modern technology has evolved from techniques invented in previous decades as a tool to solve research questions. Many of these fundamental methods are still in use in academic settings, but few are used in today's routine molecular pathology labs. Chapter 3 will further explore two groups of methods – the so-called New Generation Sequencing and Third Generation Techniques. The chapter will conclude with a section on how complex genetic data is interpreted – i.e. the field of bioinformatics.

The remaining chapters will look at the molecular pathology of different groups of diseases, taking a clinical approach. Since cancer is the consequence of chromosomal and genetic abnormalities and is the leading cause of death in the United Kingdom, it calls for three chapters. Chapter 4 serves as an introduction to cancer, and then focuses on the leading solid tumour cancer of adults, these being carcinomas. The other forms of cancer are discussed in Chapter 5, which also discusses paediatric cancer. Blood cancer is a simple grouping and is the subject of Chapter 6. The book concludes with the remaining conditions, which may variously be described as rare, inherited, metabolic, and those that fail to fall into any of these groups. However, there is considerable overlap between the groups – many metabolic diseases are both rare and inherited.

Each of these latter four chapters will explain the role of molecular pathology in the aetiology, diagnosis, and management of each particular condition.

Current Practice of Molecular Pathology

In the UK, the public health aspect of molecular pathology lies with each of the four nations: England, Northern Ireland, Scotland and Wales. Each has its own directory of genomic testing, listing diseases in various groupings, the nature of the test, target genes, technology and further eligibility criteria. These provide a part template for the layout of the chapters of this text. A second aspect of the choice of material for this book is to acknowledge the role of the UK's National Institute for Health and Care Excellence (NICE). This body offers guidance on the choice of diagnostics and treatment for various conditions, and some of the former involve gene variants, and so a role in molecular pathology.

Conflict of Interest Statement

The author has not taken any fees, stipends, hospitality, grants or any other commercial link with any of the suppliers of analysers, reagents or software described in this volume.

Acknowledgements

I am very glad to be able to acknowledge the support and contributions of Rob Dunn, Robert Baker, Jennifer Henson, Susanne Kricke, Katya Mokretar, Amy Newton, Ben Poskitt, Helen Warwicker, Isla Henderson and Richard Matthews.

1

Introduction to Molecular Pathology

A large component of human disease has a clear genetic component and can be described in a number of ways. Some of us (some say most of us) harbour abnormal genes within us that we have inherited from our parents, which may cause disease, perhaps evident at birth, perhaps developing years in the future. These are described as 'germline' or perhaps 'constitutive'. Alternatively, disease may arise from normal genes that have become abnormal due to the action of an external factor, such as ionising radiation. This may occur only within a particular organ (such as the breast or prostate), in which case the abnormalities should not be present elsewhere in the body or, indeed, in phenotypically normal tissues within the same organ. Accordingly, in certain cases, both normal and abnormal tissues may be sent for analysis and so compared. Molecular pathology is crucial in determining the cellular basis of all of these abnormalities, and in many cases, suggestions as to treatment. A third aspect of genetic analysis is the precise recognition of an infectious agent such as a bacterium or a virus.

This chapter will provide a sound and basic introduction for the chapters to follow that will address the following three factors. Section 1.1 will present a brief historical perspective of key developments in biomedical science that led to the development of molecular genetics, and then the application of this knowledge to the study of disease, i.e. molecular pathology. We then move to Section 1.2, an examination of the different types of human disease and how they come about (i.e. their aetiology). This is followed, in Section 1.3, by a view of the broad nature of the practice of molecular pathology. The chapter concludes, in Section 1.4, with an introduction to the modern practice of molecular pathology.

Molecular Pathology: A Primer for Laboratory Scientists, First Edition. Andrew Blann.

Learning Outcomes

After studying this chapter, you should confidently be able to...

- Describe major landmarks in the development of molecular pathology
- Recall the various forms of human disease and the leading causes of death
- Comment upon major features of chromosomal structure
- List common pathological alterations in sections of chromosomes
- Explain the importance of single nucleotide polymorphisms
- Describe the development of the practice of molecular pathology

1.1 A Historical Perspective

The modern practice of molecular pathology has only become possible through the slow and steady increase in our knowledge of the many facets of cell biology and deoxyribonucleic acid (DNA). We can view this as part of the development of our scientific understanding of the composition and working of the cell, the nucleus, heredity, chromosomes, genes and DNA (i.e. physiology). Only then can these be used as tools to address disease (i.e. pathology).

1.1.1 The Development of Molecular Genetics

The Nineteenth Century

Although it was well known from the Middle Ages (and probably earlier) that certain physiological features and diseases were present in families, and that these moved down the generations, perhaps one place to start is in the Victorian era with Darwin and Mendel. The former observed amendable hereditary features and postulated links with survival, publishing his seminal work 'The Origin of Species' in 1859. Mendel provided the mechanisms for the transfer of these features, which we now know as genes, and published his findings in 1865 and 1866, although it was not widely disseminated until the early 1900s. Neither was aware of each other's work.

Miescher is credited with the first description, in 1871, of a substance he named nuclein, obtained from the nuclei of white blood cells, whilst in 1882, Flemming further analysed this substance, describing it as chromatin, a material we now know to be the basis of chromosomes (Greek: coloured body). He subsequently observed the replication of chromosomes in dividing cells and concluded that these molecules were the source of information passed from generation to generation. By the end of the century, Kossel had isolated adenine, cytosine, guanine, thymine and uracil from nuclein, for which he was awarded a Nobel Prize.

The Twentieth Century – Part 1

The last century opened with Bateson's coining of the word genetics in 1906, followed by Johannsen's introduction of the term gene in 1909, emphasising the distinction between phenotypic characteristics of individuals and the material units of heredity that determine them. Levene and others extended the work of Kossel, showing that the nucleic acid component was dominated by only four of the complex molecules described by Kossel and that they were connected to each other by a sugar, the basics of nucleotide chemistry. This was followed by the recognition by Boveri, Sutton and others that chromosomes were responsible for heredity, whilst Griffith, in 1928, showed that physical characteristics of pneumococci can be transferred horizontally (i.e. not merely from parent to offspring) between different strains of the bacterium.

The 1940s were a key decade. McClintock did crucial and pioneering work on cytogenetics, such as the demonstration of transposable elements (which we now call transposons) that move from one chromosomal position to another, which was recognised by the award of a Nobel Prize in 1983. Ris and Mirsky described chromosomes as a complex system of non-histone protein, (DNA) and histone with a definite structure, whilst Avery and colleagues considered DNA a 'transforming factor' in the phenotypes of strains of pneumococcus. Beadle and Tatum published data supporting the hypothesis that genes can exert an effect on development and function through enzymes, whilst the Avery–MacLeod–McCarty experiment showed the hereditable material in chromosomes to be DNA, although some considered that heredity lay in the protein component. Smith coined the term 'genomics' to describe the total chromosomal and genetic component of an individual (the genome) in 1943, whilst decades later, in 1995, Wasinger and colleagues used 'the proteome' as a shorthand to describe the complete protein component of a genome present within a cell, a tissue, an organ, or an organism.

In 1950, Chargaff and colleagues showed that the amount of adenine in a sample of DNA was equal to the amount of thymine, and that the amounts of guanine and cytosine were also equivalent, a key component of the structure of DNA and the genes therein. In 1952, Hershey and Chase reported their now classic and crucial experiment where a virus infecting a bacterium unequivocally showed that the hereditable material was not the protein component of the chromosome but was indeed DNA. The Lederbergs did fundamental work in virology and bacterial genetics, showing the transfer of DNA between bacteria.

Although the term 'molecular biology' was introduced by Weaver in 1938, it was later developed by Astbury in 1950, whilst Butel and Melnick used the term in 1970 to describe the ability of the virus SV40 to induce in cells a transformation to a malignant phenotype. Although at the time the expression had wide usage in the exploration of any molecule, by the 1990s its use was becoming focused specifically on the gene. A key feature of molecular biology is that of the development and refinement of analytical techniques that drove the embryo discipline of biomedical science (as it was developing into) from cell-based studies to the more refined study of individual macro- (as in DNA) and micro-molecules (such as penicillin and insulin). An example of such a technique is that of X-ray crystallography, a necessity in the discovery of the structure of DNA by Watson and Crick in 1953, and of the structure of vitamin B_{12} by Hodgkin in 1955. The following year saw Tjio and Levan using cytogenetics to report the human diploid chromosome number to be 46.

The Twentieth Century – Part 2

The work of Watson and Crick led directly to knowledge of the transcription of DNA into RNA, and the translation of RNA into protein at the ribosome, with the description of ribosomal RNA in 1955 and transfer RNA in 1957. The central dogma of molecular biology was described in 1958, whilst proof of the long-hypothesised existence of messenger RNA (mRNA) was published in 1961. Other steps demonstrated the molecular basis of DNA in its four nucleotides (adenine [A], cytosine [C], guanine [G] and thymine [T]), and that a triplet sequence of these nucleotides defined an amino acid – the genetic code. For example, the triplet GGC ultimately codes (through RNA, where uracil is substituted for thymine) for glycine. Thus, a sequence of GGC-TTA-GTT in the DNA becomes GGC-UUA-GUU in the mRNA, and thereby codes for the tripeptide glycine–leucine–valine, and, by extrapolation, a protein.

The discovery in the 1950s by Luria, Bertani and colleagues of microbial enzymes able to cut specific sections of viral DNA led directly to their development in the 1960s and 1970s by Smith, Kelly, Berg and others as tools for analysing sections of DNA. These restriction endonucleases, such as EcoRI, cut DNA only at specific sites in the nucleotide sequence and provided the impetus for the major breakthrough of genetic engineering. Advances in RNA included the description of small nuclear RNA in 1968 and circular RNA in 1976 – the precursors of whole families of non-coding RNAs (ncRNAs). The 1970s also saw the discovery of methyl- groups on the DNA, a feature now known to be important in gene regulation and in cancer, and the publication in 1971 of Knudson's two-hit hypothesis of carcinogenesis, which later laid the groundwork for the description of tumour suppressors. A later development of the two-hit aspect of carcinogenesis is the escalation of a slowly developing cancer into one that is more aggressive, potentially caused by a further external factor.

Subsequent advances included DNA hybridisation, gene cloning, Southern blotting, the development of recombinant DNA techniques for vaccines (such as for the hepatitis B virus) and hormones (such as insulin and growth hormone). This phase also saw widespread use of ligating enzymes to join synthetic sections of DNA onto the ends of samples of DNA to be analysed – the process of ligation. This simple technique was to prove crucial in DNA sequencing with the development of primers.

Further steps saw the development of methods to determine the sequence of the nucleotide make-up of a DNA chain by Sanger and colleagues, and by Maxam and Gilbert (as explained in Chapter 2), used to elucidate the entire DNA sequences of a virus (1977) and a human mitochondrion (1981). This period also saw the development of fluorescence in situ hybridisation (FISH), whose origins can be traced back a decade to the use of radiolabeled probes, and the establishment of hybridising microarray technology. Much of this work was enabled by the development of several important technical methods in the 1980s, such as the polymerase chain reaction (PCR, explained in detail in Chapter 2). Other advances include the determination of repeated sections of DNA (such as dimers of T and G) described as microsatellites or tandem repeats, the number of which varied from site to site.

Technical development continued in the 1990s: Govan et al. described a method for fine-mapping DNA damage and repair in specific genome segments, whilst others developed FISH-based microarray technology, which could monitor the expression of many genes in parallel. The genomes of other organisms were subsequently reported: a bacterium (1995), a yeast (1996) and the nematode worm *Caenorhabditis elegans* (1998). This period also saw the development of DNA 'chips', often described as a 'lab on a chip', which reflected the increased use of complex analysers and robots, the technique of comparative genomic hybridisation (CGH), which was initially used to determine any gains or losses in DNA sequences, and the automated use of capillary electrophoresis in place of slab gel electrophoresis. These methodological advances are described in Chapter 2.

Key Point 1.1 The Human Genome Project effectively kick-started 'industrial' gene sequencing, and promoted the development of next-generation sequencing and bioinformatics.

A key development of this decade was the start of the Human Genome Project, which aimed to sequence an entire genome, an ambition that called for considerable international co-operation. The project demanded numerous technical innovations, one being the large-scale fragmentation of the genome into sections around 150 kbps long. These were inserted into bacterial and yeast artificial chromosomes, amplified and sequenced, according to a 'shotgun' approach, developed in 1981. This generated thousands of overlapping sequences that required the development of complex software needed for interpretation and which prompted the birth of bioinformatics. By 1997, an estimate of 90,000 human genes was postulated, and 1999 saw the landmark first report of the sequence of human chromosome 22.

The Twenty-First Century

The present century saw the publication of the DNA sequence of an insect (2000), the mouse (2002), the rat (2004), the chimpanzee (2005) and the zebrafish (2013). The mode of inheritance of eye colour was clarified in the early 2000s with studies pointing to a role for genes such as *OCA2*, although up to 50 other genes (including *HERC2*, *TYRP1*, *ASIP*, *ALC42A5*) are also implicated. The clear hereditability of height is considerably more complex, with over 12,000 very small variations implicated, although a small number dominate, and, of course, there is a considerable environmental effect of adequate nutrition.

Nomenclature Part 1 By consensus, genes are described in italics, and their protein product in standard notation. Thus, *INS* codes for insulin, and *TYRP1* for tyrosinase-related protein-1. However, in many cases, the link is not as clear as *MUTYH* codes for mutY DNA glycosylase, and some genes are named from their discovery, such as a developmental gene in the Drosophila thorax or wings, which has no parallel in our own species.

In 2001, the Human Genome Project published an initial sequencing of the genome, estimated as being composed of 30,000–40,000 genes, and the location of numerous disease genes, such as *BRCA2*, which brings a susceptibility to breast cancer. Landmarks included the sequences of chromosomes 2, 6, 7, 13, 14 and Y by 2004. Despite these advances, there remained gaps in the sequencing, and in 2022, a leading scientific journal, '*Science*', dedicated an entire issue on the completion of the genome.

The current view is that the human haploid genome consists of a little over 3000 million nucleotide base pairs with a little under 20,000 protein-coding genes. By comparison, the genome of the bacterium *Escherichia coli* has a single chromosome of some 5 million base pairs that include over 4000 protein-coding genes, whilst that of *C elegans* (commonly used in eukaryote gene research) has six chromosomes that together have around 100 million base pairs that, in turn, give rise to over 20,000 protein-coding genes. However, in our own species, the genome generates thousands of RNA molecules that do not participate in translation but seem likely to have a role in the regulation of mRNAs and DNA. These include around 14,500 species of long ncRNAs, 4800 small ncRNAs and 1700 micro-RNAs (miRNAs), to be discussed in sections to come.

Self-check 1.1 What do you consider to be the most important steps in the historical development of molecular genetics?

1.1.2 The Development of Molecular Pathology

As with the development of molecular genetics, certain steps in the development of molecular pathology are precise, whilst others are the sum of several small changes. For example, with surprising prescience, in 1902, Boveri suggested that malignant tumours may be the result of an abnormal condition of the chromosomes, a hypothesis now completely accepted.

The Fruit Fly and Cytogenetics

A key early step in the evolution of 'academic' molecular genetics into clinical molecular pathology was the work of Morgan, Muller and colleagues, who in the first half of the last century studied different phenotypes of the fruit fly *Drosophila melanogaster*. One of the first was the demonstration, in 1916, by Morgan of a link between the white eye phenotype and the X chromosome in female flies. Initially used to explore how genes are inherited, the use of X-rays greatly increased the rate of generation of different variants (mutations) of the fly. When linked to later advances in chromosome identification, this demonstrated the association between chromosomal abnormalities and variant phenotypes that was invaluable to the development of cancer genetics in our own species.

During the 1940s and 1950s, further advances in cell biology and microscopy enabled the identification of chromosomes and variants in their structure. The key event in clinical cytology was the 1959 report by Lejeune and colleagues of an extra chromosome in Down syndrome. By the end of the 1960s, cytogenetic studies had shown chromosomal abnormalities in numerous cancers, which subsequently extended to specific genes in cancer, as is explored in Chapters 3, 4 and 5.

Key Point 1.2 Early research on ionising radiation and chemical mutagens shows their effect on chromosomes in the development of cancer.

Virology

A further cause of cancer is certain viruses. In 1910, Peyton Rous reported that an avian cancer in the form of a sarcoma, a tumour of connective tissues such as muscle and bone, could be transmitted by an unknown soluble factor, smaller than a cell. Decades later, this cancer was shown to be caused by a gene in a virus, hence the Rous Sarcoma Virus (RSV). In the 1960s, Eddy and colleagues, and Huebner and Todaro, postulated viral genes as a cause of cancer (i.e. are oncogenes), with the subsequent discovery of oncogenes linked to specific tumours. The latter included *src* so named (at the time) after the **sarc**oma caused by RSV (more latterly known as *SRC*), and *myc* a transforming gene derived from the avian leukocytosis virus, causing **my**elo**c**ytomatosis.

Nomenclature Part 2 It is appropriate at this stage to discuss in more detail the systems for naming genes that often link to their product (as in part 1). There are numerous examples where, with improved knowledge, a gene is renamed, and so may be referred to by different writers in alternative forms as they are discovered. An example of this is *MYC*, originally (as *myc*) thought to be a single gene but now known to have isoforms and so a family of genes, *c-MYC*, *l-MYC* and *n-MYC*, all present on different chromosomes. Accordingly, these may be described in slightly different terms, such as *MYCN*, *N-MYC* and *n-Myc*. So, although confusing, in the current text, genes will be referred to in the shorthand used by the particular publication. With intellectual flexibility, this should not present too much of a problem in our understanding of molecular pathology.

It was later shown that certain other genes, such as *ras*, first described in a **rat** **s**arcoma virus (and later developed as Kirsten rat sarcoma, then *KRAS*, named after its discoverer, Werner Kirsten) and *TP53* (sometimes described as *tp53*), so named because its protein product was first thought to have a relative molecular mass of 53 kDa. The latter is an example of a gene whose product normally prevents the formation of cancers, i.e. are tumour suppressor genes. Thus, the loss of function of these genes permitted cancers to develop, an important step in understanding carcinogenesis. It has since been shown that the products of these genes have key roles in cell physiology, such as control of the cell cycle and apoptosis. Although Paul and Hickey pointed to the importance of molecular pathology in cancer in 1974, it was haematology that used this developing field to investigate certain clinical conditions.

Blood Cancer

During the 1960s and 1970s, cytogenetics showed abnormal chromosome structure in the most common forms of blood cancer (lymphoma and leukaemia), observations that were extended by the demonstrations of increased numbers of chromosomes and the transfer of sections between chromosomes, described as a translocation. By the 1980s, genetic analysis of one such transfer placed an oncogene next to an immunoglobulin heavy chain gene; changes were subsequently found in numerous lymphomas and leukaemias. Perhaps the leading example of this is from 1982, with the translocation of a form of *myc* (i.e. *c-myc*) next to a gene for part of an immunoglobulin molecule being linked to Burkitt lymphoma. In 1987, Weiss et al. reported the presence of the Epstein–Barr virus (EBV) in tissues of Hodgkin lymphoma, extended in 1990 by Herbst and colleagues, who showed that its DNA can be detected in paraffin-embedded tissues, both pointing to a potential causal effect and the ability to probe archived tissues. Blood cancers are further described in Chapter 6.

However, one of the most far-reaching examples of the power of molecular pathology in this setting is the demonstration of a genetic lesion in the Philadelphia chromosome. First identified using cytogenetics by Nowell and Hungerford in 1960, Rowley showed a decade later a more precise transfer of sections of chromatin between chromosome 9 and chromosome 22. The decade that followed saw the now classic report that this transfer brings together two otherwise non-pathological genes to form a new oncogene (i.e. a neo-oncogene) that causes chronic myeloid leukaemia. Subsequent research identified the protein product, and, in the decades that followed, the design of drugs to target and (in many cases) effect a cure. These observations will be revisited in more detail in Chapter 6.

Genetic analysis of other leukaemias followed. The Philadelphia chromosome was also demonstrated in cases of acute lymphoid leukaemia (as it was then described), and in chronic myeloid leukaemia. The abnormality was detected using leukaemic-specific mRNA sequences amplified *in vitro*. *c-fms* (with origins in a feline leukaemia virus, and since renamed *CSF1R*) was confirmed as a molecular marker for certain types of acute myeloid leukaemia in 1987. The presence of oncogene *ETS2* on chromosome 21, which when translocated to chromosome 8, was found to be associated with a type of acute myeloid leukaemia, whilst others reported deletions in part of chromosome 7 in myelodysplasia and other myeloid disorders. In 1990, Becher et al. reported a case series of duplication of part of chromosome 17 in leukaemia, and by the middle of the decade, all forms of blood cancer had been linked to a genetic lesion.

By the 1990s, advances in molecular genetics were being used in routine clinical practice. An example of this is the use of PCR technology to probe for the presence of any instances of the neo-oncogene in the Philadelphia chromosome being present in suspected relapsed leukaemia after transplantation, for minimal residual disease in the treatment of acute lymphoblastic leukaemia, and the use of FISH for the presence of the Philadelphia chromosome on glass slide smears.

Self-check 1.2 Give examples of the value of virology in the development of molecular pathology.

Solid Organ Cancers

Studies from the 1960s and earlier demonstrating familial breast cancer strongly suggested a genetic component, ultimately proven. The 1980s also saw the expansion of molecular pathology into cancer of a number of diverse organs, with the development of techniques such as PCR permitting further analyses. Dubeau and colleagues showed that re-arranged genes can be detected within tissues that are embedded in paraffin wax. In 1984, others provided firm evidence that the human papillomavirus was the causal factor in many genital organ cancers, especially the cervix. Subsequent research used a DNA probe to find viral DNA in oral tissues and certain dermal lesions.

A key advance in molecular pathology in 1987 was the finding that several different types of malignant cells produce increased levels of one of the alternative forms of basic fibroblast growth factor to stimulate their own growth and that of other cells. Subsequent work showed upregulation of its gene (*FGF2*) in transformed cells and its interaction with a cell surface receptor. The late 1980s also saw the cloning and subsequent analysis of sections of chromosome 22, showing the presence of genes that had the potential to be an oncogene, i.e. proto-oncogenes.

Examples include *SIS* and *EWSR1*, the latter linked to many cases of Ewing sarcoma.

In 1990, research into liver and ovarian cancer was enhanced by reports of, respectively, the detection of hepatitis B virus DNA and PCR for *c-myc* in formalin-fixed paraffin wax-embedded tissue. In the same year, Rodenhuis et al. reported activation of a *ras* oncogene in non-small cell lung cancer. By the end of the decade, genetic abnormalities had been reported in all major cancers, with many instances of several genes linked to the same disease. Furthermore, *TP53* was shown to be the most frequently mutated gene in human cancer, underlining the importance of tumour suppressor genes, explained in more detail in section 2.3.2. In 1994, several groups reported using the technique of CGH to show altered DNA sequence copy numbers and chromosomes in breast cancer tumours and cell lines, in primary small cell lung cancers, renal cell carcinomas and malignant gliomas. *BRCA1* and *BRCA2* were reported, in 1994 and 1995, to be responsible for hereditary breast and ovarian cancer, and subsequently mapped to chromosome 17 and chromosome 13, respectively, whilst in 1996, DeRisi and colleagues used a microarray of over 1000 DNA elements to analyse tumour suppressor gene expression in melanoma.

The molecular pathology of solid organ cancers is further discussed in Chapters 4 and 5.

Other Diseases

The list of diseases with a genetic component (some being hereditary) is considerable. As is discussed in detail in Section 1.3.4 and in Chapter 7, the various forms of haemoglobinopathy constitute the leading global hereditary disease. During the 1960s and 1970s, Weatherall, Orkin and others showed that the disease, known to be caused by an abnormal amino acid sequence in a haemoglobin molecule, was subsequently shown, in turn, to be due to mutations in the genes for this molecule. Wexler and colleagues were amongst the first to point to the location of a gene for a particular disease – that being Huntington disease on chromosome 4, and named *HTT*, coding for huntingtin. The leading form of familial hypercholesterolaemia (FH), in its heterozygous form, and more so in its homozygous form, brings high levels of serum cholesterol and a markedly increased risk of premature cardiovascular disease. Goldstein and Brown, and their colleagues, showed this condition to be due to defects in the receptor for low-density lipoprotein cholesterol, and in 1984 assigned the gene for this molecule to chromosome 19. The presence of mutations in this gene (*LDLR)* unequivocally defines the disease. Further research showed the effects of other genes that interact with the low-density lipoprotein cholesterol receptor, and which cause increased plasma cholesterol. Retinoblastoma is the leading ocular cancer, molecular pathology showing most cases resulting from a loss of function, or deletion, of *RB1*, found on chromosome 13. Further advances in virology included the sequencing of numerous pathogenic viruses, setting the way for the development of rapid diagnosis kits for use in routine diagnostic pathology (Table 1.1).

In 1987, Robakis and colleagues cloned and sequenced the gene for amyloid and localised it to the long arm of chromosome 21, thus providing a huge leap forward in our understanding of Alzheimer disease. The year 1990 saw a report of *in vitro* amplification of DNA for *CYP21B*, coding for 21-hydroxylase, in the prenatal diagnosis of congenital adrenal hyperplasia, and that Hunter syndrome, a form of mucopolysaccharidosis, is caused by mutations in the gene for iduronate-2-sulphatase on the X chromosome.

Table 1.1 Selected focused analytical kits.

Pathology of interest	Examples of focused kit (pathogen or genes determined)
Infectious diseases	• Influenza panel: influenza A and B, pandemic H1 influenza virus, H3 influenza virus, respiratory syncytial virus. • Neuro panel: enterovirus, EBV, simplex virus, human adenovirus, human cytomegalovirus, human herpesvirus 6 and 7, human parechovirus, human parvovirus B19, Varicella Zoster virus.
Oncology	• Myeloproliferative neoplasms: *BCR*::*ABL*1, *JAK2, CALR, MPL, ABL1.* • Acute lymphoblastic leukaemia: *BCR*:: *ABL1, TEL*::*AML1, E2A*::*PBX1, MLL*::*AF4, MLL*::*AF9, MLL*::*ENL, ABL1.*
Coagulation/thrombophilia	• Prothrombin (*G2021A*), Factor V Leiden (*FVL*), MTHFR (*C677T* and *A1298C*).

Advances in molecular genetics were not confined to the diseases in man. In 1990, Golemboski et al. used genetic engineering in botany with their report of inducing resistance to tobacco mosaic virus into *Nicotiana tabacum*, whilst Whetter et al. brought the technology to the veterinarian world with the cloning of an equine infectious anaemia virus.

Epigenetics

Mutations are not only found in the DNA nucleotide sequence: epigenetics considers other changes to chromosomes that do not directly involve the nucleotide sequence. In many cases, there are changes to the exposed, outermost section of DNA, often a methylation, and in others there are changes to histone proteins within the chromatin, mostly acetylations. Enzymes that cause these changes (e.g. DNA methyltransferase) can be inhibited with drugs such as 5-azacitidine and decitabine. These (and several other) drugs have been licenced for use in diseases that include myelodysplastic syndromes and acute myeloid leukaemia, whilst inhibitors of histone acetylation, such as vorinostat and pracinostat, have been approved for cutaneous T cell lymphoma and acute myeloid leukaemia, respectively. Other targets for pharmacotherapy include lysine-specific demethylase-1, which acts on histones and may be an effective therapy in prostate cancer.

Summary

The previous section has provided an outline of certain events likely to be recognised as representative of importance in the early days of the development of molecular pathology. By the mid-1990s, the sub-discipline of molecular diagnostics was becoming established, as were the publications of dedicated learned journals in which new findings could be disseminated. In the early 2000s, roles for miRNAs in cancer were becoming recognised, opening a vast new field of investigation. As Figure 1.1 shows, using key words of 'Molecular Biology' and 'Molecular Pathology' as cited by 'PubMed', the depository of the National Library of Medicine of the USA, the number of research papers increased exponentially from around 1990 to 2024, and is likely to continue.

> **Key Point 1.3** Molecular genetics, in the form of molecular pathology, has enabled scientists to link abnormal genes with a mechanism likely to point to the causes of a particular disease and so to potential treatments.

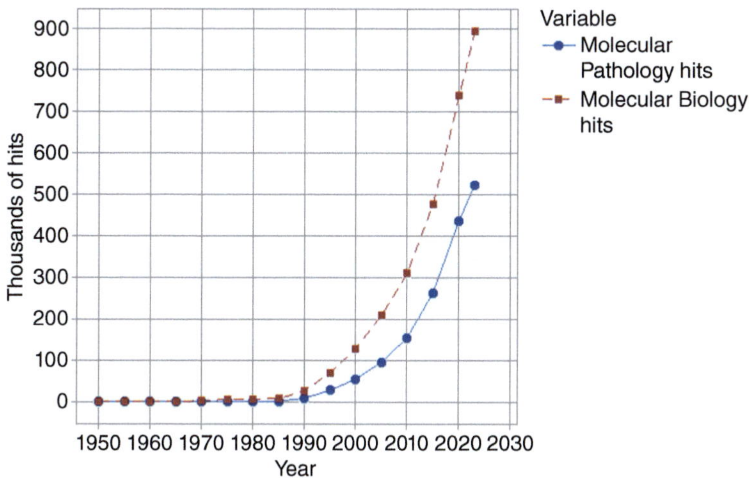

Figure 1.1 PubMed hits for molecular biology and molecular pathology. Squares: molecular biology, circles: molecular pathology. *Source*: National Library of Medicine / Public domain / https://pubmed.ncbi.nlm.nih.gov/.

1.2 The Scope of Human Disease

In order to study the various forms of human disease, we must understand the basis of their classification. They may be defined as those that are evident at, or shortly after, birth, in which case they are described as congenital. Of the remainder, some may arise at any time, and these may also be classified as those developing from within the individual, and those linked to external factors. Many diseases can also be inherited. In all of these groupings, molecular pathology can be very useful in establishing the nature of the disease.

1.2.1 The Classification of Disease

The International Classification of Disease (ICD) provides a system by which diseases can be collected into sections, based mostly on organs and organ systems, as shown in Table 1.2.

Some diseases have both an internal and external precipitant. An example of the latter being cardiovascular disease, where many cases result from the effects of risk factors, such as smoking, hypertension and

Table 1.2 The ICD-11 classification of diseases.

Disease group	Examples
Infectious and parasitic diseases	Tuberculosis, bacterial diseases, viral infections (viral hepatitis and HIV), mycoses and protozoal diseases
Cancer (neoplasms)	Melanoma, cancer of the breast, prostate, blood, brain and other parts of the central nervous system
Blood and blood-forming organs	Nutritional, haemolytic, aplastic and other anaemias, coagulation defects
The immune system	Primary immunodeficiencies, autoinflammatory disorders and diseases of the thymus
Endocrine, nutritional and metabolic diseases	Diabetes mellitus, disorders of the thyroid and pancreas, nutritional deficiencies, obesity
Mental and behavioural disorders	Schizophrenia and delusional disorders, mood disorders, neuroses, disorders of personality
The nervous system	Inflammatory and atrophic diseases of the central nervous system, movement disorders, stroke
The eye	Disorders of the conjunctiva, sclera, cornea, iris and ciliary body, lens, choroid and retina
The ear	Diseases of external, middle and inner ear
The circulatory (cardiovascular) system	Hypertensive and ischaemic heart diseases, pulmonary heart disease
The respiratory system	Acute upper respiratory infections, influenza and pneumonia, chronic lower respiratory diseases
The digestive system	Diseases of the salivary glands oesophagus, stomach and duodenum, liver and gallbladder
The skin	Infections of the skin and subcutaneous tissue, bullous disorders, dermatitis, eczema, urticaria
The musculoskeletal system	Arthropathies, systemic connective tissue disorders, soft tissue disorders, osteopathies
The genitourinary system	Renal diseases, urolithiasis, ureter, bladder, male and female genital organs and breast disorders
The perinatal period	Foetus and newborn affected by maternal factors, complications of pregnancy and delivery
Developmental anomalies	Congenital and chromosomal abnormalities (duplications and trisomies)

Adapted from https://www.who.int/standards/classifications/classification-of-diseases.

diabetes, whilst hypercholesterolaemia may have a genetic component that is frequently inherited. Other causes of disease that arises within the body include autoimmune diseases (although some may be linked to an external stimulus) and cancer.

However, there are numerous examples of malignancy being the consequence of an external factor, such as the hepatitis B virus and the hepatitis C virus being linked to liver cancer, and radiation causing blood cancer. Nevertheless, many cancers appear without a clear scientific precipitant and are due to seemingly spontaneous genetic aberration.

1.2.2 Morbidity and Mortality

The preceding section and the ICD system make no comment on the extent to which disease brings an increased likelihood of mortality. For example, although relatively common in an elderly population, osteoarthritis is a major morbidity, but is rarely directly fatal, but may be indirectly. Similarly, asthma is relatively common, as are ocular problems, with a high proportion of people wearing reading and/or long-distance spectacles, but these are infrequently listed on death certificates.

One of the leading UK government data-collecting bodies is the Office for National Statistics (ONS). This organisation publishes a set of data of the various forms of death, which includes those from accidents and crime as well as from disease. The reporting of the types of death follows that of ICD and shows that the most common cause of death is cancer (Table 1.3).

Whilst there can be many causes of cancer, the link between this disease and its clinical development (generally, a tumour) is an abnormality in the DNA sequence of a particular cell within the 'host' organ involved. This therefore provides the justification for the investigation of the DNA in these cancers, and so the development of clinical aspects of molecular pathology.

Key Point 1.4 The ICD provides a globally recognised common standard for the investigation and reporting of disease.

Despite the focus on cancer, there are a great number of other diseases that have a chromosomal or genetic component, and which are not necessarily fatal but bring a considerable burden of morbidity, an example being Down syndrome (as discussed in Chapter 7). Knowledge of an individual's

Table 1.3 Leading causes of death in England and Wales, 2023.

Cause	Number of deaths	Percentage (%)
All causes	580,108	100
Cancer	152,418	26.3
Disease of the circulatory system	139,984	24.1
Respiratory disease	71,144	12.3
Mental and behavioral disorders	44,000	7.4
Nervous system disease	42,709	7.4
Digestive system disease	30,858	5.3
COVID-19[a]	11,818	2.0
Endocrine, nutritional and metabolic disease	10,706	1.8
Genitourinary disease	9049	1.6
Infectious and parasitic disease	7695	1.3
Musculoskeletal system disease	4360	0.8
Disease of the skin and subcutaneous tissue	2544	0.4
Congenital malformations, deformations and chromosomal abnormalities	1516	0.2

Figures do not sum to 100% as only selected categories are presented.
[a] In 2021, this caused 67,057 (11.5%) of total deaths and in 2022, was linked 22,396 deaths (3.9%) and may be linked to fewer deaths in 2024. *Source*: Adapted from Nomis web / https://www.ons.gov.uk/peoplepopulationandcommunity/birthsdeathsandmarriages/deaths/methodologies/mortalitystatisticsinenglandandwalesqmi / Last accessed at 08/09/2024.

genetic profile can predict ill health and can be useful in focusing on the treatment of those with the greatest risk of developing disease, as in the case of Huntington disease. The presence of the mutant Factor V Leiden, a variant of 'normal' coagulation factor V, may bring the need for anticoagulation in certain circumstances. Thus, the knowledge of potential or actual health issues represents a major step forward in medical care.

Self-check 1.3 What are the leading causes of death in England and Wales? – Which of these can be prevented, and how can this be achieved?

1.3 Chromosomal Abnormalities and Disease

The astute laboratory scientist working in molecular pathology will appreciate numerous aspects of the physiology and pathology of the cell, and of the nucleus in detail. The objective of this section is to summarise gross concepts in both chromosomal and genetic disease, and in doing so, prepare the path for chapters to come. The key laboratory method used to determine the number of chromosomes and certain chromosomal abnormalities is cytogenetics. The practical aspects are described in Chapter 2.

1.3.1 Chromosomes in Health and Disease

Normal Chromosome Structure and Function

The leading component of chromosomes, chromatin, is a complex of DNA and histones, the latter a collection of five forms (H1–H5) of a globular protein. An octamer core complex of H2–H4 that forms a template around which around 146 base pairs of DNA can coil is a nucleosome, whilst H1 and H5 link adjacent nucleosomes. This system allows for more efficient packaging of the 2 m of DNA in our nucleus, and also its unwinding for transcription and replication.

Each chromosome is numbered by size from 1 to 22 (the autosomes), and in alphabetic groups (such as group E for chromosomes 16, 17 and 18), whilst the sex chromosomes are named X and Y. Each chromosome has a different density of histones and frequency of nucleotides whose pattern can be visualised by staining with dyes, such as quinacrine and Giemsa, hence Q-bands and G-bands. The latter is more commonly used: dark staining G-bands are more likely to localise dense chromatin enriched in thymine and adenine, whilst sections with a high component of cytosine and guanine, and which are loosely packed, stain lighter. However, before staining, chromosomes must be freed from the tight constraints of the nucleus, best done by inducing an in-vitro culture of cells to undergo mitosis, and then arresting the process before the new cells are formed. Careful manipulation of these 'poisoned' cells allows the cell to burst and release its 46 chromosomes, which can then be stained and visualised as a karyogram (Figure 1.2).

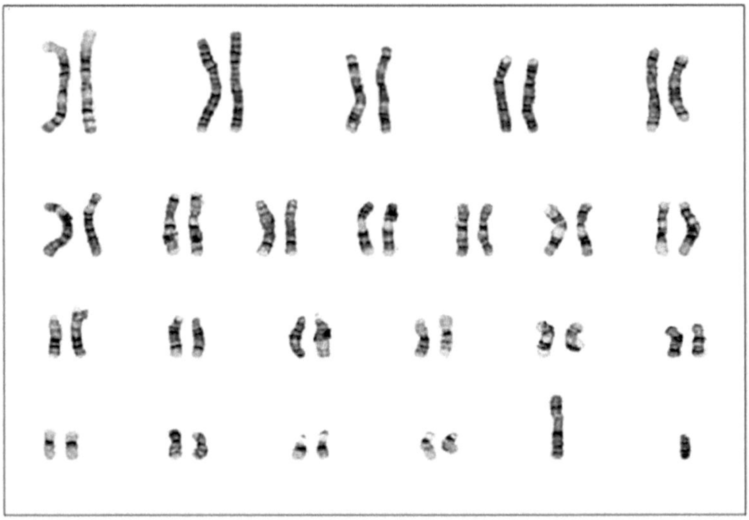

Figure 1.2 Karyogram of the human chromosome complement. The numbering system goes from left to right and top to bottom, so the top row is pairs of chromosomes 1–5, one each from the mother and father. The bottom row is chromosomes 19–22, with, at the bottom right, the unpaired X and Y chromosomes.

Not all the DNA in a chromosome codes for a functioning protein. Two such regions are the telomere (at the ends of each chromosome) and centromere. The former are composed of up to 800 repeats of the nucleotide sequence TTAGGG, which protect the two ends of the chromosome from degradation each time the cell replicates. The centromere is composed of densely packed heterochromatin and is involved in sister chromatid binding in mitosis and is the site of the attachment of kinetophores. The exact position of the centromere varies between chromosomes and allows a classification:

- Where the centromere lies roughly in the middle of the molecule, the chromosome is described as metacentric.
- Should the centromere be slightly offset, the chromosome is submetacentric.
- An acrocentric chromosome has its centromere markedly offset.

The positioning of the centrosome leads therefore to two chromosomal arms. The shorter being given the label p (after the French for

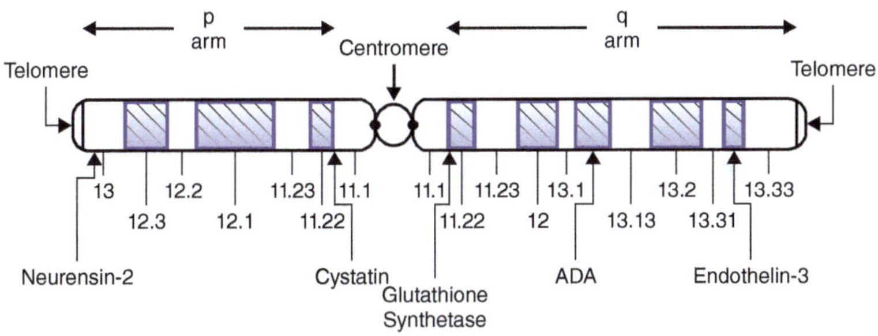

Figure 1.3 Structure and banding of chromosome 20 showing the location of certain genes. The numbers refer to a particular band. The size and position of the bands is for illustrative purposes only. The position of genes is approximate. ADA = adenosine deaminase, the *ADA* being present at 20q13.12.

petite), the longer arm being named q (as q follows p in the alphabet). The centromere is also important as it is the starting point for the numbering of the dyed G-bands of the chromosome – which numerate from the centromere towards each of the telomeres, in relatively well-defined blocks of light and dark staining. This numbering system and the p/q labels allow us to locate the position of specific genes along the chromosome (Figure 1.3). For example, the location for adenosine deaminase, an enzyme involved in the synthesis of haemoglobin, is at 20q13.12.

DNA That Fails to Code for Proteins

Apart from the centromere and telomere, many other sections of chromosomes fail to code for protein. Indeed, it has long been established that much of the entire DNA genome fails to code for mRNA, tRNA or rRNA, once described as 'junk' DNA. A gene may contain alternate sections of DNA that do (exons) or do not (introns – from intergenic region) code for mRNA, whilst longer sections of the genome that fail to code for proteins may be described as gene deserts. An additional form of non-coding DNA are thousands of 'satellites' – repetitive sequences of up to six base pairs (such as ATGATGATGATG) often repeated up to 50 times and which may also be described as short tandem repeats. Pseudogenes are sections of DNA that resemble functioning genes but have an intrinsic defect rendering them non-functional.

Certain sections of DNA code for RNA species that themselves do not code for proteins. These ncRNAs may be classified into one of two major families.

- Housekeeping ncRNAs include ribosomal RNA (rRNA, involved in translation), transfer RNA (tRNA, transporting amino acids), small nuclear RNA (snRNA, involved in RNA processing), small nucleolar RNA (snoRNA, with roles in the modification of RNA), and telomeric RNA (TRNA, whose function include participation in the synthesis of telomeres).
- Regulatory ncRNAs include miRNAs, with functions in translational control and RNA stability), lncRNA (with roles in nuclear structure, epigenetics and imprinting), circular RNA (circRNA, the inhibition of miRNA activity, often described as an RNA sponge) and small interfering RNA (siRNA, often involved in the degradation of RNA).

As many of these ncRNAs have roles in the regulation of DNA and other RNA molecules, some are also linked to cancer and cardiovascular disease. However, as these fascinating molecules are not (as yet) part of the routine clinical practice of molecular pathology, any discussion will be limited.

Classification of Chromosomes and Their Genes

Table 1.4 summarises data on the chromosomes, and the number of protein-coding genes, non-protein-coding genes and pseudogenes present in each chromosome. The data sum to a genome size of some 3.1 billion base pairs, 19,785 protein-coding genes, 16,158 ncRNA genes and 14,671 pseudogenes, a total of 50,614 genes.

Despite the validity of this data, interpretation must be cautious as, for example, there are instances where a sequence of DNA can transcribe more than one species of RNA, and certain ncRNAs can prevent the translation of mRNAs. The size of the chromosome correlates significantly with the number of protein-coding genes (correlation coefficient [r] 0.63, $p = 0.001$), the number of ncRNA genes ($r = 0.84$, $p < 0001$), but most significantly with number of pseudogenes ($r = 0.93$, $p < 0.001$), and so, overall, with the total number of genes ($r = 0.81$, $p < 0.001$).

The Mitochondrion

Mitochondria have their own DNA (mtDNA), arranged as a ring of 16,569 base pairs. Of the 37 genes, 13 code for subunits of ATP synthase, cytochrome C oxidase and NADH (nicotinamide adenine dinucleotide – hydrogen)

Table 1.4 Selected features of chromosomes.

Number	Size (million base pairs)	Protein-coding genes	Non-coding RNA genes	Pseudogenes	Total gene count
Acrocentric chromosomes					
13	113.6	323	510	441	1274
14	101.2	667	693	543	1903
15	99.7	591	679	512	1782
21	45.1	234	315	195	744
22	51.3	460	356	333	1149
Y	53.8	58	95	391	544
Metacentric chromosomes					
1	248.4	2031	1474	1254	4759
3	201.1	1060	916	808	2784
16	96.3	842	650	443	1935
19	61.7	1413	599	485	2497
20	66.2	536	424	292	1252
Submetacentric chromosomes					
2	242.7	1247	1165	1055	3467
4	193.6	748	735	726	2209
5	182.0	854	848	688	2390
6	172.1	1041	865	816	2722
7	160.6	922	708	842	2472
8	146.3	679	714	612	2005
9	150.6	779	621	664	2064
10	134.8	729	656	612	1997
11	135.1	1285	730	772	1627
12	133.3	1017	788	618	2423
17	84.3	1162	760	520	2442
18	80.6	270	393	264	927
X	152.5	837	464	785	2086

Data on genes is pooled from different sources and is approximate. Adapted from https://www.ncbi.nlm.nih.gov/assembly/GCF_009914755.1/#/st.

dehydrogenase, 22 code for a transfer RNA, and 2 code for the large and small subunit ribosomal RNA. Associated with the large subunit is humanin, a 21 amino acid peptide believed to have a role in protecting from Alzheimer disease and having anti-apoptotic activity. miRNAs have been localised within mitochondria, and so are presumed to have a role in the physiology of this organelle.

A Caveat The number of genes quoted here is brought together from different sources, and so is very likely to differ from data published elsewhere. Furthermore, advances in technology may also change these numbers, as the role of some genes may be revised.

Self-check 1.4 From Figure 1.3, describe the location of the gene for cystatin.

Key Point 1.5 Chromosomes are classified by their size (and so a group), the position of the centrosome and whether of autosome or sex nature.

Having described the structure of the chromosome, we will next address the pathology of chromosomes and their genes according to a size-based scale from large to small. This will first be an examination of differences in the number of chromosomes, and then large-scale abnormalities within chromosomes, and will conclude with micro-changes to the genome.

1.3.2 Errors of Chromosome Numbers

Quality control of the process of cell replication generally ensures that the two sister chromatids lining up on the metaphase plate separate cleanly, with one each being drawn to opposite poles of the cell by microtubules. Thus, each new 'daughter' cell has its correct number of chromosomes (46: euploidy). However, should this mechanism fail, then both sister chromatids may be drawn to one pole, and none to the opposing pole. Consequently, one new cell will have an extra chromosome (47: hyperploidy) and the other one fewer chromosome (45: hypoploidy). Both these deviations in chromosome segregation are described as aneuploidy, the process itself being non-disjunction. Should non-disjunction occur shortly after fertilisation, all cells will be affected, but should it occur later in embryonic development, only some cells may be aneuploid, other euploid, a state known as mosaicism, and is relatively common in cancer.

Aneugens (factors causing aneuploidy) include industrial chemicals (such as benzene and carbamly compounds), pesticides, toxins in tobacco smoke and X-rays. These factors also cause general damage to cells (i.e. are cytotoxic).

Hypoploidy

Absence of one chromosome is a monosomy. Although most consider this lesion to be present when an entire chromosome is missing, others describe deletion of an entire p or q arm of chromosomes as a monosomy. Perhaps the only major example of a complete monosomy is Turner syndrome, where an entire X chromosome is missing, so that the karyotype is 45,XO. However, mosaics for Turner syndrome are known, and the phenotype can also be present if part of an X chromosome has been deleted. A leading phenotype alteration is short stature: mosaics have a height between that of women who are full monosomics and euploid women. Notably, many cancer cells preferentially lose small chromosomes (which may contain tumour suppressor genes) and are therefore hypoploidic.

Hyperploidy

In contrast to hypoploidy, there are numerous forms of hyperploidy, by far the most frequent being trisomy. An extra copy of chromosome 21 produces Down syndrome, and with a frequency of around 1150 per million live births, it is the most common chromosomal condition, even though 40–80% of conceptions are estimated to be lost during pregnancy. The karyotype is 47,XX or 47,XY, although there can be mosaics, and the leading cause is an increased age of the mother, where the risk of this trisomy rises 40-fold from maternal age 20 to 45. Patau syndrome and Edwards syndrome are trisomies of chromosomes 13 and 18, respectively, both with a frequency of around 200 per million; others include trisomy of chromosomes 7, 8, 9, 18 and 22. However, as with Down syndrome, many refer to partial duplication of either arm of a particular chromosome, and mosaics may be present. Polysomy of up to eight extra chromosomes has been described.

There are numerous sex chromosome hyperploidys:

- Klinefelter syndrome results from an additional X chromosome, and so, karyotype 47,XXY. Estimates of presenting frequency vary around 1000 per million newborn males (i.e. 500 per million total births), although the diagnosis may be missed.
- Jacobs syndrome is the product of an extra Y chromosome in males, hence 47,XYY, and is cited as having a frequency of 980 per million male births.
- Triple X syndrome, with a karyotype of 47,XXX, presents with a frequency of around 1000 per million females.
- Additional X chromosomes are present in tetrasomy X (karyotype 48,XXXX), with a frequency of 20 per million females.
- Extra copies of both chromosomes, i.e. 48,XXYY, present with a frequency of around 40 per million male births.
- Pentasomy 49,XXXXX presents at around 8 per million females.

Self-check 1.5 What are the leading clinical conditions linked to hypoploidy and hyperploidy?

1.3.3 Errors of Sections of Chromosomes

This section describes abnormal changes to the structure of chromosomes, of which there are several forms, most being translocations, inversions, duplications, deletions and insertions. All these alterations are found in, and in many cases are the cause of, various forms of solid organ tumours (Chapters 4 and 5), blood cancer (Chapter 6), and other diseases (Chapter 7). Ring-like, circular, chromosomes are present at a frequency of around 20 per million and may also be found linked to certain diseases. Ring chromosomes 14 and 20 are linked to epilepsy, whilst examples of other ring chromosomes include 1, 2, 3, 4, 5, 13, 16, 18 and the X chromosome (with Turner syndrome), many of which are linked to other forms of pathology.

The International System for Human Cytogenetic Nomenclature

This system explains the nature of the chromosomal lesion. Consider a gene on chromosome 5 at position p12 and a second gene on chromosome 10 at position q20. A reciprocal translocation occurring between these chromosomes is defined as t(5;10)(p12;q20). Note that the chromosome with the lower number is placed first, followed by the position on each respective chromosomal arm. In inversions and duplications, only one chromosome is named, followed by the positions of the changes, such as inv6(q11;q22). Note also that the locations both have the same q notation, so this lesion is within the q arm and does not cross the centromere, although others do, as in 6(p12;q12). Ring chromosomes such as r(6)(p23;q25) tell us that the break and re-join position is between p23 and q25.

Translocations

Most translocations can be classified into one of two groups. A reciprocal translocation, the most common, is characterised by two chromosomes exchanging sections of DNA, whilst a Robertsonian translocation is where large q arms of two chromosomes are joined together at the centrosome. In a homologous translocation, the two chromosomes can be of the same type, e.g. one copy of chromosome 15 joining with the other copy of chromosome 15. By contrast, a heterologous translocation is formed between two different types, e.g. one copy of chromosome 15 and one copy of chromosome 14. In these examples, the q arms fuse, whilst the remaining small p arms are consumed within the cell (Figure 1.4).

There are numerous examples of these translocation errors in pathology. We have already discussed the formation of the Philadelphia chromosome, where parts of chromosomes 9 and 22 transpose, so causing chronic myeloid leukaemia. The precise location of the break/join region is position 34.1 on the q arm of chromosome 9, and position 11.2 on the q arm of chromosome 22. Hence, the new, hybrid chromosome is designated t(9:22)(q34;q11). Heterologous Robertsonian translocations are present in around 3% of cases of Down syndrome and are written as 47XY,t(14q21q), where the q arm of chromosome 21 is translocated onto the q arm of chromosome 14. The reciprocal translocation of Ewing sarcoma brings together parts of chromosome 11 and chromosome 22 to form t(11;22)(q24;q12), merging *EWSR1* with *FLI1*, which can produce a new EWS-FLI1 fusion protein. A homologous (of the same chromosome) Robertsonian translocation involving chromosome 17 (hence isochromosome i(17q)) has been reported in cases of leukaemia.

Inversions

Inversions come in two forms: pericentric (where the centromere is not involved) and paracentric (where the centromere is at the centre of the inversion (Figure 1.5). The latter has been reported in cases of Down

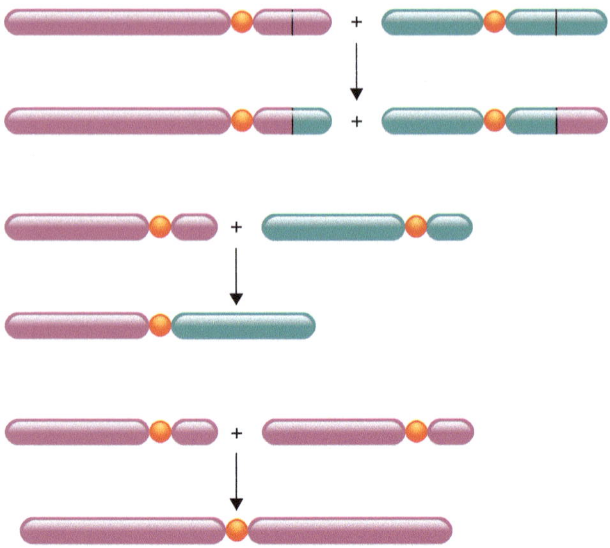

Figure 1.4 Chromosome translocations. Upper panel: a reciprocal translocation, where sections of chromosomes are swapped between different chromosomes (i.e. the terminal short sections of the 'pink' and 'cyan' chromosomes). Middle panel: a heterologous Robertsonian translocation where the q arms of different chromosomes (pink and cyan) fuse – the p arms are lost. Lower panel: a homologous Robertsonian translocation, where q arms of the same (pink) chromosomes fuse, and the small p arms are again lost. This latter translocation produces an isochromosome (iso = the same).

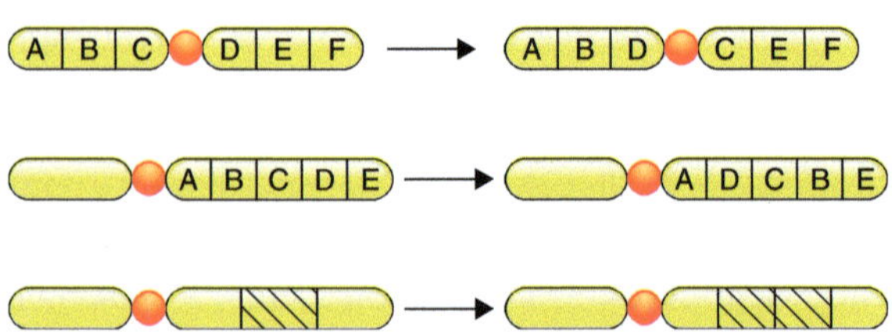

Figure 1.5 Chromosome inversions and duplications. Upper panels: inversions in the nucleotide sequence ABCDE to ADCBE including (acrocentric) and excluding (pericentric) the centromere. Lower panel: duplication of a section of the q arm of a chromosome.

syndrome (9(q32q34)), Klinefelter syndrome (11(q21q23), 12(q15q24) and 5(q21q32)), acute leukaemia 7(q11q22)) and three inversions in cases of mental retardation (1(p22p36), 3(p13p25) and 13(q12q22)). The pericentric inversion [inv(16)(p13q22)] brings together two otherwise unrelated genes on different arms of chromosome 16, the fused product likely to be responsible for acute myeloid leukaemia. Inversions in a chromosome 22 intron (denoted I21I) of *F8* (coding for coagulation factor VIII) have been reported in haemophilia.

Duplications

Duplications are characterised by a section of a chromosome repeating itself further down its arm (Figure 1.5), and once more there are numerous examples – a second *MYC* is present in many cases of breast, cervical, oesophageal and ovarian cancer. A duplication in Xq27.1 has been linked to growth hormone deficiency and intellectual disability. This region contains *SOX3*, mutations in which have been associated with X-linked hypopituitarism. Duplications in 1q21.1 in schizophrenia have been described and are present in around 10% of cases of Duchenne muscular dystrophy (generally in exons 2–10 of *DMD* at Xp21). Micro-duplications at Xq26.3 encompassing *GPR101* are associated with X-linked acrogigantism. However, there are numerous instances of the physiological duplication of genes, an excellent example being those of the different forms of haemoglobin, which undoubtedly arose from a single ancestral gene.

An allied aspect of duplication is that of copy number variation, or variants (CNV), which have developed into a major analytical tool. As discussed earlier, not only varying sections of chromosomes can be duplicated, such as entire genes, but a small number of nucleotides, of varying length, can also be duplicated. An example of this is the trinucleotide CAG, repeated many times in Huntington disease, although CNVs may represent normal phenotypic variation, and represent as much as 10% of the human genome. Nevertheless, the total CNV load in schizophrenia exceeds that of controls. Although large CNVs can be benign in nature, one study pointed to 8% of the population carrying a CNV >500 kbps compared to 25% of those with intellectual disability.

Conversely, increased CNVs may be beneficial, and example being *AMY1* at 1p21.1, coding for amylase, whose substrate is starch. It has been suggested that an increased CNV provides a more active species of amylase, and so a selective advantage of an improved ability to digest starchy foods.

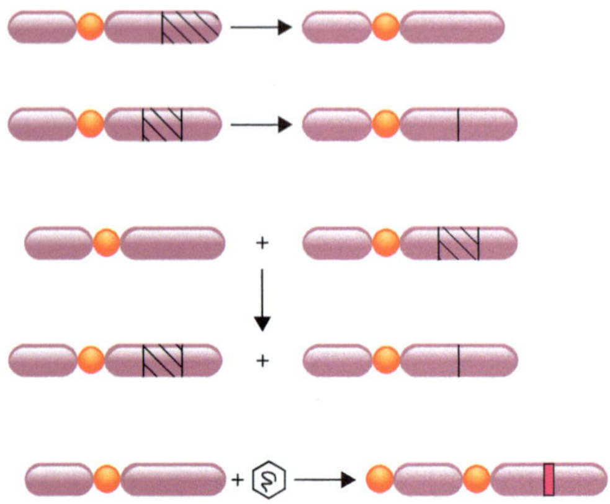

Figure 1.6 Chromosome deletions and insertions. Upper panel: deletions of the terminal and a within-arm section of chromosomes. Middle panel: insertion of a section from one chromosome to another. It follows that the 'donating' event from the chromosomes on the right is a deletion. Lower panel: insertion of viral DNA.

Deletions

These lesions can range from a single nucleotide to large kilobase sections, and even entire p or q arms, many of them rare, and lead to hundreds, if not thousands, of diseases (Figure 1.6). Examples of pathological deletions include:

- Prader–Willi syndrome can be caused by a 15q11q13 deletion and so loss of several small nucleolar RNAs, leading to the characteristic short stature and obesity.
- Angelman syndrome also results from a deletion in 15q11q13, with the loss of *UBE3A*, whose enzyme product targets proteins for intracellular digestion.
- The multi-system Williams syndrome is caused by a 1.5–1.8 Mb deletion at 7q11.23, leading to developmental delay, a hoarse voice and microdontia (small teeth).
- The deletion 22q11.2 causes the relatively common DiGeorge syndrome (250 per million births), the loss of 30–40 genes producing numerous pathological changes.

- Micro-deletions in 16p11.2, with an incidence of 100 per million, are characterised by language/growth retardation, obesity and seizures.
- Around 60% of cases of Duchenne muscular dystrophy are caused by deletions in Xp21.2-21.1, the location of *DMD* that codes for dystrophin.

Other deletions in their respective genes have been shown to cause growth factor deficiency, anti-thrombin deficiency, thalassaemia, elliptocytosis, Lesch–Nyhan syndrome, osteogenesis imperfecta, retinoblastoma and Wilms tumour. Further details of the pathology of these genes are explained in chapters to come.

Insertions

Genetic material inserted into a chromosome can be from another chromosome or from a retrovirus (Figure 1.6). An example of the latter is the insertion of a section of the genome of the hepatitis B virus into the X chromosome of hepatocytes, causing hepatocellular carcinoma. The precise mechanism for the viral insertion provides a potential method for the insertion of normal genes, possibly as a plasmid or a viral vector, to replace genes lost by a deletion or another mechanism. Several diseases are characterised by the insertion of a variable number of sets of three nucleotides (typically CAG, CTG, CGG or GAA) that range from 39 to several thousand copies and are described as repeat expansion disorders.

Together, insertions and deletions may be described as indels. Should any particular genetic lesion have a size greater than 50 base pairs, it may be described as a structural variation.

> **Key Point 1.6** Molecular pathologists test for differences in the number of chromosomes and for differences within chromosomes.

> **Self-check 1.6** Explain the difference between a reciprocal translocation and a Robertsonian translocation.

1.3.4 Micro-changes to the Genome

The previous section has summarised major abnormalities of chromosome structure. We now move from macro-changes to entire chromosomes and large sections of chromosomes, to micro-changes to the sequence of nucleotides in the DNA, and how they are linked to a disease. The genetic code, based on the nucleotide sequence, is a blueprint for the production of proteins. A major triumph of molecular genetics has been to show how alterations in the nucleotide sequence lead directly to disease. Perhaps the greatest earliest example of this was the demonstration of the cause of the major form of sickle cell disease, discussed in the following section, and in Chapter 7. Some changes are described as 'nonsense mutations', where the gene becomes non-functional, and causes disease such as deficiency in coagulation factors VIII and IX. Disease may also be caused by mutations in the processing of RNA, and include some thalassaemias and some cases of phenylketonuria.

Single-nucleotide Polymorphisms

Whilst the aberrations in chromosome number and structure described earlier are interesting, they quantitatively pale before the most common form of genetic variation, the single nucleotide polymorphism (SNP). Some of these SNPs may be seen as a substitution (of one nucleotide for another, as in sickle cell disease), others an insertion of a particular nucleotide, these previously being described as a point mutation. An alternative descriptor for an SNP is a single nucleotide variant, used by some commentators.

An early estimate from the Human Genome Project in 2001 put the number of SNPs in the genome at around 1.4 million, later updated to 10 million, hence >1% of the genome, leading to the figure of one SNP per 1250 base pairs. However, further updates suggest SNPs to be present at a frequency of 1 per 300 base pairs, which by extrapolation points to 11 million SNPs in the genome. Also, others cite data of over 84.7 million SNPs, with 64 million having a frequency of <0.5% of the population, ~12 million having a frequency between 0.5 and 5%, and ~8 million having a frequency of >5%. The average number of SNPs per chromosome has been reported as around 59,000, ranging from 130,000 in chromosome 1 to 4200 in the Y chromosome.

Reference Point Wikipedia carries a series of four lists of 19,259 genes, with links to further details of each of them:

https://en.wikipedia.org/wiki/List_of_human_protein-coding_genes_1: A1BG–EPCIP

https://en.wikipedia.org/wiki/List_of_human_protein-coding_genes_2: EPDR1–MTMR1

https://en.wikipedia.org/wiki/List_of_human_protein-coding_genes_3: MTMR2–SLC17A7

https://en.wikipedia.org/wiki/List_of_human_protein-coding_genes_4: SLC17A8–ZZZ3

Implications of SNPs

The consequence of an SNP is often the generation of an alternative form of the protein. A nucleotide may be added into a sequence following incorrect repair by polymerases, leading to changes such as a frameshift mutant. For example, an SNP insertion of an adenosine (A) residue into ATC/GGC/ATA/GGA gives ATC/GGA/CAT/AGG, and so an alternative amino acid sequence) or a stop codon (such as TAG, TAA or TGA). In some cases, an SNP will not result in a change in protein structure (as, for example, the codons AAA and AAG both code for lysine), and so will be pathogenically irrelevant, a change described as synonymous. However, in others, the SNP (such as leading to codon AAT) will indeed result in a different amino acid (asparagine), and so potentially a different protein product that may have altered functional activity, or be non-functional, and so pathogenic. Alternatively, an SNP may be in the form of a substitution. It was long known that sickle cell disease was due to a change in the amino acid sequence of a section of the haemoglobin molecule (from glycine to valine), and accordingly, the SNP (at the time described as a point mutation) causing this change (triplet GAG to GTG) was amongst the first demonstrations of the importance of this transition.

Not only are SNPs the most frequent change in the genome, but they are also responsible for the most wide-ranging forms of disease, most of which are cancer, to be discussed in Chapters 4–6. Other SNPs are found in non-malignant disease, such as in glucose-6-phosphate dehydrogenase deficiency, α1-antitrypsin deficiency, hereditary amyloidosis and FH, as will be discussed in Chapter 7.

Normal and Abnormal Genes

From a philosophical perspective, it could be argued that since the gene for brown eyes is the most common (i.e. is 'normal'), then the less common gene for blue eyes is 'abnormal'. However, we would most likely define an abnormal gene as a variant of a normal gene that brings a risk of, or actual, disease. Examples of these include those that lead to FH and to sickle cell disease, both caused by an SNP. Perhaps the true role of molecular pathology is to identify these abnormal genes, and so provide the practitioner with additional information that may lead to targeted management, often with drugs.

Table 1.3 tells us the leading causes of death are cancer and cardiovascular disease, and in turn, most cases of the latter are linked to other risk factors, as discussed in Chapter 5. Many cancers also have risk factors, but in several instances, these act on a proto-oncogene, converting them to true oncogenes. These are discussed in more detail in Chapter 3.

Developing Analyses

Molecular genetics continues to provide new knowledge and so opportunities to determine diagnosis and direct management. An example of this is the growing family of ncRNAs (as introduced in Section 1.3.1), many of which have regulatory roles in translating mRNA and in acting on DNA. As many clinical trials have shown, these molecules may enter routine practice. An example of this is the small interfering RNA zilebesiran, which specifically targets angiotensinogen synthesis in the liver, leading to a reduction in blood pressure, and so a potential treatment for hypertension. It may be predicted that technical advances will see the development of point-of-care molecular pathology testing that may be used in general practice.

> **Self-check 1.7** Which type of error within a chromosome is the cause of chronic myeloid leukaemia?

1.4 The Practice of Molecular Pathology

A key feature of molecular pathology is that it has an impact on every sub-discipline within pathology and within clinical medicine, as outlined in the figure in the preface. A major driver of this was the Human

Genome Project, an international collaboration, which by 2003 had sequenced around 85% of the genome, rising to 100% in 2022, reporting 3.12 billion haploid base pairs. These provided an unparalleled opportunity to locate aberrant genes. A further initiative was the development of next-generation sequencing (described in Section 3.1), also described by some as massive parallel sequencing. With these and other analyses came considerable data on the presence of pathological alterations to the genome, often demanding interpretation by the newly developed discipline of bioinformatics (Section 3.3). A consequence of this is the understanding of an impressive mass of detail regarding the working of the body and how the disease comes about.

1.4.1 Personalised Medicine

This expression (also described as precision medicine) refers to the treatment of the patient as an individual, as opposed to treatment of their condition or disease. An example of the latter is the prescription of aspirin after a myocardial infarction or stroke, where, in the United Kingdom, a dose of 75 mg a day is taken by all, regardless of the overall effect of the drug. However, in other countries, that daily dose may be 87 mg or 100 mg. Similarly, certain direct-acting oral anti-coagulants are given in a single dose to all those needing this treatment. A disadvantage of this approach is that, as defined by the natural distribution, in some the dose will be suboptimal (and so less effective) and in others supra-optimal (possibly leading to adverse side effects). Personalised medicine adjusts the treatment precisely, so it is effective for each individual. Examples of this include the anticoagulant warfarin, where the effective appropriate dose in some may be 1 mg daily, in others 20 mg daily, and the cholesterol-lowering drug atorvastatin, where the required dose may be 20 mg, 40 mg, 60 mg or 80 mg daily, dependent on the needs of the patient.

Much of the variability of drug effects (pharmacokinetics and pharmacodynamics) can be explained in terms of genetics, and accordingly, knowledge of the presence of certain genes involved in the metabolism of a particular drug can guide treatment. An example of this is in the use of the drug gefitinib in lung cancer – it is effective only in those whose genome carries a certain mutation coding for the receptor for epidermal growth factor. A second example is of the enzyme thiopurine methyltransferase, which contributes to the processing of thiopurine drugs, such as 6-thiopurine and azathioprine, used in cancer chemotherapy. As the enzyme effectively neutralises the drugs, low levels of the enzyme lead to enhanced activity of the drug, which can then be damaging to the bone marrow and perhaps lethally toxic. The leading cause of the enzyme deficiency is a loss-of-function mutation in the enzyme's gene and can be assessed with relative simplicity. The presence of these mutations can only be determined by molecular pathology, and in oncology, this procedure is described as cancer pharmacogenomics.

More recent advances include obtaining a sample of a patient's tumour and exposing it *in vitro* to a variety of drugs to see which is most effective, and so a targeted treatment. A further example of personalised medicine is in generating chimaeric antigen receptors on T-lymphocytes (CAR-T). In this complex procedure, a major step is the transfection of a patient's own cells with engineered DNA specific for a particular antigen, which are then returned to the host where they seek and destroy cells bearing the antigen. CAR-T therapy is well-established in a number of blood cancers and may also be so for solid organ tumours. Another procedure is where genetic engineers/editors may attempt to deliberately insert semi-artificial DNA by site-directed mutagenesis and CRISPR/Cas-9 technology into, for example, a stem cell in order to correct a genetic disease.

1.4.2 National Developments in Molecular Pathology

Driven by the inevitable fall in the cost of DNA sequencing an entire genome (Figure 1.7), itself the product of automation, which in 2021–2022 plateaued at a little under $1000, the number of uses for molecular pathology is increasing. This, and other developments, has brought a revolution in our view of health and disease, and as links between genes and disease can be determined with more precision, aberrant genes can be identified so that the individual can be counselled regarding their future. We will discuss these features in chapters to come.

> **Key Point 1.7** Advances in technology have driven down the cost of molecular analysis, permitting the investigation of many rare diseases not possible using older techniques.

With these and other advances, the demand from clinicians also increased as they appreciated the value of determining their patient's

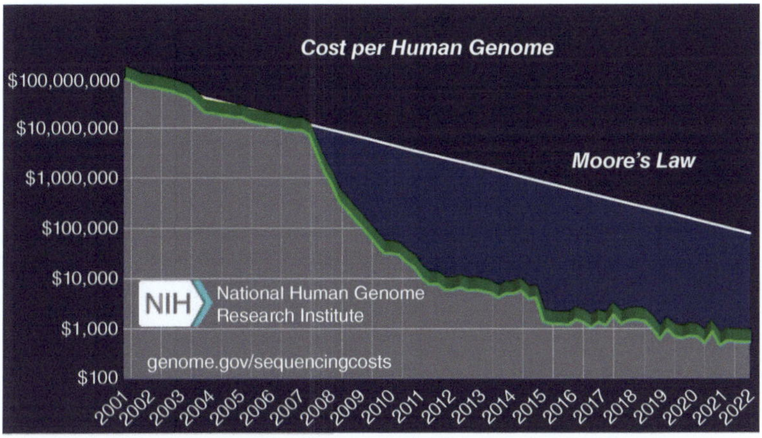

Figure 1.7 Falling cost of sequencing the human genome over time. *Source:* National Library of Medicine / Public domain/ https://www.genome.gov/about-genomics/fact-sheets/Sequencing-Human-Genome-cost.

molecular genetics. As this demand grew, the highly technical and specialised nature of the investigation could not be justified within a general hospital or even a university teaching hospital. Thus, the laboratory work became centralised in regional laboratories, then re-organised as Genomic Laboratory Hubs, focusing on work and expertise in dedicated labs, a far more efficient use of resources, both in analysers and scientists. In the UK, these centres of excellence were established in Birmingham, Cambridge, Manchester, Bristol, Newcastle, Aberdeen, Dundee, Glasgow and Cardiff, with three sites in Edinburgh and London (Great Ormond Street, Guy's and St Thomas's and St Bartholomew's Hospitals). Quality assurance of the work in these labs is overseen by the United Kingdom Accreditation Service. Devolution of health care to the UK's nations has been the development of guidelines regarding those genetic tests that are available (see further reading for websites). Of these guidelines, the most comprehensive is that of NHS in England, and accordingly, this resource will form a major part of this textbook.

Despite the high costs of complex analyses, the falling costs of certain of these, and the required analysers, alongside improvement in quality control, saw the marketing of simple kits for the most common clinical cases that could be performed outside the regional Genomic hubs, i.e. in General Hospitals. An example of this is the development of a simple method for determining SNPs in *EGFR*, coding the epidermal growth factor receptor, to help diagnose and manage certain lung cancers. Further advances include small machines to confirm or deny a bacterial or viral infection from a swab taken from a prospective patient. These analysers require minimal scientific expertise and will be discussed in Chapter 2.

Self-check 1.8 Why did it become necessary to establish regional centres for molecular pathology?

1.4.3 A Move Back to the General Hospital

The natural history of molecular pathology may be viewed in a number of stages.

- First was the transformation of certain aspects of research-based molecular genetics into hospital-based molecular pathology.
- As the clinical value of molecular pathology was developing, and becoming recognised, the techniques moved out of the research labs into those of routine pathology departments.
- However, the complexity of many of these analyses called for the establishment of dedicated centres of excellence – the Genetic Hubs described earlier.
- With the development of analysers and reagents in kit form focusing on specific and relatively simple investigations, and which called for a minimum of technical expertise, molecular pathology moved back into the general hospital.

This last step describes the typical development of a biomedical diagnostic process, moving from the research lab to routine practice. Many commercial suppliers offer kit packages containing all the required reagents where the rapid analysis of common genetic abnormalities and the presence/absence of certain pathogenic microorganisms can be undertaken. Many of these kits focus on a particular gene or disease, whilst other kits offer a panel allowing the parallel analysis of several genes, often in a semi-specific pathological area (Table 1.1). Further aspects of these focused panels will be detailed in chapters to come. Technology is advancing so rapidly that perhaps one day molecular pathology may be available in a point-of-care setting and perhaps in primary care.

Self-check 1.9 What are the key factors in moving molecular pathology to the General Hospital?

1.5 Conclusion and a View Forwards

In this chapter, we have explored several aspects of molecular pathology, beginning with its development from molecular genetics, and the nature of the diseases it can address. Table 1.5 offers a selected chronology of certain steps in both these disciplines, and as early as 1987, Weatherall had identified a wide variety of diseases that could be ascribed to a single gene defect (Table 1.6). Naturally, the current list would be considerably longer. A full understanding of the structure, classification and function of chromosomes is essential, as is knowledge of the various forms of both large-scale and small-scale abnormalities that bring about these diseases.

A notable statistic, 864, is the number of different genes that are referred to in this book. This number represents around 4.4% of the 19,785 protein-coding genes in our genome. Of those 12 genes cited most frequently in the text, the joint leaders are *TP53* and *BRAF*, followed in sequence by *KRAS*, jointly *NTRK1* and *NTRK3*, jointly by *ALK, BRCA1* and *BRCA2*, jointly by *ABL1* and *RUNX1* and then by *BCR, EGFR, BCOR, NRAS* and *RB1*. Of further note is the frequency of genes in a particular family. Of these, the most commonly described are those of *NTRK* (with some 120 referrals) and *MYC* (around 75), although these are both single referrals and those where the gene was fused to another.

Many genes are named according to their function – *SLC* (with 22 referrals) describes a solute carrier, such as of sodium, zinc or carnitine (such as *SLC4A1*), whilst *COL* (20 entries) prefixes genes linked to collagen formation (such as *COL3A1*), and *KCN* (14 entries) prefixes many genes coding for parts of a potassium channel (such as *KCNA2*). Of the 37 mitochondrial genes, prefixed by *MT*, 17 are described in this book. One hesitates to suggest that the increased frequencies of these genes imply some form of a rank order of a degree of importance for the practice of molecular pathology.

Molecular genetics and molecular pathology have now evolved to such a degree that separate (mostly) non-overlapping strands can be described.

Table 1.5 Selected chronology of molecular genetics and molecular pathology.

	Molecular genetics	Molecular pathology
Nineteenth century	Darwin, Mendel, nuclein, chromatin, chromosomes, mitosis and meiosis.	Scrotal cancer linked to paraffin and tar in chimney sweeps.
1900–1920	Genetics, gene, the maternal and paternal nature of chromosomes. Eye colour as a model of dominant and recessive gene inheritance.	A filterable factor causing cancer in chickens (later discovered to be the Rous sarcoma virus).
1921–1940	Chromosomes and heredity Molecular biology.	X-rays cause genetic mutations in *Drosophila.*
1941–1950	Genes and enzymes Genomics.	Xeroderma pigmentosa is hypothesised to be due to an abnormal gene.
1951–1955	Structure of DNA, ribosomal RNA and transfer RNA. DNA is the hereditable material.	Subtypes of sickle cell disease hypothesised to have a genetic basis.
1956–1960	Humans have 46 chromosomes. Microbial enzymes cutting DNA. Discovery of DNA polymerase.	Trisomy of chromosome 21 in Down syndrome. Loss of X chromosome in Turner syndrome. Philadelphia chromosome.
1961–1965	Discovery of mRNA and the triplet genetic code. Birth of genetic engineering. A tRNA sequenced.	The viral oncogene. Radiation causes chromosome damage.
1966–1970	RNA-dependent DNA polymerase. Wide spectrum of restriction endonucleases. In situ hybridisation with probes.	EBV causes lymphoma. Knudson's two-hit hypothesis of carcinogenesis.
1971–1975	Construction of an artificial plasmid. Construction of an artificial gene. Development of Southern blotting with a hybridising probe.	Carcinogens, such as benzopyrene, are mutagenic towards DNA. Retinoblastoma linked to chromosome 13.
1976–1980	*Thermus aquaticus* DNA polymerase. Sanger sequencing. Sequencing of a viral genome. Fluorescence in situ hybridisation (FISH). Direct transfer of cloned genes from bacterial to mammalian cells.	Tumour suppressor genes. Characterisation of *p53.* 9q + /22q-translocation causes CML EBV sequenced.

	Molecular genetics	Molecular pathology
1981–1985	PCR process optimised. Sequencing of a mitochondrial genome. Microarray technology developed. Establishment of bioinformatics.	The genetic basis for Lynch syndrome as a risk factor for colorectal cancer. Cystic fibrosis caused by lesions at 7q31.2.
1985–1990	Approval of a recombinant DNA Hepatitis B vaccine. Restriction fragment length polymorphisms. Development of multiplexing.	Early onset familial breast cancer linked to 17q21. Retinoblastoma gene *RB1* located at 13q14. *p53* named a tumour suppressor gene.
1991–1995	Sequencing of a bacterial genome. Start of the Human Genome Project. Establishment of biobanks.	Aneuploidy in breast cancer detected by FISH. Discovery of *BRCA1* at 17q21 and *BRCA2* at 13q12-13.
1996–2000	Optimisation of real-time PCR. Sequencing of a yeast genome. Sequencing of chromosome 22. Development of gene/DNA chips, pyrosequencing and nanopore sequencing.	The gene causing Chediak–Higashi syndrome mapped to 1q42-43. Imatinib licenced to treat *BCR-ABL1* positive leukaemias.
2001–2005	Sequencing of a mouse, a rat and a chimpanzee. Completion of the Human Genome Project. Development of multiplex ligation-dependent probe amplification. 'Next-generation sequencing (NGS)' described.	Progression of myelodysplasia linked to deletion in chromosome 1p. Osteoarthritis linked to lesions in chromosomes 4, 6 and 16.
2006–2010	Development of genome-wide association studies and of NGS RNA sequencing. Concept of genetic personalised medicine.	Mitochondrial DNA deletion syndromes mapped. Genetic basis of Asperger disorder.
2011–2014	Development of droplet digital PCR, and of sequencing by ligation.	Susceptibility to bipolar disorder linked to 7q34. Lesions in 2q24.2-24.3 linked to epilepsy.

Chronology deliberately censored 10 years prior to the publication of this text.

Table 1.6 Examples of molecular pathology of single genes.

Nature of gene abnormality	Linked disease
Single nucleotide polymorphism	α- and β-thalassaemia, sickle cell disease, glucose-6-phosphatase deficiency, α1-antitrypsin deficiency, hereditary amyloidosis, FH, coagulation factor VIII and IX deficiency.
Deletion	α-, β- and δβ-thalassaemia, growth hormone deficiency, anti-thrombin deficiency, elliptocytosis, Lesch–Nyhan syndrome, Duchenne muscular dystrophy, osteogenesis imperfecta, chorionic somatomammotropin deficiency, retinoblastoma, Wilms tumour.
Inversion	δβ-Thalassaemia.
Gene fusions	Haemoglobin variants, thalassaemia, red/green colour blindness.
RNA processing mutations	α- and β-thalassaemia, phenylketonuria, haemoglobin E, Christmas disease.

Source: Modified from Weatherall (1987).

Molecular genetics maintains its position as a leading research tool for dissecting how our bodies work, and so the consequences of errors (i.e. disease). This work feeds directly into the health care industry with its needs for tools in diagnosis and management, i.e. molecular pathology. In turn, this can be viewed as complex investigations (perhaps for a number of genetic analyses) as may be undertaken in a reference or Hub setting, and the move to simple investigations for a single gene in a routine NHS setting. The purpose of this book is to enable the reader to operate confidentially in any of these fields.

The purpose of this chapter is to equip the reader with a basic vocabulary of molecular pathology that will be needed in the chapters to follow. The next two chapters describe laboratory methods in molecular pathology.

Self-check 1.10 Confirm with yourself that you have completed the learning objective at the beginning of this chapter.

Bibliography

Akhoundova D, Feng FY, Pritchard CC, et al. Molecular genetics of prostate cancer and role of genomic testing. Surg Pathol Clin. 2022;15(4):617–628.

Domagala-Kulawik J. New frontiers for molecular pathology. Front Med. 2019;6:284. doi:10.3389/fmed.2019.00284.

Nurk S, Koren S, Rhie A, et al. The complete sequence of a human genome. Science. 2022;376(6588):44–53.

Pertea M, Shumate A, Pertea G, et al. Thousands of large-scale RNA sequencing experiments yield a comprehensive new human gene list and reveal extensive transcriptional noise. bioRxiv. https://doi.org/10.1101/332825.

Pujol N, Heeke S, Bontoux J, et al. Molecular profiling in non-squamous non-small cell lung carcinoma: Towards a switch to next-generation sequencing reflex testing. J Pers Med. 2022;12:1684.

Ren L, Yang Y, Li W, et al. Recent advances in epigenetic anticancer therapeutics and future perspectives. Front Genet. 2023;13:1085391. doi:10.3389/fgene.2022.1085391.

Weatherall DJ. Molecular pathology of single gene disorders. J Clin Pathol. 1987;40:959–970.

Websites

The National Institute for Health and Care Excellence: https://www.nice.org.uk/

Genomic directories of the four nations of the UK: https://www.england.nhs.uk/publication/national-genomic-test-directories/

https://www.nss.nhs.scot/publications/scottish-molecular-pathology-test-directory/

https://medicalgenomicswales.co.uk/index.php/download-services

https://belfasttrust.hscni.net/service/laboratory-services/clinical-genetics/requesting-genetic-testing/

2

Methods in Molecular Pathology 1

Basic Concepts and First-generation Techniques

Perhaps unsurprisingly, advances in the clinical aspects of the practice of molecular pathology have followed the development of new techniques. The latter have evolved from those based on cell biology, cytology and chemistry, to the use of electrophoresis and enzymology, chromatography and nucleotide chain sequencing. The almost simultaneous development of a number of novel and analyser-based techniques early in the 2000s, collectively described as 'next-generation sequencing' (NGS) was a major advance. The complexity and importance of these methods, and those that followed, 'third-generation sequencing', demand their own chapter, which will follow. Therefore, for convenience, these techniques in this section are described as 'first-generation techniques'.

Focusing initially on some clinical aspects, the opening section of this chapter on first-generation techniques will focus on the development of these developing analyses. Accordingly, we begin with the individual person under investigation, who is likely to become (if not already) a patient, and from whom a sample of tissue must be obtained. Sections that follow will consider the numerous forms of analysis of this sample, beginning in Section 2.1 with a consideration of the steps necessary to process the sample. This is often described as pre-analytical considerations and focuses on the nature of the clinical sample – in most cases blood or tissues from a particular organ.

Section 2.2 will cover cytogenetics and karyotyping, which focuses on detecting gross changes in chromosome number and structure. An extension of this is to use fluorescence in situ hybridisation (FISH) to

Molecular Pathology: A Primer for Laboratory Scientists, First Edition. Andrew Blann.

detect gross abnormalities in chromosome structure, such as deletions of part, or all, of a p or q arm, and other changes, as discussed in Section 2.3. The next level of analysis is to search for abnormalities in specific genes, which begins with the preparation of nucleic acids from a clinical sample in Section 2.4, followed by methods of sequencing sections of DNA in Section 2.5, and its development with microarrays in Section 2.6.

Despite the numerous platforms, there are certain instances where these broad sections overlap, and where this is relevant, they will be discussed.

Learning Outcomes

After studying this chapter, you should confidently be able to...

- Describe key aspects of sample acquisition and preparation
- Comment on the need for chromosomal analysis
- Explain how certain abnormalities can be determined by cytogenetics
- Understand the process of fluorescence in situ hybridisation
- Outline key steps in a polymerase chain reaction procedure
- Grasp the concepts underlying Sanger sequencing
- Describe the principles of microarray analysis

2.1 Sample Acquisition and Preparation

2.1.1 General Considerations

Almost all materials for clinical genetic analysis come from blood, bone marrow, tissues from specific organs, fine needle aspirates of soft tissues such as the thyroid, or cell samples from body cavities, such as the pleura or peritoneum. Analysis may also be performed on tissues that have already been processed for routine cell pathology (such as in a block of paraffin wax) or, in some circumstances, on other tissues that are dried onto a glass slide. An important area is the microbiological determination of an infection, such as by SARS-CoV-2, from a throat swab, or a swab from an infected wound or ulcer, where the presumed pathogen (if present) can be transferred from the swab to a liquid medium, which is then processed as if it were plasma or serum. A sputum sample may also be analysable.

A leading component of sample acquisition and analysis is quality assurance, the most important being the avoidance of contamination during collection and subsequent processing. All chromosome and gene work must ensure the integrity of the sample, as contamination with some normal tissues may bring a false negative. All laboratories will have their own specific processes, such as standard operating procedures (SOPs) in this respect.

An important consideration is the nature of the sample. It may be expected that the DNA within the cells of a tumour carries the pathogenic genetic abnormality causing the particular malignancy. However, this does not answer the question of whether the particular abnormal gene (or genes) is present in tissues that appear to be normal. Should a sample of phenotypically normal tissue (i.e. may be described as germline) carry the inactivated abnormality, then the same malformed gene in the paired tumour must have been activated by a process yet to be determined. Conversely, should the abnormal gene in a tumour sample be found to be normal within a sample of phenotypically normal tissue, it may be assumed that the pathogenic process operated on that localised tissue alone. This has implications for carcinogenesis, population screening and the future health of an individual.

2.1.2 Blood and Soft Tissue

By far, the most common soft tissue analyses are performed on blood, generally obtained through a standard venepuncture into a vacutainer with the anti-coagulant EDTA, which provides a germline sample (unless there is a blood cancer). A bone marrow aspirate, often requested in the investigation of blood cancer, will also be anticoagulated. A blood film will generally be made of the sample, and stained as per routine methods, for light microscopy, and a sample may also be analysed by fluorescence flow cytometry for standard markers of malignancy. The next step is to obtain mononuclear white blood cells (predominantly lymphocytes) by density centrifugation over a solution such as the Ficoll, a polysaccharide. The enriched leukocyte population can then be analysed by cytogenetics, FISH (elaborated in Section 2.3), and used as a source of nucleic acids. As discussed earlier, clinical samples for microbiological analysis are also in a fluid phase.

2.1.3 Tissues from Solid Organs

In most cases, 'fresh' tissues arrive in the laboratory, perhaps directly from the operating theatre. If so, it may be necessary to micro-dissect to remove other tissues (such as fat and connective tissues) to ensure, as much as possible, a high-quality sample. In certain circumstances, these tissues may be homogenised for analysis, which will therefore destroy any anatomy and process as if they were soft tissue samples.

Tissues may also be sent for analysis from the routine cell pathology department, in which case they will probably already be embedded in paraffin wax. As aforementioned, the tissues can be processed in the molecular pathology laboratory, but in either cases, a section of the tissue may be placed on a glass slide. If this is done, an appropriate section of tissue can be removed to provide an analysable sample, possibly for FISH studies, but certainly for the harvesting of nucleic acids. Figure 2.1 summarises these early steps in practical molecular pathology.

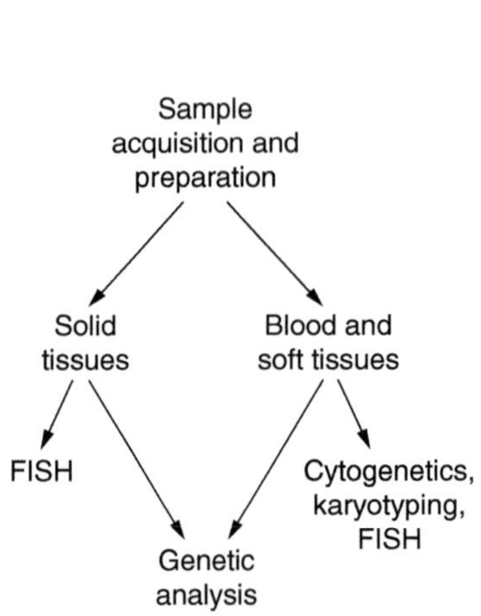

Figure 2.1 Early steps in practical molecular pathology. FISH, fluorescence in situ hybridisation.

Methodological Note The laboratory scientist must adhere to the local SOP, specific to their laboratory. It is most likely that such documents will vary in small degrees from laboratory to laboratory.

Self-check 2.1 What is the most important pre-analytical feature required in order to minimise false negative results?

2.2 Cytogenetics and Karyotyping

Strictly speaking, the descriptor cytogenetics, as applied to cytological analyses of the last century, is in almost all cases a misnomer as genes themselves were not (yet) analysable, although technical development in the last part of 20th century did indeed permit genetic testing, as described in Section 2.3.

The number and structure of chromosomes may be determined by cytogenetics and are best illustrated in a karyogram (Figure 1.2). As outlined in Chapter 1, the normal human diploid karyotype is 46XX or 46XY, formed from 22 pairs of autosomes and two sex chromosomes, this pattern being described as euploidy.

2.2.1 Practical Considerations

The earliest and most widely used technique for studying chromosome structure in a clinical setting was (and can still be today) to culture a suitable cell, often a leukocyte, with mitogens such as phytohemagglutinin or concanavalin A to induce that cell into mitosis. To the culture is then added colchicine, a chemical that inhibits microtubule formation so that the duplicated chromosomes align at the metaphase plate and remain there.

A small collection of cells, perhaps 50 μL, may then be dropped onto a glass slide, the impact of which will burst the cell and release the duplicated chromosomes. An alternative method lyses the cells in hypotonic potassium chloride. The slide can then be dried and stained with dyes so that it can be viewed by light microscopy (Figure 2.2). This provides the classic double chromosome shape, where two sister

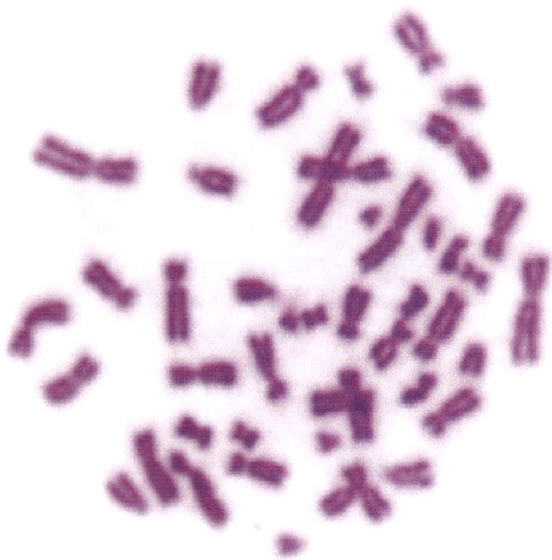

Figure 2.2 A metaphase spread. *Source:* Aldrieny EA, et al. J Microscopy Ultrastructure. 2013;1:89–95.

chromatids are joined at the centromere. A common error is to consider this 'X' shape to be that of a chromosome whilst in the nucleus, whereas within that organelle, chromosomes are packaged as a single chromatid strand.

Further processing can separate the sister chromatids, and so each individual chromosome can be visualised and collected together as a karyogram. Use of different dyes such as Giemsa and quinacrine can highlight light and dark bands in the chromosomes to aid identification, as discussed in Section 1.3.1 and shown in Figures 1.2 and 1.3.

Key Point 2.1 In many laboratories, the ability to generate and comment on a karyogram is an essential skill.

2.2.2 Historical Perspective

Hsu, in a 1954 publication, used an in situ hybridisation method to show extensive fragmentation of chromosomes in a carcinoma, whilst Bender showed that X-irradiation of epithelial cells *in vitro* causes breaks in the chromosome. In 1959, two key papers using this method were published. Lejeune and colleagues reported a series of nine children with Down syndrome whose karyotype included three copies of chromosome 21 (hence trisomy), whilst Ford et al. reported a case study of a girl with Turner syndrome whose karyotype had only one X chromosome (hence monosomy), thus opening the field of clinical cytogenetics. Within a few years, cytogenetics had demonstrated additional trisomy abnormalities, deletions and a ring chromosome.

Subsequent analyses in diseases such as lymphoma, leukaemia, retinoblastoma, Waldenstrom macroglobulinaemia and cervical cancer reported hyperploidy, further instances of monosomy and the loss of one of the arms of a chromosome (a deletion – subsequently found to be common in numerous cancers). In 1969, Buchanon and Croft reported a case of Down syndrome and acute leukaemia with numerous chromosomal abnormalities, an early finding of a later-established firm link between the syndrome and the malignancy.

As introduced in Chapter 1, a major and widely significant finding was reported by Nowell and Hungerford in 1960 of an abnormal chromosome in cases of chronic myelocytic leukaemia, subsequently named the Philadelphia chromosome (Figure 2.3). In 1971, Rowley reported the use of quinacrine fluorescence to show the loss of the Y chromosome in three cases of myelodysplasia, and later used the same technique to show the Philadelphia chromosome to be a translocation between chromosomes 9 and 22 (i.e. t(9;22); Figure 2.3). By the mid-1980s, radiolabelled probes and hybridising Southern blotting were used to demonstrate the

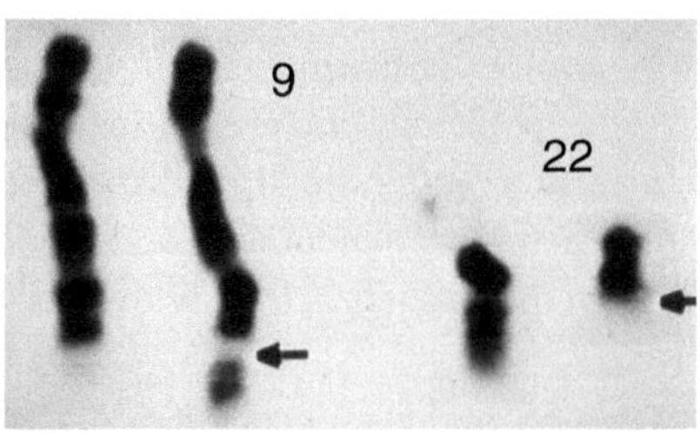

Figure 2.3 The Philadelphia chromosome. From left to right: a normal chromosome 9, an extended chromosome 9 with a translocated part of chromosome 22, a normal chromosome 22 and the remainder of chromosome 22 where a section has translocated to chromosome 9. *Source:* https://www.nature.com/scitable/content/the-philadelphia-chromosome-25070/.

amplification of oncogenes such as *c-myc* and *KRAS* (described by some as *c-Myc* and *k-Ras*, respectively) in lung and other cancers.

2.2.3 Modern Clinical Applications

Whilst the Philadelphia chromosome can be detected by karyotyping, this method has been superseded by other techniques, principally FISH, which is discussed in the following sections. Nevertheless, conventional cytogenetic analysis remains of value in certain settings, although it does demand a high degree of expertise in interpretation of the karyogram.

Aneupoidy

Determination of an abnormal chromosome component can come in several forms of analysis – for complex karyotypes, hyperdiploidy, a monosomal karyotype, near haploidy, low hypoploidy and common aneuploidy testing. As described in Section 1.3.2, by far the most common hypoploidy is Turner syndrome (XO), another being monosomy of chromosome 5. However, there are numerous forms of trisomy, the most common being Down syndrome (an extra chromosome 21), and others being Klinefelter syndrome (XXY), Jacobs syndrome (XYY) and trisomy of chromosome 8, chromosome 13 (Patau syndrome), chromosome 18 (Edwards syndrome) and chromosome 9. More recently, software can determine common abnormalities in a karyogram with sensitivities and specificities that are close to 100%.

Deletions

Cytogenetics can also demonstrate the loss of part of, or all, one of the arms of a chromosome, i.e. a deletion, or the need to investigate specific sections within a certain chromosome. The principal chromosomal lesions of this nature, and its particular pathology, are shown in Table 2.1. However, in many cases, the ability to detect often small abnormalities calls on considerable skill, so for more assurance, many of these deletions are detected by more complex genetic methods, such as NGS as discussed in Section 2.5.

> **Key Point 2.2** Although many aspects of cytogenetic karyotyping and the assessment of chromosome integrity have been replaced by genetic testing, it remains an important fundamental tool of the molecular pathologist.

Table 2.1 Common cytogenetic deletion investigations.

Chromosome (arm)	Pathology
1p36	1p36 deletion syndrome: can be variable with 1.3–30 mbps lost, and may cause a severe multisystem disorder.
1p/19q	The co-deletion of parts of chromosome 1 and chromosome 19 is found in a small number of gliomas.
5q	Deletion from q33 is linked to myelodysplasia and potentially a leukaemia.
6p	Deletion from p24-25 brings many congenital abnormalities, whilst schizophrenia may be linked to abnormalities in p24-22.
7q	Deletion from 7q is present in certain lymphomas and is a cause of Williams syndrome.
9q34	Deletion causes Kleefstra syndrome, a very rare condition with features such as autism and poor psychological development.
11q23	Deletion causes Jacobsen syndrome; severity depends on the precise deletions but often includes intellectual disability, and immunodeficiency.
12p	There can be various deletions, e.g. from p1-11, p11-13, p13 and the distal zone of 12p. Features include autism, and abnormalities of the fingers and toes.
13q	Several deletions are known, e.g. the centromere to 13q32 and from 13q32. Characteristics include intellectual disability, low birth weight and cleft palate.
17p11	Deletion of ~3.7 Mb from p11.2 causes the Smith–Magenis syndrome, whilst duplication of this section leads to Potocki–Lupski syndrome.
17q12	Microdeletion of this section links to development problems with the urinary system, and the potential to develop a form of diabetes.
19p13	Features of this deletion include a tall stature, intellectual disability and an unusually large head.
22q11	A deletion from q11.2 leads to DiGeorge syndrome.
22q13	Deletion of this section leads to a variety of pathological features, dependent on the exact nature of the lesion, which can vary from 130 000 bases to 9 mbps.

This list does not intend to be exhaustive but is illustrative of which analyses are considered common and/or important. Mbps, million base pairs.

Self-check 2.2 Describe the importance of the Philadelphia chromosome in molecular pathology.

2.2.4 Paediatrics and Cytogenetics

A major part of the specialism of paediatrics is of malignancy and congenital disease, and as such whilst much of the previous text has applied to those of all ages, special mention of this discipline is called for. This is evidenced by the fact that there are many specialist hospitals addressing only on diseases of children, and so their laboratories will also focus on these conditions.

In many cases, cytogenetics is called upon to confirm a preliminary diagnosis based on signs and symptoms. Some such diagnoses will already be relatively firm, such as Down syndrome, which may not even need cytogenetics, although a differential diagnosis is Edwards syndrome, with which it shares several features. Numerous congenital malformations, such as in Turner syndrome, may be noted soon after birth, and clinical cytogenetics may not immediately be called for. But the precise nature of the lesion, such as loss of a complete X chromosome, or a deletion of a section of chromatin, may be helpful. However, in many cases, diagnosis in cases of non-specific features such as failure to thrive, a depressed growth curve and altered mental development may require cytogenetics. In other cases, other features may appear (or fail to appear), during or after puberty, such as in Klinefelter syndrome, a diagnosis that cytogenetics will confirm or refute.

Further details of paediatrics are presented in Chapters 4, 6 and 7.

2.3 Fluorescence In situ Hybridisation

Classical microscope-based cytogenetics can determine only the most gross changes to chromosome structure, being unable to determine microdeletions and other minor changes. However, the technique can be extended with the use of FISH, where fluorochromes are linked to section-specific and gene-specific probes. It can be used on metaphase spreads (as aforementioned) or on whole, intact cells during interphase.

2.3.1 The Evolution of FISH

The 1970s also saw the development of this further key technique. Watkins used a complementary RNA from the virus SV40, labelled with radioactive tritium, to hybridise with, and thus show the presence of, viral DNA in kidney cells that had been previously infected with the virus. In 1977, Rigby and colleagues published the nick-translation technique for generating radiolabeled DNA probes, whilst Cheung, Tishler and colleagues refined the method by the binding of a fluorescent microsphere to a strand of a mRNA probe to show the localisation of specific complementary haemoglobin α- and β-genes in chromosomes 4 and 5 (subsequently shown to be incorrect – the genes lie on chromosomes 16 and 11, respectively) – hence the birth of FISH.

The advances reported by Singer and Ward, and by Angerer and Angerer, and others, in the early eighties were to detect mRNA using radiolabeled and biotinated polynucleotide probes. Whilst radiolabeled probes were still in use in the early-mid 1980s, these ultimately gave way to fluorochromes, although in 1984 Renz and Kurz reported a colourimetric method using enzymes. In 1986, Pinkel and colleagues used a Y-specific probe linked to fluorescein-labelled avidin and biotin to show the presence of two Y chromosomes in metaphase spreads and whole cells derived from the amniotic fluid, and in 1988, they used FISH to detect trisomy 21 and translocations of chromosome 4. In the same year, Leonard and colleagues used high-resolution fluorescence-activated chromosome sorting and in situ hybridising complementary DNA probes to show that the interleukin-2 receptor gene is to be found at 10p14-15. A year later, Hopman and colleagues reported a method for the simultaneous detection of two DNA target sequences using fluorescence and enzymology. The ability to investigate more than one abnormality at a time is described as multiplexing, a technique to be widely used in gene analysis, as described in Section 2.4.5. At the end of the eighties, Nederlof and colleagues reported simultaneous three-colour (red, blue and green fluorochromes, each detectable at different wavelengths) FISH linked to probes specific for chromosomes 1, 7 and 15.

Comparative Genomic Hybridisation

By the 1990s, advances in FISH technology led to the development of this method that could determine the presence of multiple copies of the same sequence of DNA, i.e. copy number variation (CNV), in a sample without the need for cell culture. Typically, tumour DNA labelled with one fluorochrome and normal DNA labelled with another is applied to a metaphase spread, and difference in binding is noted. Variants included the detection of complete and partial chromosome gains and losses, extra-chromosomal gene amplification, and the detection of abnormalities in formalin-fixed paraffin-embedded solid tumours. Lawrence et al. used comparative genomic hybridisation (CGH) to detect HIV integrated into the genome, whilst others used it to define the carrier status of female carriers of the deletion in Xp21 that leads to Duchenne–Becker muscular dystrophy; and to demonstrate silencing translocations and deletion in 17p13.3 that lead to the Miller–Dieker syndrome. Reid et al. used the method to report a multicolour FISH for the simultaneous detection of five chromosomes, and Kallioniemi et al. used it to show that two-thirds of primary breast cancers had an increase in CNVs mostly in 17q22-24 and 20q13. The general concept of CGH has been expanded into other analyses and platforms, as we will see in the chapters and sections to come.

With these and other technical developments, numerous variants of FISH appeared, and many are in use in the present setting to detect a wide range of genetic abnormalities. Figure 2.4 illustrates the principles of FISH and some common applications with chromosomal lesions.

2.3.2 Uses of FISH Variants in Pathology

Translocations

The principles of translocations are described in Section 1.3.3 and Figure 1.4. Probably the most common variants of FISH are dual-coloured, where one probe is linked to a fluorochrome giving a red colour, and the second probe is linked to a fluorochrome giving a blue colour. In one variant, probes are complementary to regions of DNA that include genes of interest, such as A and B. In a normal setting, the resultant picture of the sample will see the two sets of each colours far apart: four separate sets of colour will be seen as each gene is present on two chromosomes,

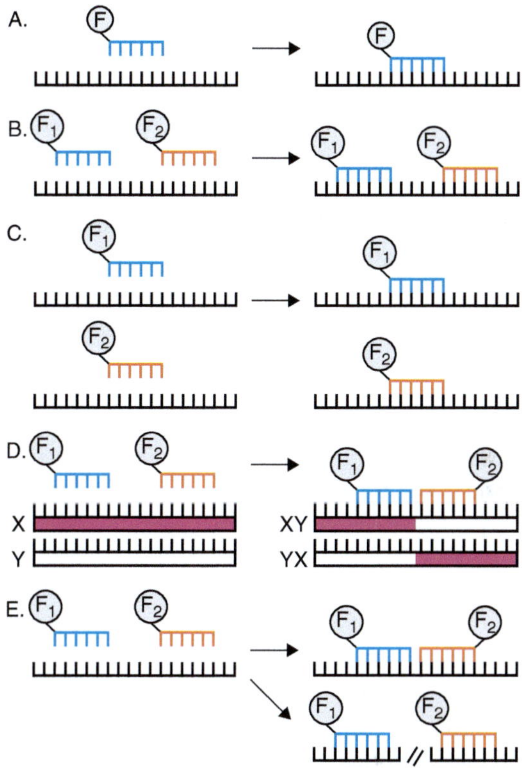

Figure 2.4 Variants of fluorescence in situ hybridisation. (A) illustrates the basic principle of FISH, whereby a nucleic acid probe (typically 20–50 nucleotides linking to a 40–50 base pair section of the sequence) linked to a fluorochrome (F) locates its complementary nucleic acid strand in the sample. (B) shows two different probes (F_1, F_2) designed to match with different sections of nucleic acids on the sample. The probes may be on different chromosomes within the sample (C) and so may be identified in separate parts of the slide. An extension of the latter is illustrated in (D), where a translocation had brought together two genes that would normally be far apart, perhaps on different chromosomes; here, the translocated gene can be identified by the two colours being adjacent to each other. The modification (E) of this dual FISH sees probes for different parts of the same gene that in health will co-locate (as in E1), but in translocation and certain other lesions, the gene breaks apart, so that the colours will appear distant from each other (E2).

i.e. A, A, B, B. However, should there be a translocation bringing together the genes of interest, the two spots of colour will be adjacent, i.e. A, AB and B. This model is often described as dual fusion FISH. An alternative method is to probe only for the hybrid gene (AB) formed from the two individual genes, thus requiring only one large probe that is complementary of a part of A and a part of B.

In a break-apart FISH, two probes that are complementary to either side of the gene in question (such as the 3' end and the 5' end) will co-locate if that gene is intact. However, should the gene be broken apart, as often in a translocation, the two spots of distinct colours will be separate.

Table 2.2 shows a selected collection of translocations, the genes involved and a related pathology that can be determined by FISH, whilst Figure 2.5 shows FISH in the context of a translocation. The role of this powerful method in cancer will be developed in Chapters 3, 4 and 5.

Deletions

The diploid nature of the genome includes two copies of each chromosome and thus two copies of each gene. A common genetic lesion in this respect is loss of heterozygosity (LOH), where instead of two copies of the same gene being present, one on each homologous chromosome, only one is detected, implying a deletion in the other. However, in many cases, this LOH deletion may not be pathogenic should the gene of the sister chromosome be functional, although this is not always the case. An example of this is in Williams syndrome, the critical gene being *ELN* at 7q11.23, coding for elastin. Using a probe to this region, the finding of only one of the two fluorochrome signals to be expected implies the LOH

Table 2.2 Selected translocations, their genes and linked pathology.

Translocation	Genes	Linked pathology
t(11;14)(q13;q32)	*CCND1::IGH*	Mantle cell lymphoma
t(2;3)(q13;p25)	*PAX8::PARγ1*	Follicular thyroid cancer
t(12;16)(q13;p11)	*DDIT3::FUS*	Myxoid liposarcoma
t(9;22)(q34;q11)	*BCR::ABL1*	Chronic myeloid leukaemia
t(11;22)(q24;q11)	*FLI1::EWSR1*	Ewing sarcoma

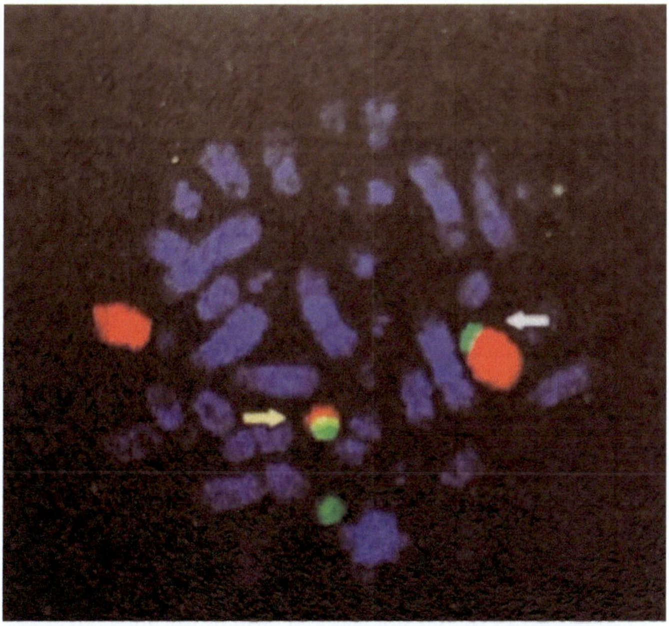

Figure 2.5 FISH and a translocation. Red and green fluorochromes are linked to probes for different genes. Normally, the genes binding the two probes would be on different chromosomes. However, the arrows highlight two chromosomes with both colours, thus defining a reciprocal translocation. Some of the 'red' genes have moved to the 'green' chromosome, and vice versa. *Source:* Hoffbrand et al. (2011). Figure 11.12C, p.161, with permission from John Wiley & Sons, Ltd.

deletion in a section of around 27 genes, thereby making the diagnosis. It follows that the normal functioning of the second copy of *ELN* is not enough to prevent the disease.

Taking the 19p13 deletion syndrome as a further example, FISH and subsequent sequencing have identified a large number of deletions, ranging from micro-deletion of some 700 kbps, to extensive loss of some 3000 kbps, each leading to their own particular phenotype. p13 has 5 bands, named from the centromeres as 13.11, 13.12, 13.13, 13.2 and 13.3, which together host around 935 genes. *COMP*, at p13.11, codes for cartilage oligomeric matrix protein: deletions in this gene lead to pseudoachondroplasia and multiple epiphyseal dysplasia, a form of connective tissue disease.

GCDH, at 19p13.13, codes for glutaryl-CoA dehydrogenase: an 18 bp deletion coding for six amino acids leads to glutaric aciduria type 1, and the congenital sign of macrocephaly, with the subsequent development of neurological disease. Larger deletions in 19p13.3, of around 290 kbps, can see the loss of several genes (such as *APC2, PLK5* and *MBD3*), which results in complex phenotypes with features such as attention deficit, speech delay and hypotonia.

Deletions and Tumour Suppressors

Experimental *in vitro* cell biology in the 1980s showed that phenotypically malignant cells could be 'rescued', and their normal phenotype restored by the transfer of certain chromosomes. Other data showed that loss of only certain exact sections of a chromosome was linked to malignancy. This process led to the concept of 'tumour suppressor genes' (amply demonstrated at present in a clinical setting), by the use of a radiolabeled *ras* probe, of the partial deletion of 11p in breast cancer that correlated with the size of the primary tumour.

Other leading tumour suppressor genes detected by FISH include *RB1* at 13q14.2, coding for the retinoblastoma protein Rb, and *p53* (or perhaps *TP53*), at 17p13.1, coding the protein p53. Loss of these genes caused by a deletion, therefore, leads to a failure of production of their protein products, which would normally suppress tumour growth. Hence, the deletion of these tumour suppressor loci (or, indeed, any such gene) is linked to the development of cancer. Later, it was shown that microdeletions or SNPs in tumour suppressor genes, leading to loss of function, would also lead to malignancy.

Key Point 2.3 The concept of a tumour suppressor gene was a major breakthrough in our understanding of cancer. The crucial importance of this discovery is discussed in more detail in Chapter 3 and will be re-examined many more times throughout this volume.

Other Genetic Lesions

Use of FISH is helpful in determining lesions such as inv(2)(p21p23), involving *EML4* and *ALK*, present in 4–8% of non-small cell lung cancers. Similarly, amplifications can be demonstrated by multiple instances of the binding of a section-specific FISH probe (Figure 2.6).

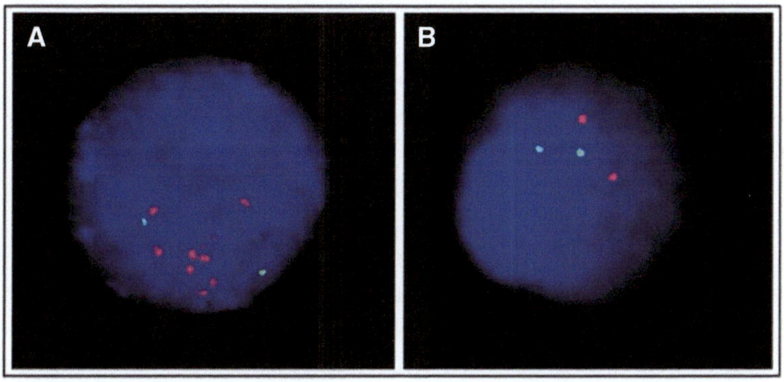

Figure 2.6 Gene amplification detected by FISH. An interphase spread probed by FISH for *RUNX1*, located at 21q22.12. (A) Eight pink dots are present on the patient's sample, proving amplification, and (B) only two on the healthy control. The blue dots are a control probe. *Source:* Garcia et al. (2013).

In certain circumstances, this can be described as a form of duplication, a phenomenon first described using FISH, where 16–50 copies of the oncogene *myc* were described in cancer cell lines. Duplication of other oncogenes has since been described in breast cancer (e.g. *FGFR1*), colorectal cancer (*KRAS*), ovarian cancer (*AKT2*) and many other malignancies. A FISH-defined duplication in sections of 19p13.3 has been linked to male infertility, and FISH was used to show human papilloma virus DNA inserted into 8q24.1 in a small cell carcinoma of the cervix. In a clinical setting, gene duplications such as these are now often identified by NGS methods, as discussed in Chapter 3.

2.3.3 Developments in FISH

As described earlier, CGH emerged in the 1990s and may be used to determine the frequency of the CNV of certain sections of DNA in different samples. The method was originally developed to detect regions of loss or gain of DNA sequences between a presumed abnormal sample and a known normal sample, but it has since evolved into other formats, such as microarray. At the chromosomal level, it can observe amplifications, duplications and deletions. Examples of this include prenatal diagnosis of a trisomy, and in a study of monosomy of chromosome 3 in melanoma, 57% of patients with this lesion suffered with metastatic disease.

Subsequent developments include that of a microarray, as discussed in Section 2.6, and are extended for use in a clinical setting. For example, the importance of CNV may be illustrated by Charcot–Marie–Tooth disease type 1A, where in early research, a dual-colour FISH reported a duplication in 17p11.2 that spanned at least 1100 kbps, the responsible gene being *PMP22*. Subsequent work on over 10,000 individuals with peripheral neuropathy showed around 80% of diagnostic alleles to be CNVs at this locus. Other developments include a single-nucleus analysis pipeline for multiplexed DNA FISH (hence SnapFISH), which can be used to investigate chromatin loops, disruptions that have been linked to a variety of malignant, developmental and psychiatric conditions.

Key Point 2.4 FISH is a very important tool for detecting a number of chromosomal and genetic abnormalities in a wide variety of malignant and non-malignant diseases, although many such abnormalities can also be detected by other methods such as NGS.

Self-check 2.3 Describe the basis of the two-colour FISH for the investigation of a potential translocation.

2.4 Preparation of Nucleic Acids for Analysis

We now move from the analysis of chromosomes and substantial changes to genes, to micro-changes within genes. To do this, a sample of nucleic acid is required.

2.4.1 Obtaining DNA

DNA can be obtained directly from the clinical material, or from a sample of RNA, as is described in Section 2.4.2. If using clinical material (as opposed to a purified collection of cells from tissue culture) to obtain a sample of DNA, step one is to clarify and concentrate the test material, as described in Section 2.1. Step two is to part-digest the sample with a detergent such as Triton X-100 or sodium dodecyl sulphate, followed by use of an enzyme, such as proteinase K, to remove the proteins and thus release the DNA from the chromatin. In step three, the DNA is purified by one of several methods, although two dominate: a solid phase extraction and a precipitation extraction.

- Solid phase extraction involves passing the DNA solution over a solid matrix such as silica, often in a column, followed by washing with a neutral buffer, after which the DNA is removed with an eluting buffer.
- Unwanted proteins can be precipitated out of the reaction solution with a mixture of phenol and chloroform, and the precipitate removed by centrifugation. An alternative to these potentially hazardous solvents is to use a saturated sodium chloride solution. Once free of contaminant protein residues, the DNA can then be precipitated out of solution in ethanol.

Quality control of the purification can be determined using an ultraviolet spectrophotometer by checking absorbance at 260 and 280 nm; a ratio of 1.7 to 1.9 is considered acceptable. However, in more complex analysis, or where the mass of nucleic acid is small, this may be unsuitable. In such cases, an alternative method uses a fluorescent dye specific for nucleic acid, such as ethidium bromide or SYBR Green, to be added to the sample, and the degree of fluorescence quantified. Commercial kits are available, which provide all the necessary reagents, such as those produced by the biotech company Qiagen as their DNeasy kit. Once obtained, 'raw' DNA must be stored at $-20^{\circ}C$ to $-80^{\circ}C$.

Chromatin Immunoprecipitation

This procedure is useful in determining those proteins, such as transcription and other regulatory factors, that are associated with a particular section of DNA. It therefore calls for a specific antibody for the protein in question. Once immunoprecipitated, the proteins can be stripped from the DNA so that the latter can be sequenced. Examples include sections binding to histone H3, *c-Myc*, β-catenin, *RUNX3* and *GATA3*.

2.4.2 Obtaining RNA

In several circumstances, a sample of RNA may be called for, such as for direct analysis or to generate DNA, or in a clinical setting to demonstrate an abnormal gene such as BCR::ABL1. Cytoplasmic RNA is an important product, as it proves the effectiveness of the transcription process, the sum of all RNA in a cell being the transcriptome. RNA may also be obtained from plasma or serum, where it is found contained in protein complexes or extracellular vesicles (e.g. exosomes) from which it must be released, such as by proteinase digestion or organic extraction.

The process for obtaining RNA has much in common with that for DNA, but a specific problem is the presence of naturally occurring enzymes that denature RNA (i.e. RNases). A further issue is the heterogeneity of this cellular RNA, estimated at around 20–40 pg, of which 0.5–1.0 pg is mRNA, the remainder being tRNA, rRNA and ncRNA (which include miRNAs, a species that can also be analysed by these techniques). A single mammalian cell contains typically 300,000 individual mRNA molecules, but of course this number can fluctuate widely. A further requirement is to eliminate any DNA in the sample with DNAses.

As mRNA has a tail of many adenine molecules (hence poly-A), it can be purified by exploiting this feature, binding to a strand of the oligonucleotide (oligo = a few, often shortened, for convenience, to oligo) poly-thymine. The latter can be linked to a solid phase, such as cellulose beads, often in a mini-affinity column. Purification of other forms of RNA, which lack the poly-A tail, is more complex and relies on processes such as size selection and precipitation. Commercial kits are available for the easy preparation of RNA, which generally include a mini-column, RNase-free DNA, carrier RNA, and RNase-free reagents and buffers, examples being Qiagen's RNeasy kit and ThermoFisher's MagMAX kit.

With the exception of the ribosome, RNA does not generally associate with other intracellular proteins. Accordingly, proteinase treatment may be called for, although a useful step is a guanidinium thiocyanate solution, which will also precipitate the DNA present. Once the unwanted material is removed, purified RNA can again be precipitated in ethanol. Storage is as for DNA. Small RNAs, such as miRNAs, can be obtained by size separation in a gel electrophoresis, where the RNAs are eluted from the gel. An alternative method is to graft a poly-A tail onto the miRNA with a ligase, and then to proceed as for mRNA. Kits that provide reagents for synthesis of a complementary DNA (cDNA) from miRNA are also available.

Once purified, RNA can be the substrate for the generation of cDNA, as DNAs are generally required for automated gene analysis. As mRNAs have a poly-A tail, a poly-thymine primer can be used to start the generation of the cDNA, given an RNA polymerase (a reverse transcriptase), primers and nucleotides. This will provide a single strand of DNA that can be the basis of a polymerase chain reaction (PCR; see Section 2.4.4) to generate double-stranded copies. In a modification of this process, a fluorochrome can be added to the reaction mixture, so the cDNA product can be identified by its fluorescence and potentially quantified. A typical use of cDNA is in a microarray, as explained in Section 2.6.

The upscaling of RNA to DNA sequencing was developed on an industrial scale by the NGS method of RNA-Seq, described in Section 3.1.6.

2.4.3 The Library

There are various methods for collecting sections, or fragments, of DNA, from which extracts can be obtained and analysed.

The Genomic Library

In the early years of molecular genetics, analysis of the sheer mass of DNA in a nucleus demanded radical new techniques, one being the development of DNA libraries using methods borrowed from microbiology. This required a series of complex steps, where DNA was non-specifically broken down into sections of varying lengths, a common method being digestion with restriction endonucleases, another being sonication. This has been described as a 'shotgun' approach, as fragments are heterologous and unselected. These fragments were then (generally) inserted into plasmid vectors, which themselves were transfected into a bacterium where they proliferated clonally. Ideally, each bacterial colony, of which there would be thousands, would host a clone of a single plasmid and thus would generate multiple copies of the same single section of DNA. An alternative is to bypass the viral stage and insert the DNA directly into bacterial chromosomes, which later developed into bacterial artificial chromosomes, of perhaps 300,000 base pairs, being independent of the bacterial chromosome.

The next step involved transferring a representative sample of each bacterial colony onto a solid substratum such as nitrocellulose where they could be interrogated with a specific probe. This allows the identification of those colonies hosting a plasmid binding to the hybridising DNA to then be harvested and the DNA collected for analysis. A library is thus created from the product of all the cloned DNA and is described as genomic as it represents all the DNA in the nucleus, regardless of whether it transcribes (i.e. is an exon) or does not transcribe (i.e. is a non-coding intron).

The Exome Library

Just as an exon is that part of a section of DNA that can be transcribed as mRNA, the exome becomes the sum of all DNA sequences that code for mRNA. This therefore differs from the transcriptome, which is the mRNA itself. The advantage of the exome is that it, by definition, has the potential to translate to proteins, and that its preparation does not call for manipulation of the nucleus. A further expression of note in this respect is therefore 'whole exome sequencing' (WES), which is distinct from 'whole genome sequencing' (WGS).

The cDNA Library

An alternative form of a library is that described earlier, i.e. of cDNA that has been generated from mRNA. This therefore differs from a genomic library as it has transcribed DNA from which introns have been removed. Thus, a cDNA library will contain a far smaller collection of DNA sections.

Sequencing a Library

Current routine practice in molecular diagnosis allows DNA in both types of library to be sequenced by NGS methods, which is to be discussed in Chapter 3. However, before we progress further, we must first understand some basic principles of molecular genetics.

2.4.4 Amplification of DNA: The Polymerase Chain Reaction

It is probable that the steps described earlier will not provide sufficient material for a robust analysis, and accordingly, the mass of nucleic acids must be increased. This is achieved by the revolutionary process of the PCR, the precise procedure being freely available from many sites online. A very simplified protocol for a PCR reaction (which will vary according to circumstances) is as follows:

1. Creation of a cocktail (or master mix) of the required constituents in a small plastic vial:
    a. A Taq polymerase (originally from the bacterium *Thermus aquaticus*) to generate the copies of the required sample DNA.
    b. Two DNA primers (forward and reverse, perhaps 20 base pairs long) for specific sections of the nucleotide sequence of the

DNA to be amplified (and which therefore demands knowledge of that nucleotide sequence).

   c. The sample of single-stranded DNA from the subject that is to be copied.
   d. Building blocks of the four deoxynucleotides (A, T, G and C, although some protocols use uracil in place of thymine) as triphosphates.
   e. An appropriate reaction mixture with magnesium chloride, buffer/diluent (in some cases, distilled water), all of which must be free of enzymes capable of digesting DNA/RNA. Some protocols call for additives such as formamide, dimethyl sulphoxide, ammonium sulphate and detergents.

2. The amplification itself, that proceeds in a device called thermal cycler, into which the particular temperatures and times for each step are programmed, such as
   a. 95 °C for denaturation of the sample DNA;
   b. 55–65 °C for the annealing of the primers to the sample DNA;
   c. 72 °C for the extension steps where the nucleotides build the new DNA; and
   d. the number of cycles (i.e. steps a–c), typically 25–30 or so, which can generate a billion or more copies, depending on the size of the sample DNA to be copied.
3. The reaction vial is placed in the cycler and set for the programme to run.
4. Once complete, quality control is performed on the resultant solution (e.g. by gel electrophoresis). If not used immediately in further analyses, the amplified nucleic acid must be stored, perhaps in aliquots, at −70 °C.

Sections of DNA or RNA that are the source and/or products of amplification are referred to as amplicons. Production of DNA or RNA is summarised in Figure 2.7, and although RNA is now rarely analysed directly, it remains a useful research tool.

During the 1980s, the industry launched machines to rapidly synthesise artificial genes, probes and primers, with comprehensive reagents kits, and the first automatic sequencing machine based on capillary electrophoresis. With updates, these advances opened the door to additional technologies that would lead to uses in clinical practice.

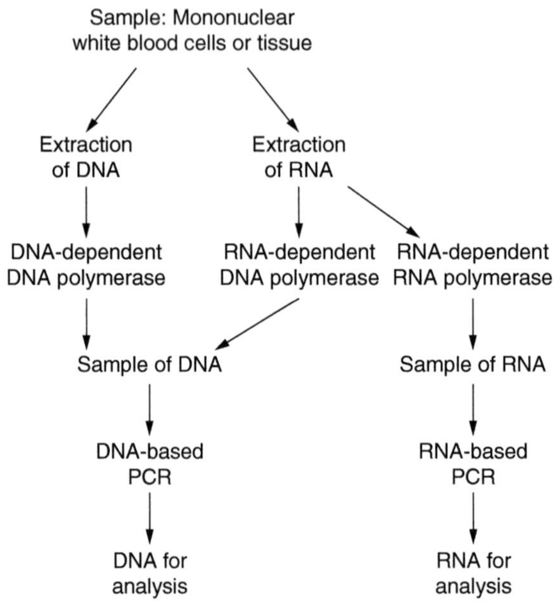

Figure 2.7 Production of DNA and RNA.

Self-check 2.4 What is the key difference between a genomic DNA library and a cDNA library?

Key Point 2.5 The PCR is a crucial process in molecular genetics, for which its inventor, Kary Mullis, was awarded a Nobel Prize.

2.4.5 Developments in PCR

Unsurprisingly, there are numerous variants of this standard original PCR methodology. What follows is a selection.

Real-time PCR

The 'standard' PCR, as developed, was a perfectly good method of sequencing a chain of DNA and, although slow and labour-intensive, was, in its time, an excellent research tool. However, technological advances saw the introduction of more rapid, automated analysers that proved very popular in clinical settings. Thermal cyclers were soon adapted to show how the PCR reaction is progressing cycle by cycle, hence real-time PCR (rtPCR: not to be confused with reverse transcriptase, RT-PCR), also described as quantitative PCR (qPCR).

Quantification of the reaction can be assessed by increasing levels of a fluorescent marker in the reaction vessel. One approach is to include a fluorescent dye such as SYBR Green, which intercalates into double-stranded DNA, such that increasing signal strength implies an increasing mass of DNA. An alternative approach is to use a TaqMan© probe, formed from a probe that has a fluorescence marker at one end and a quencher molecule at the other, the latter required to literally quench a nearby fluorescence signal by a process called fluorescence resonance energy transfer (Figure 2.8). Therefore, in both cases, the increase in

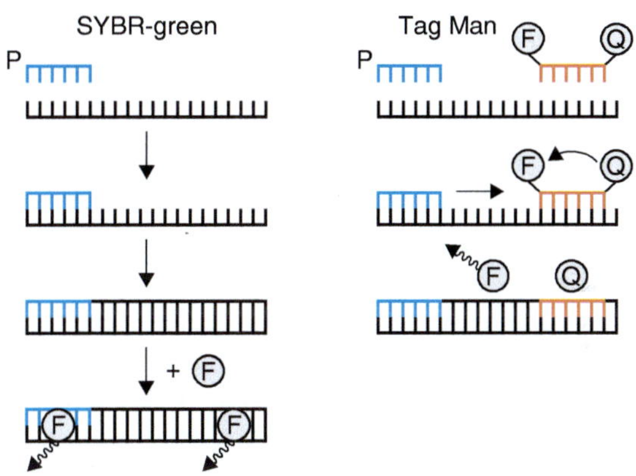

Figure 2.8 Quantifying DNA amplification. On the left, the SYBR Green method proceeds as a standard PCR, where a primer (P) binds ssDNA and initiates the formation of a complementary DNA (cDNA) strand by the polymerase, thereby generating a dsDNA. The fluorochrome SYBR Green (F) will then intercalate into the DNA and emit light when stimulated. On the right, the TaqMan probe, with a fluorochrome at the 5' end and a quencher (Q) at the 3' end, binds a part of the ssDNA remote from the primer, which also initiates formation of a complementary strand. The close proximity of the quencher to the fluorochrome ensures the latter does not emit a signal when stimulated. The polymerase will cleave the fluorochrome from the probe, and when liberated from the effect of effects of the quencher, the fluorochrome will emit a signal when stimulated.

fluorescence is a surrogate marker of the mass of DNA in the reaction vessel, generally quantified by indices such as the cycle threshold (Ct). Both methods have their pros and cons.

Multiplexing

Soon after the original 'one-channel' PCR technique was developed, it was amended to permit the simultaneous amplification of several sequences of DNA. Early methods from the late 1980s described the use of five different fluorochromes, including fluorescein and rhodamine linked to primers designed to hybridise to different parts of the target gene. Thus, if only the cDNA linked to the fluorescein probe is amplified, the product will be fluorescent in the green part of the spectrum. Likewise, amplification of the DNA linked to the rhodamine will produce a red product. Should both be amplified, the product will be yellow, that being the product of red and green. This two-colour system was subsequently used by other methods.

Clinical developments started to appear in the literature in the 1990s. The procedure was used to diagnose and family-trace gene lesions in Lesch–Nyhan syndrome, in Duchene–Becker muscular dystrophies, and in cystic fibrosis, whilst others reported a method for the simultaneous screening of the prothrombin G20210A mutation and Factor V Leiden. As discussed in Chapter 1, the tumour suppressor gene *p53* is implicated in the aetiology of 40–75% of cancers. A multiplex PCR, primers being used to amplify different exons, was developed to specifically search for deletions in the *p53* locus, finding that one such deletion is present in around a third of all breast cancer tumours.

As elsewhere in biomedical science, industry has responded to the need for a simplification of gene analysis so it can be used in routine clinical practice by providing kits that include all the reagents for a given analysis. One such was used to detect the presence of the EBV, cytomegalovirus and herpes viruses in the saliva and serum of patients with HIV (as introduced in Table 1.4). The degree of methylation of DNA has important consequences for the regulation of transcription and carcinogenesis. A methylation-specific multiplex PCR was used to probe for four methylation marker abnormalities in 781 women with breast cancer, finding the panel outperformed existing clinical diagnostic methods in the detection of minimal and early-stage tumours.

This multiplex PCR method is faster and can be used in a wide range of investigations, such as linkage analysis, pathogen detection and genetic disease diagnosis. The general principle of multiple and parallel assays was further developed into techniques such as microarray technology (Section 2.6) and pyrosequencing (Section 2.7.1). The rolling development of an rtPCR analysis of four genes (A–D) is illustrated in Figure 2.9.

Points 1–4 show that the samples have cycle threshold values of around 14, 19, 22 and 26, respectively, suggesting that sample A had the greatest mass of DNA in it, sample D the least. Indeed, the latter has not reached a plateau phase, suggesting either an extremely small amount of DNA, or a technical or contamination anomaly that requires investigation.

Figure 2.10 shows a typical thermal cycler, with the plot of the increase in fluorescence of time being shown on a screen at the top face of the device. Although the primary metric, an increased in fluorescence, being

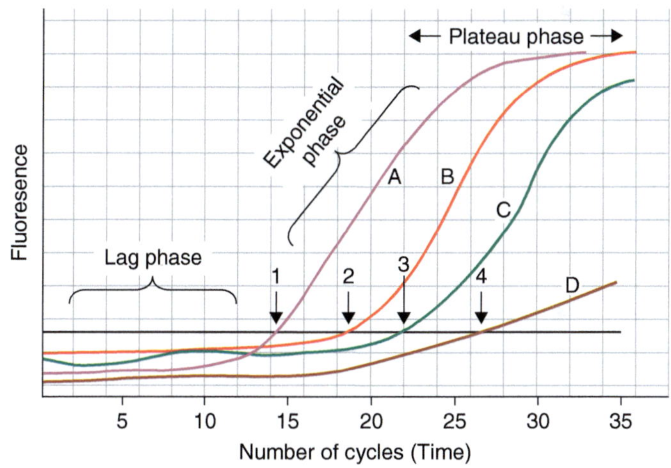

Figure 2.9 Real-time increase in fluorescence in an rtPCR reaction. The four lines A, B, C and D represent the amplification of different genes, each with a probe linked to a fluorochrome. Each fluorochrome is linked to a specific wavelength that does not overlap with other fluorochromes. The reaction can be visualised in three phases. In the first or lag phase, the signal from the growing mass of amplified DNA (or more correctly, the linked fluorescence) does not yet exceed the background, or threshold, level. As the reaction proceeds, the fluorescence signal rises above the threshold, entering the exponential phase. This proceeds up to the plateau phase, where the reaction is complete. The key metrics are shown as 1–4, the points where the fluorescence signal rises above threshold, representing the cycle threshold (Ct).

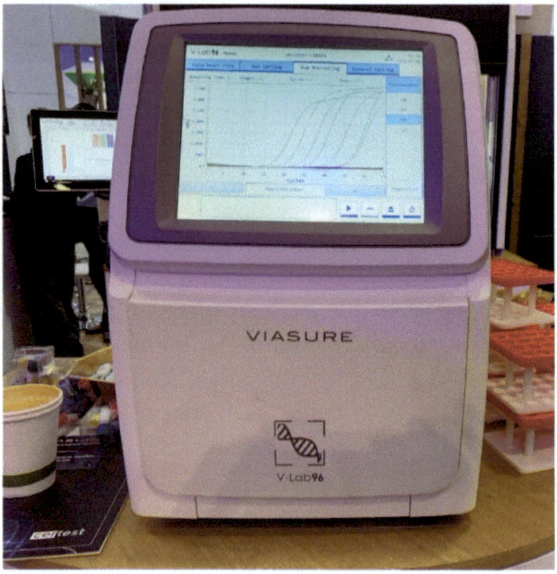

Viasure

Figure 2.10 A thermal cycler showing rtPCR sigmoidal curves. The display shows seven curves of the increase in signal. The cycle threshold can be determined from this plot, and that on the far left shows the most rapid generation of signal, reflecting a large amount of DNA in the sample. Conversely, the curve on the far right increases much later and therefore indicates a small amount of signal. The extent to which these can therefore be defined as positive or negative is a matter of interpretation.

an analogue estimate, is often inaccurate: digital techniques provide better accuracy (see Section 2.2.4). Nevertheless, rtPCR has since developed into a mainstream method used in a routine setting for a number of clinical indications, such as the presence of viral DNA.

> **Rapid Diagnostic Molecular Pathology** A multiplex PCR procedure can be used to detect the relative amounts of different sections of DNA in a given sample. An example of this is the focused detection of certain named respiratory viruses in a nasal or throat sample, providing proof of a particular infection (e.g. whether the patient's symptoms are due to an influenza virus, SARS-CoV-2 or a different pathogen such as a bacterium).

Multiplex Ligation-dependent Probe Amplification

This procedure was developed in the early 2000s, but as with almost all methods, it has biochemical roots going back decades. Of the various functions of ligase enzymes is their ability to repair damaged single-strand DNA chains by restoring the phosphodiesterase link between adjacent nucleotide chains. It follows that loss-of-function mutants leading to reduced ligase activity are linked to certain disease (such as ataxia telangiectasia and xeroderma pigmentosum) where damaged DNA is not repaired. Nevertheless, the ability of ligase to join two strands of DNA is exploited in multiplex ligation-dependent probe amplification (MLPA).

The procedure differs from standard PCR in that it is not the target DNA that is amplified, but an extended oligonucleotide formed from two smaller probes, each complementary to a section of the gene or region of interest, and that are merged by the ligase. It follows that the method requires knowledge of the nucleotide sequence of the section of the sample to be amplified. Each probe is linked to a primer, and one probe, often the smaller, has a unique 'stuffer' insert that simply extends the length of the probe, allowing products to be identified. The fine detail is illustrated in Figure 2.11.

As up to 40 probes can be used simultaneously, the procedure is both fast and efficient. MLPA was first applied clinically to investigate deletions in *MLH1* and *MSH2* in colorectal cancer and Lynch syndrome, in *BRCA1* in breast cancer, and in *CMT1A* and *HNPP* in Charcot–Marie–Tooth disease. Further analyses include prenatal aneuploidy screening for abnormalities in chromosome 13, chromosome 18, chromosome 21, the X chromosome and the Y chromosome, in the investigation of haemoglobin gene deletions in thalassaemia, Duchenne–Becker muscular dystrophy, antithrombin deficiency, chromosome 1p/19q co-deletion, and in the relative CNVs of *CDKN2* in bladder cancer. It is also used to determine the number of CNVs in malignancies, methylation changes and SNP detection. In 2013, the method was amended to be able to amplify sections of miRNA, using ribonucleotide-modified DNA probes, whilst in 2020, Tamminen and colleagues published a digital MLPA and used it to screen for antibiotic resistance.

Droplet Digital PCR

Also known simply as digital PCR, or sometimes emulsion PCR, this variant was first reported in 1999, but was developed for general

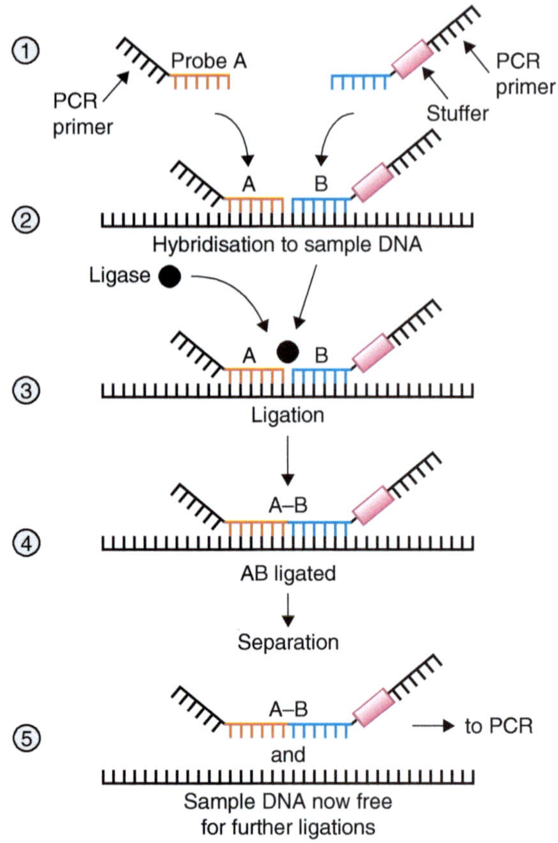

Figure 2.11 Principle of multiplex ligation-dependent probe amplification. In step one, primers to adjacent parts of the gene or region of interest are hybridised with the sample DNA, often overnight. Step two sees the application of the ligase, which will join the two primers to generate one long extended probe, with primers that are matched to the target. If the DNA in the target fails to bind to either probe, no extended probe will be generated. The final step is to dissociate the extended probe and the target DNA, and then amplify the former in a standard PCR, as one of the probes is fluorescently labelled.

laboratory use with a chip in the mid-2000s and is based on an oil/water emulsion of tens of thousands of microdroplets. Each droplet may contain any number of library DNA fragments partitioned in its own micro-reaction vessel, which are then amplified in a PCR. The digital aspect refers to a analysis where each droplet is defined as positive or negative. The problem that arises is that a droplet may contain a varying number of positive results; fortunately, this can be corrected by Poisson mathematics. Advantages of droplet digital PCR (ddPCR) include absolute quantification ability, low cross-contamination, low reagent consumption, high sensitivity and high throughput.

One of the first clinical uses of ddPCR was in determining the presence of HIV, the presence of *HER2* and its copy number, in formalin-fixed paraffin-embedded sections of breast cancer tumours, and in the detection of donor DNA in transplant recipients. Subsequent reports showed its ability to evaluate the value of serum lncRNA in hepatocellular carcinoma, to detect tumour DNA in the plasma of patients with urothelial carcinoma and to determine BCR::ABL1 levels in patients with chronic myeloid leukaemia who received tyrosine kinase inhibitor therapy. The latter research points to a potential value of ddPCR in the routine testing for measurable residual disease in blood cancers, suggesting it could be a possible replacement for rtPCR, currently the most commonly used technique.

Without doubt, additional innovations in this technique will appear.

Self-check 2.5 How does MLPA differ from a standard PCR?

2.5 DNA Sequencing

The greater part of the routine molecular pathology laboratory is to confirm or exclude the presence of a particular gene likely to have pathological significance. We have already looked at the potential of FISH in the detection of a gene in a metaphase/interphase spread, but in other instances, a different approach is required.

As molecular pathology grew into a discipline of its own, certain experimental strands could be identified. There was (and is) a continuing need to explore potential alterations in suspect genes in specific clinical pathological situations and to determine the genomes of pathogenic organisms, a classic example being that of SARS-CoV-2. A second area was the growing need to determine the presence of a specific abnormal gene in a patient, or in an individual before they become a patient. With knowledge of the genetic basis of a disease, clinicians can tailor their treatment options, perhaps with specific drugs designed to combat a given genetic problem.

2.5.1 Sanger Sequencing

Almost simultaneously in 1977, two groups published methods for sequencing DNA. Maxam and Gilbert's method involved specific chemical modification of each nucleotide and the use of radioactive phosphorus, with subsequent analysis of that modified nucleotide. The alternative, from a group led by Fred Sanger (Figure 2.12), updated and simplified an earlier paper, where a DNA polymerase (from the bacterium *Escherichia coli*) produces a modified nucleotide chain when provided with an alternative nucleotide (a di-deoxynucleotide – i.e. those without hydroxyl groups on ribose carbon atoms 2 and 3), subtly different from the standard deoxynucleotides where the hydroxyl group is present on carbon atom 3. This results in a termination of the elongation of the chain, so that the sequential products can be determined by electrophoresis.

After some years of scientific natural selection, the sequencing community came down on the side of the Sanger procedure. With knowledge of the nucleotide sequence, artificial complementary gene probes could be synthesised to enable the further identification of particular genes *in situ*. In 1980, Sanger was awarded his second Nobel Prize for developing this method of sequencing.

Figure 2.12 Fred Sanger.

Key Point 2.6 The Sanger procedure was undoubtedly the single most important scientific development in molecular genetics and molecular pathology, with over 30,000 PubMed citations to date. Although the basic technique is rarely used in a modern routine molecular pathology laboratory, it is nonetheless an essential analytical tool and is frequently referred to in a variety of analyses.

2.5.2 Technical Advances

Sanger's original process had two major drawbacks – the use of radioactivity and the nature of the slab gel electrophoresis for demonstrating the different fragments of DNA generated. The former was solved by using fluorochromes, which can be induced to emit light at one wavelength when stimulated by light of a different wavelength. An extension of this principle was rapidly developed with the use of four different fluorochromes, one linked to each of the nucleotides (Figure 2.13.).

Although a relatively simple method, it could not (at the time) be easily scaled up to the 'industrial' level required for the sequence of large sections of DNA. The second problem was addressed by the development of capillary electrophoresis. These two developments came together in 1986 with the first rapid automated sequencer, the Applied Biosystems AB370 analyser, which could detect 96 bases in one run and 500,000 in a day. Subsequent machines extended this to a batch of hundreds of samples and read up to a million bases in a day. A third major advance was the development of improved DNA polymerases, some of which were engineered, often tailored for a specific process.

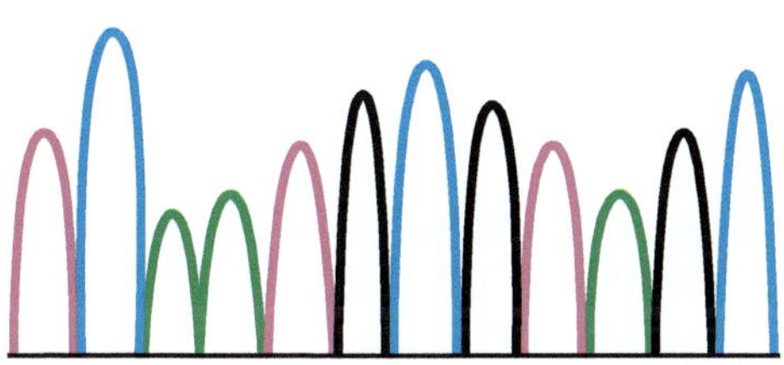

Figure 2.13 Fluorochrome printout of a Sanger sequence. Each of the four nucleotides is linked to a separate fluorochrome, shown here as black = G, green = A, red = T and blue = C. The following different colours give the DNA sequence of the sample, so this example is TCAATGCGTAGC.

As technologies such as robotics and other methods advanced, so did the ability to detect first the most common genetic abnormalities and then those present at a smaller and smaller frequency. This therefore brought the ability to provide a particular diagnosis with greater confidence. Table 1.2 in Chapter 1 has shown that the leading cause of death in England and Wales is cancer. Accordingly, molecular pathology focused on this form of disease, reporting strong links between genetics and clinical conditions, as discussed in Chapters 4, 5 and 6.

2.5.3 The Human Genome Project

As introduced in Chapter 1, the Human Genome Project boldly set out, with the benefit of automatic sequencing machines, in the early 1990s to sequence the entire three billion base pairs of the genome, then estimated to include some 50,000–100,000 genes. By 2001, this estimate had fallen to 30,000–40,000 genes, compared to a little under 20,000 at present. Using scaled-up modifications, and the conventional Sanger sequencing method, the project took 3–4 years to complete at the cost of some $300 million, which by 2003 had sequenced 92% of the genome. This enormous scientific achievement spawned a revolution in molecular genetics and so molecular pathology. It was based on new methods, collectively described by some as NGS, by others as massive parallel sequencing, or high-throughput sequencing. These are described in Section 2.7.

Self-check 2.6 What do you consider to be the major biochemical advance in the Sanger method?

2.6 Microarrays

2.6.1 The Theory

Microarrays represent a natural next step in technology from PCR and FISH (Section 2.3). Although first described as a research tool in the 1980s, with developments such as processing on a 96-well microtitre plate or on a glass slide, they soon became a mainstream technique in the mid-1990s. An early method placed 256 different sequences of eight nucleotides onto a solid phase 1.28 x 1.28 cm array, subsequently described as a chip, a process further developed by introducing robotics. A two-colour fluorescence karyotype array on a 1.8 x 1.8 cm chip was used to detect alterations in different chromosomes, whilst others used a 1.0 x 1.0 cm chip to investigate changes in the expression of genes linked to heat shock proteins and inflammatory markers such as interleukins and tumour necrosis factor.

The microarray principle relies on the fixation of short oligonucleotide sections of DNA (originally described as spots) with a known genetic sequence placed robotically onto a solid substratum such as glass or a sheet of nylon. This, of course, first requires knowledge of the nucleotide sequence of the particular gene being fixed, the ability to artificially synthesise that gene and the procedure for attaching it to its substratum. The process works most efficiently with a sample of purified mRNA from the patient, from which cDNA may be generated; the product being linked to a visualising marker, probably a fluorochrome, which is then applied to the solid substratum bearing the fixed genes. An alternative is to extract the RNA from a cell sample, attach a fluorescent label and then apply it directly to the solid phase oligonucleotide array. Should there be an annealing match, the labelled patient's

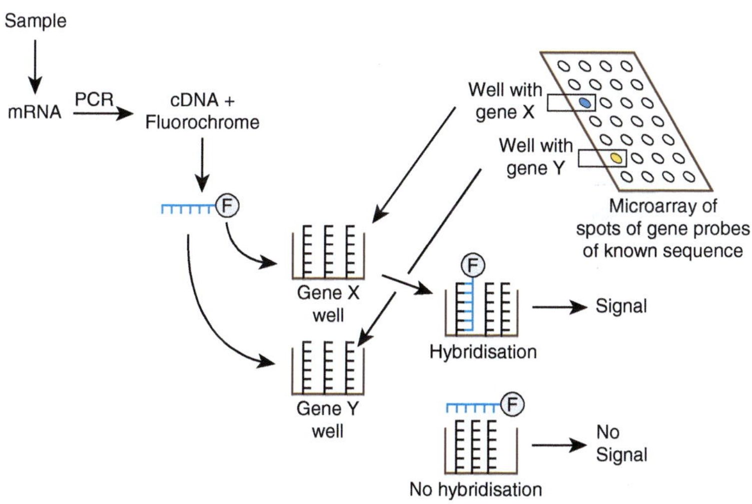

Figure 2.14 Basics of microarray analysis. (a) A simple representation of key steps. In a single sample analysis, the cDNA is labelled with a fluorochrome that may hybridise with the fixed oligonucleotide to gene X on the microarray spot. Alternatively, if gene Y is not present in the cDNA library, there will be no hybridisation and therefore no signal, so that only gene X is present in the sample.

(Continued)

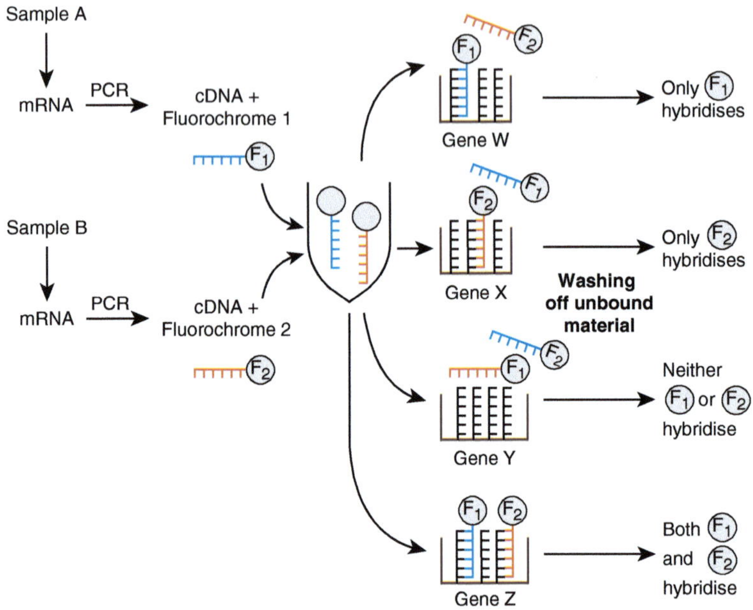

Figure 2.14 (Cont'd) (b) An extension of the process sees two (or more) samples analysed in parallel. These can be, for example, from a tumour and adjacent normal tissues, patients with two different conditions, or a patient and a healthy control. In this illustration, both cDNAs are placed in a single reaction vial, and aliquots applied to wells or spots with gene-specific oligonucleotides, shown here as genes W–Z, although in practice there can be many more. Should there be a hybridising match between a gene in the sample and the solid phase probe, the fluorochrome can be detected. The different fluorochromes allow either, both or neither hybridising event to be recorded.

sample will hybridise with the oligonucleotide, and once unhybridised sample material is washed off, the product can be determined by fluorescence laser scanning (Figure 2.14).

2.6.2 The Practice

A major advance in microarray design was the reduction of the solid phase to a small cassette, or chip, linked to the ability to assess multiple samples at the same time, i.e. multiplexing. This, when applied to an

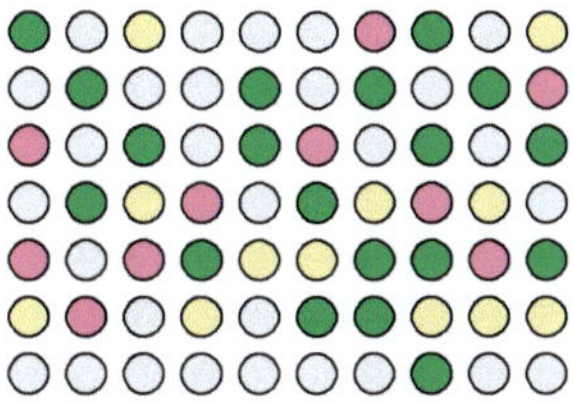

Figure 2.14 (Cont'd) (c) Representation of a microarray when two different samples of cDNA, each with a different fluorochrome, are added to an array where each spot carries oligonucleotides of a different gene, as in part (b) above. Should the sample with the red fluorochrome hybridise to DNA fixed to the spot, the laser detector will record a red spot, and similarly where the sample with the green fluorochrome hybridises, the resultant spot will be green. If the samples do not contain the gene in question, no fluorochrome binding will be detected and the spot will not be coloured. However, if both samples contain the same gene, both colours will be present and so, in this example, the spot will appear yellow. This particular chip has 70 known genes: in practice there may be hundreds.

appropriate analyser, enabled, and therefore provided, a much greater range of genes that could be identified at a lower unit cost and with greater speed, thereby bringing the technique into the routine clinical pathology setting and the true birth of practical molecular pathology. An early example of this is the detection and characterisation of mutations in *CFTR* that cause cystic fibrosis, whilst others used microarray technology in an early pharmacogenomic setting in this disease to probe for genes whose activity is linked to the xanthine drug CPX, a potential treatment. Neurofibromatosis is an autosomal dominant condition often caused by loss of function or deletion of tumour suppressor *NF2* at 22q12: use of a microarray form of CGH found 24 deletions in this locus in a cohort of 116 patients.

The technique is admirably suitable for the detection of genes in different populations of people (i.e. patients and healthy controls) and different populations of tissues (malignant and healthy controls). An example of the latter from 1999 compared tissues from 532 renal tumours with those from 6 normal kidneys and reported 89 differentially expressed genes in the malignant tissues. Another study from 2001 compared prostate cancer tissues, tissues with benign disease, and normal tissues, finding increased expression of a number of genes in the malignant tissue, suggesting their roles in carcinogenesis and potential as diagnostic markers and targets for therapy. Figure 2.15. shows a typical microarray analysis for the determination of leukaemia. High-resolution SNP genotyping microarrays were used in a 2007 study of 119 patients with low-risk myelodysplasia, finding a uniparental disomy duplications in 46% of cases, deletions in 10% and amplifications in 8%.

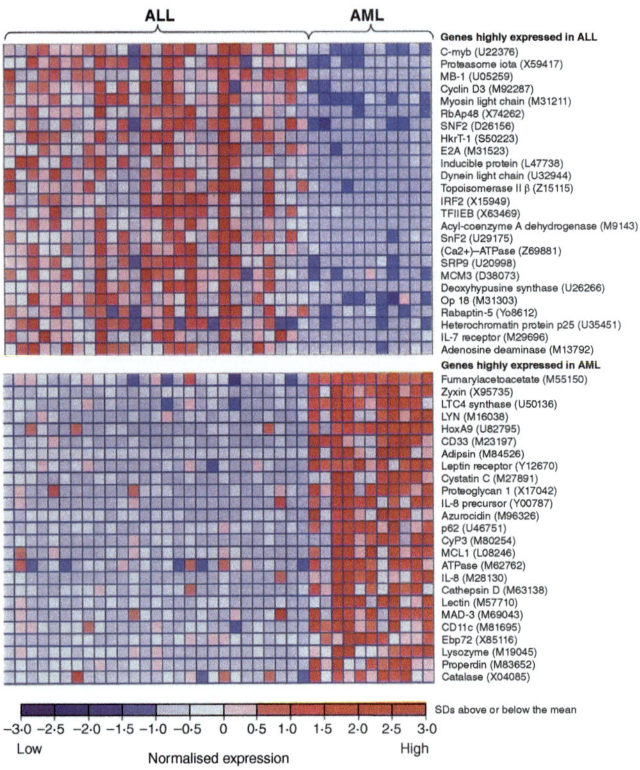

Figure 2.15 Microarray analysis. The upper panel shows (in red) the binding of the patient's sample genes, consistent with a diagnosis of acute lymphocytic leukaemia, collected on the left of the array. In the lower panel, the patient's sample binds to a different pattern of fixed oligonucleotides, with red being dominant on the right side, suggesting a diagnosis of acute myeloid leukaemia. Each column represents analysis from a single patient, and each row corresponds to a particular gene. *Source:* Hoffbrand et al. (2011). Figure 11.13B, p.162, with permission from John Wiley & Sons.

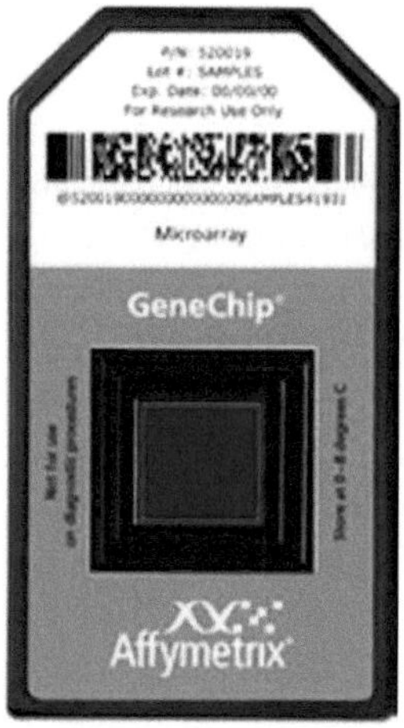

Figure 2.16 An Affymetrix GeneChip.

These abnormalities were linked to patient outcomes, thus identifying those in need of early therapeutic interventions. Microarray analysis has shown a link between the CNV of certain genes in certain breast and other cancers, whilst the multiple sampling nature of microarray can be amended for CGH, where a test sample (often pathological) is compared to a reference sample (generally healthy).

Affymetrix GeneChip (Figure 2.16.) microarrays rapidly became the market leader and have been used to investigate up- or down-regulated mRNAs and miRNAs in early and advanced atherosclerotic plaque from coronary arteries, and to determine the presence of miRNAs in epithelial ovarian cancer. The microarray method was also used to identify upregulated genes (such as code for matrix metalloproteinases) in the synovium of patients with early inflammatory arthritis and to identify *ADM, ANGPTL4, USP8, EDN, NFIL3, MSR1* and *CEBPD* as candidate genes for hypertension.

2.6.3 Disease-linked Gene Panels

One of the earliest large-scale uses of microarray in cancer was an analysis of 371 primary lung adenocarcinoma tumours, of which 242 had matched normal samples. Large-scale deletions (>50% of an arm) were found in 14 chromosomes, large-scale amplification (>50% of an arm) in 9, focal deletions (such as in 5q11.2) in 7 and focal amplifications in 21 (such as in 14q13.3). In many cases, known tumour suppressor genes were detected, including *MDM2, MYC, CDK4, KRAS, CCNE1, TERT, ARNT, CDKN2A/CDKN2B* and *PTEN*, whilst new candidate genes *NKX2-1, MBIP, VEGFA, PTPRD, PDE4D* and *AUTS2* were reported.

Similarly, 206 samples of glioblastoma tumour (of which 143 had matched normal peripheral blood) were probed by a number of methods, including microarray. High-level focal amplifications were found in (in order) *EGFR* (in >80% of samples), *CDK4, PDGFRA, MDM2, MDM4, MET, MYCN, CCND2, PIK3CA* and *AKT3*. The most common gene to have a homozygous deletion was *CDKN2A/B*, followed by *PTEN, CDKN2C, RB1, PARK2*, and *NF1*, whilst those genes with the most common somatic mutations include (in order) *TP53, PTEN, NF1, EGFR, ERBB2, RB1, PIK3R1* and *PIK3CA*.

The importance of these reports is that they added to the general view of a focused panel of genes most likely to have a role in diagnosis of a particular disease, a view that was to develop in the future with the establishment of specific panels for the investigation of specific disorders. These points are developed in the clinical chapters to come.

Self-check 2.7 What do you consider to be the major biochemical advance in the development of microarray technology?

Key Point 2.7 Microarray technology was the first to allow true high-throughput parallel analysis that could be brought to routine practice.

2.7 Summary of First-generation Techniques

The first aspect of molecular pathology (and, indeed, any biomedical analysis) is to ensure the sample is in its correct format, adequately labelled and uncontaminated, as described in Section 2.1. This was followed in Section 2.2 with a discussion of cytogenetics and karyotyping, which is performed on whole cells, and focuses on chromosomes, where changes such as the number of chromosomes, and gross changes within chromosomes, may be sought. Section 2.3 examined an extension of analysis of this nature – FISH – to look for specific changes within chromosomes, such as translocations and deletions.

The second part of this chapter looked at methods for determining changes in the number or structure of genes. To do this, nucleic acids must first be extracted from cellular material and then amplified to provide sufficient material for further analysis, as described in Section 2.4. The leading process of DNA amplification is the PCR. Once a purified sample of nucleic acid has been obtained, an important analysis is to determine the nucleotide sequence, which provides information regarding the presence or absence of an abnormality that may be pathological. The leading method for this is that of Sanger, as described in Section 2.5, and knowledge of the technique is essential. The scientific aspect of the chapter concludes in Section 2.6 with an examination of microarray technology.

2.7.1 Moving On

The impetus of the Human Genome Project led to the industrial production of numerous techniques, many of which became suitable for routine clinical practice – the NGS methods these were subsequently overtaken by other platforms that came to be known as third-generation techniques. Close attention to the present chapter will provide the reader with a firm grounding for a discussion of these techniques.

Self-check 2.8 Confirm to yourself that you have completed the learning objectives at the beginning of this chapter.

Bibliography

Aparna GM, Tetala KKR. Recent progress in development and application of DNA, protein, peptide, glycan, antibody, and aptamer microarrays. Biomolecules. 2023;13:602.

Cusenza VY, Bisagni A, Rinaldini M, et al. Copy Number variation and rearrangements assessment in cancer: Comparison of droplet digital PCR with the current approaches. Int J Mol Sci. 2021;22:4732.

Dunbar SA. Nucleic acid sample preparation techniques for bead-based suspension arrays. Methods. 2023;219:22–29.

Garcia DR, Arancibia AM, Ribeiro RC, et al. Intrachromosomal amplification of chromosome 21 (iAMP21) detected by ETV6/RUNX1 FISH screening in childhood acute lymphoblastic leukemia: A case report. Rev Bras Hematol Hemoter. 2013;35:369–371.

Hoose A, Vellacott R, Storch M, et al. DNA synthesis technologies to close the gene writing gap. Nat Rev Chem. 2023;7:144–161.

Hughes RA, Ellington AD. Synthetic DNA synthesis and assembly: Putting the synthetic in synthetic biology. Cold Spring Harb Perspect Biol. 2017;9:a023812.

Liu X, Liu S, Wang H, et al. Potentials and challenges of chromosomal analysis in prenatal diagnosis. Front Genet. 2022;13:938183.

Maxam AM, Gilbert W. A new method for sequencing DNA. Proc Natl Acad Sci USA. 1977;74:560– 564.

Sanger F, Air G, Barrell B, et al. Nucleotide sequence of bacteriophage φX174 DNA. Nature. 1977;265:687–695.

Sanger F, Nicklen S, Coulson AR. DNA sequencing with chain-terminating inhibitors. Proc Natl Acad Sci USA. 1977;74:5463– 5467.

3

Methods in Molecular Pathology 2

Next-generation Sequencing and Third-generation Sequencing

In Chapter 2, we looked at the development of a series of techniques that provided the basis of molecular pathology. Industry responded to the demands from researchers and practitioners, producing analytical platforms that came to dominate practice. Together, they have markedly driven down both the cost of an analysis and the turnaround time compared to even the most sophisticated Sanger methods, bringing this nature of analysis into routine clinical medicine.

A corner was turned in molecular genetics in the mid to late 90s with the co-development of a number of post-Sanger methods, many of which overlap and include the use of high-density nylon as a solid phase, biochips high-throughput techniques, automatic sequencing, microsphere beads, the use of lasers, robotics and advances in recognition systems, alongside complex computational software required for the terabytes of information, i.e. bioinformatics. Together, many of these were, in 2005, described as next-generation sequencing (NGS), and by 2008, the expression was in common usage as an umbrella for a number of different technologies.

Alongside further developments, some refer to NGS as second-generation sequencing, recognising additional new advances, generally from 2010, as third-generation sequencing (TGS). These platforms are described in Section 3.2, although some consider certain methods to lie in either group. Nevertheless, the immense amount of data generated by NGS, and subsequently TGS, demanded new methods of analysis, which spurred the development of bioinformatics, as explained in Section 3.3.

Molecular Pathology: A Primer for Laboratory Scientists, First Edition. Andrew Blann.

Learning Outcomes

After studying this chapter, you should confidently be able to...

- Describe the principles of pyrosequencing
- Comment on the science underlying Illumina sequencing
- Explain the basis of SOLiD sequencing
- Understand the key points of the Luminex xMAP platform
- Provide examples of the wider applications of NGS
- Grasp the concepts of nanopore technology
- Describe key features of the PacBio platform
- Explain the processes underlying Ion Torrent sequencing
- Appreciate the advances made by next-generation sequencing
- Summarise leading third-generation sequencing techniques

3.1 Next-generation Sequencing

This section will describe a number of methods or platforms in rough order of their appearance. These more complex analyses enabled scientists to determine precise differences in the genome, as any two copies of the genome differ by around one change per 1000 bases (i.e. 0.1%), by far the most common being a SNP. These occur at a frequency of around one per 300 bases, and so there may be as many as 10 million SNPs, the consequence of which makes up 90% of an individual's genetic fingerprint.

3.1.1 Read Length

The early Sanger methods all required determination of the sequence from an electrophoresis plot by eye, sequentially working through the position of each nucleotide. This manual read length was bounded only by the length of the electrophoresis gel apparatus and could resolve details of up to 100 nucleotides, although analysers later extended this to reads of 500–1000 base pair (bp) fragments, with up to 96 kbps per run.

A key feature of all high-throughput techniques is the length of DNA that can be sequenced or investigated by the read length in each round of analysis. This 'eyeballing' method became impossible due to the large number of sections of DNA being sequenced by automatic analysis, interpretation of which required complex software, i.e. bioinformatics. A further feature of NGS automation was of a finite length of DNA correctly sequenced in each analysis, so that the read length was reduced. Fortunately, the methods generate thousands (later, millions) of overlapping short/small reads, from which an overall picture of the complete sequence could be built, and the analysis of short reads such as 36 bps can be enhanced by the use of mass spectroscopy. The advanced technology of TGS enables longer read lengths (Section 3.2).

3.1.2 Pyrosequencing

This procedure, whose scientific roots go back to the 1990s, was amongst the first to be developed as a commercial large-scale automated technique. One of the most commonly used methods calls for a random shotgun rendering of the genome to form a crude library, with fragments then ligated to adaptor linker sequences that permit the binding of a primer. A further advance is to add a unique tag to each primer before PCR amplification, a method known as barcoding, which can be applied to other library preparations. The primers, in turn, allow the hybridisation of an individual DNA molecule on to the surface of a microbead in droplets of an emulsion. Notably, the method does not rely on radioactive or fluorescent markers, or oligonucleotides. It was brought to the market by 454 Life Sciences/Roche in 2005, whilst Qiagen also produced a pyrosequencer with dedicated reagent kits.

The Theory

The chemistry of this 'sequencing by synthesis', as described in 1998, relies on the process of the construction of a growing DNA chain by a DNA polymerase, and the release of a pyrophosphate molecule ($P_2O_7^{4-}$, often abbreviated as PPi) from one of the nucleotides, as shown in Figure 3.1.

The chemistry of the pyrophosphate reaction is as follows:

1. The initial reaction involves the template of the immobilised unknown strand of library DNA to be sequenced, a primer, DNA polymerase and a nucleotide triphosphate.
2. Should polymerase incorporate the nucleotide into the strand, a molecule of pyrophosphate is released as a by-product (Figures 3.1 and 3.2).

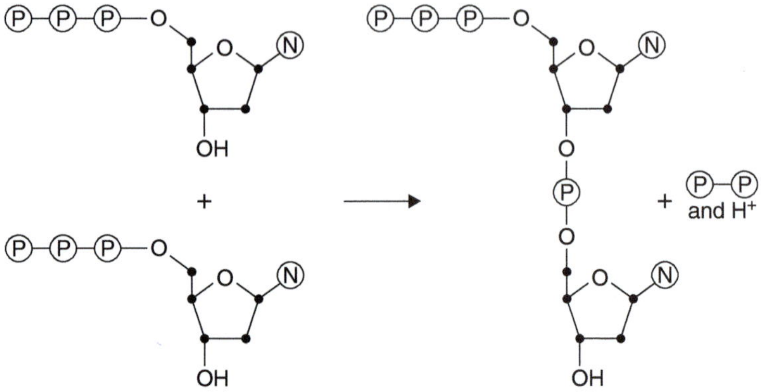

Figure 3.1 Nucleotide chemistry. The two nucleotides on the left form a dimer on the right via a phosphodiesterase bond. This generates a two-atom molecule of pyrophosphate and a hydrogen ion (far right). The latter is a crucial part of the Ion Torrent method described in Section 3.1.2. N = a base, OH = a hydroxide group, the heavy black spots represent a carbon atom. Numerous bonded hydrogen ions are omitted for clarity. Incidentally, removal of the oxygen atom from the hydroxide converts the deoxyribose to di-deoxyribose, the basis of Sanger sequencing (Section 2.5.1).

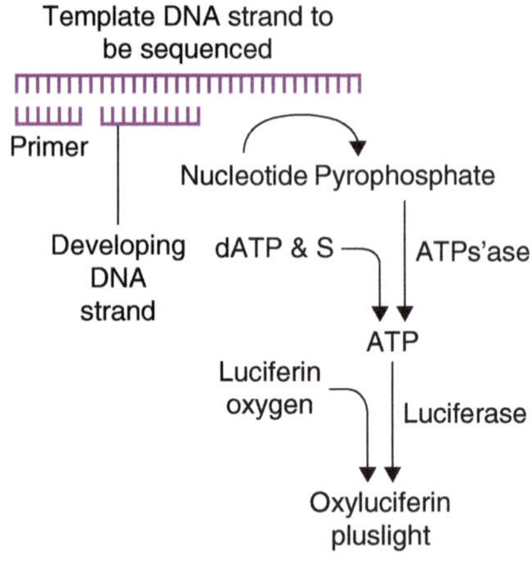

Figure 3.2 The basis of pyrosequencing. A nucleotide is added to the developing DNA strand by a polymerase, which releases the pyrophosphate molecule. This, with the ATP variant, is converted to ATP, which is, in turn, a substrate for luciferase in the generation of a photon of light. ATPs'ase = ATP sulphurylase.

3. The pyrophosphate is a part-substrate for ATP sulphurylase, which generates a molecule of ATP, which is, in turn, the substrate for luciferase.
4. When provided with its other substrates (luciferin and oxygen), the luciferase generates the product oxyluciferin and also releases a photon of light, easily picked up by a detector such as a charge-coupled device.
5. Unincorporated nucleotide triphosphates are degraded by apyrase.
6. The cycle returns to step 1 with another nucleotide added to the reaction vessel.

If the nucleotide is not incorporated into the growing strand, no pyrophosphate is generated, and thus no photon of light is emitted (Figure 3.2). An important point is that standard ATP cannot be used in step 1, as it is also a substrate for luciferase. Instead, the variant dATPαS is used, as it recognised as a substrate by the DNA polymerase, but not by the luciferase.

The resultant data can be visualised and is not dissimilar to that of Sanger sequencing shown in Figure 2.14, whilst a pyrogram is shown in Figure 3.3. A considerable advantage of this method is that it avoids the need for labelled nucleotides and electrophoresis, but its limitations include the requirement for complex interpretation. Furthermore, the length of the sequenced nucleotide is relatively short, early reports

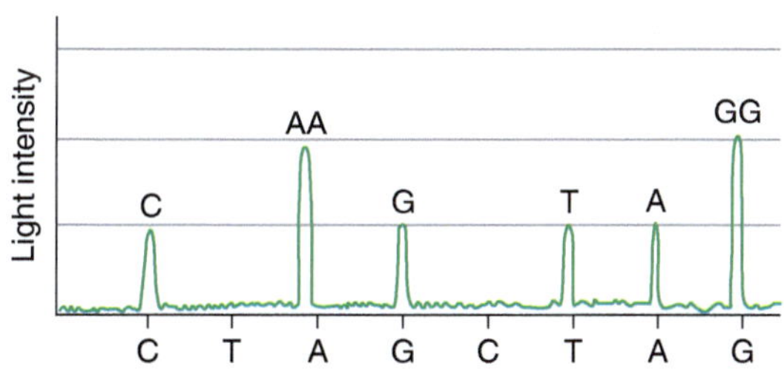

Figure 3.3 A pyrogram. The vertical axis is light intensity, and the horizontal axis represents the order in which nucleotides are offered to the reaction mixture. The first peak on the far left represents light emitted when a cytosine is added to the developing chain. The next to be offered is a thymine, which is not added to the chain, and thus there is no signal. The enlarged peak for the adenosine variant indicates that it is added twice, followed by a single guanine. The pattern continues, so this sequence is CAAGTAGG, which converts to a sequence of TGGACGAA in the sample DNA. A peak higher than those illustrated points to a triple nucleotide.

describing a read length of 70–130 bps, whilst later research has sequenced 300–450, and even 800, bp reads, and a hugely increased 80–120 mbps per run. However, given the ease of automation of the process, it is therefore not only fast and efficient but also flexible, accurate and amenable to parallel processing. Of further note is the ability of pyrosequencing to detect methylations, principally of cytosine residues.

The Practice

By the early 2000s, pyrosequencing was reporting data of clinical interest, such as the genetic basis of rifampicin resistance in *Mycobacterium tuberculosis*; the report of polymorphisms in *XRCC1* at 19q13.31, coding for a DNA repair protein, in colorectal cancer; and the report of 15 mutations in proto-oncogene *BRAF* at 7q34, coding a signal transduction kinase, in melanoma cell lines. It was also used to analyse the tumour suppressor *TP53* and in HLA typing. Perhaps the earliest landmark use of 454-pyrosequencing was the publication of the genome of James D Watson, a task that cost around a hundredth of that of traditional capillary electrophoresis methods and was completed in two months. Compared to reference genomes (explained in Section 3.1.8), Dr Watson's genome contains 3.3 million SNPs causing over 10000 amino acid substitutions, and over 22000 insertions or deletions ranging from 2 to almost 39000 bps. Pyrosequencing has also been used to explore variants in DNA methylation in colorectal cancer, glioblastoma and in other conditions.

> **Self-check 3.1** Pyrosequencing relies on phosphate groups. Where do these come from?

3.1.3 Ion Torrent Technology

The Theory

The chemical basis of this process is shown in Figure 3.4, which bears a striking similarity to Figure 3.1, in that it has much in common with pyrophosphate sequencing. Analysis starts with library fragmentation of the sample DNA (as described in Section 2.4.3); each fragment is linked to a bead via an adaptor, which is then subjected to PCR to produce multiple copies on each bead. These are then loaded onto a chip with thousands of

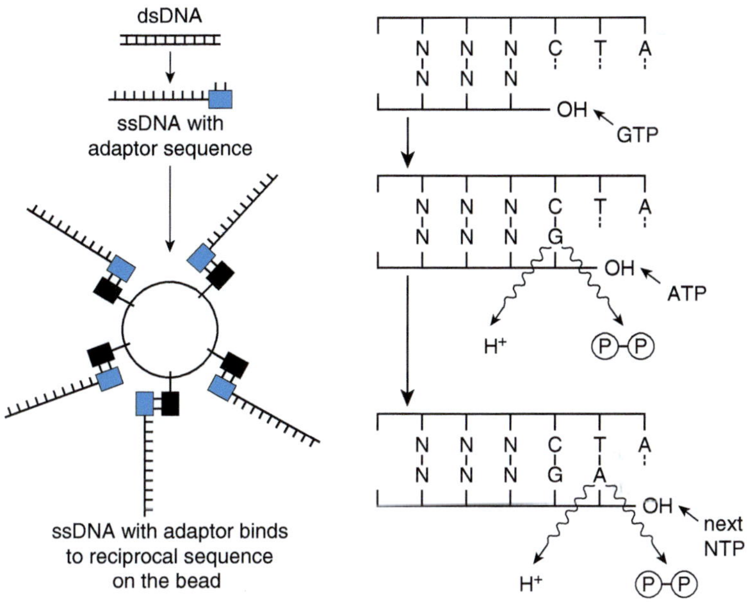

Figure 3.4 Ion Torrent sequencing. On the left, the sample dsDNA is stripped to ssDNA and an adaptor sequence is added. Multiple copies of the reciprocal sequence to this adaptor are present on a bead, to which the sample DNA binds, and multiple copies of the sample ssDNA are generated. On the right, the upper sample strand forms the template for the reciprocal strand to be synthesised. Should a C residue appear on the sample strand, it will attach a molecule of GTP, which the DNA polymerase will link to the terminal –OH group, generating a pyrophosphate group (PP) and a hydrogen ion. Should the next nucleotide in the sample template be T, it will bind a molecule of ATP, which will ligate to the terminal –OH group of the G, generating further PP and hydrogen ions, and so on. In the next round, should the well be flooded with CTP, it will fail to bind to the A and so there will be no product, and no signal detected.

nanowells, so that each nanowell has only one bead, perhaps thousands of copies of the same fragment, which are flooded sequentially with nucleotides. Each time a nucleotide is added to a developing DNA chain, a hydrogen ion is released from the hydroxyl group that links adjacent nucleotides. With this happening simultaneously on each of the multitude of developing chains, a large number of protons will be released.

The hydrogen ion pool is detected by an exquisitely sensitive ion sensor under the nanowell, which is a field-effect transistor, and thus effectively functions as a form of a pH meter. This gives rise to ion semiconduction sequencing, a principle exploited by the 'Ion Torrent' range of analysers marketed by Thermo Fisher Scientific from 2011. With a large number of nanowells, the process can be seen as massive parallel sequencing, with the ability to sequence entire genomes or to specifically target particular genes. Read lengths are of the order of 200 bps, and the method can be enhanced by using ChIP (Section 2.4.1) for studies of histone modifications in tumour tissues.

The Practice

The technique has been used by microbiologists to type meningococcal isolates, to sequence drug resistance mutation in *M. tuberculosis*, to characterise human papillomavirus isotypes, and for an investigation of the biology of the protozoan parasite *Trypanosoma brucei*. One of the first clinical uses of this method was in a public health investigation to characterise the genome of potentially lethal *E. coli* isolates, to determine those mutations in *CFTR* that cause cystic fibrosis, to sequence *BRCA1* and *BRCA2*, to sequence the mitochondrial genome, and to sequence cancer genes in formalin-fixed paraffin-embedded tissues and in fine-needle aspirates.

More recent work has also followed these trends. The complex in-hospital transmission of a norovirus, with different strains on different wards, was revealed by deep Ion Torrent sequencing, whilst the method was used to point to different cervical and vaginal microbiomes in early miscarriage, where *Gardnerella vaginalis* and *Mycoplasma girerdii* were present. Further applications included publication of the genome of the rabies virus, development of a method for the detection of the susceptibility of *M. tuberculosis* to 14 drugs, and an exploration of HIV-1 drug resistance.

3.1.4 Illumina Sequencing

The scientific roots of this technique go back to the mid-1990s and were brought to market by the biotech company Solexa in the mid-2000s. The company and its technology were subsequently acquired by Illumina, which came to dominate the industry, hence the name.

The Theory

The technical advance made by Solexa's method is the use of ligation to attach two different fixed synthetic DNA stands to each end of the library DNA to be analysed. In a pre-analytical step, the sample DNA has two adaptors ligated to each end of the strand. This allows a bridge or loop to be formed when the adaptors find their reciprocal oligonucleotides attached to a solid phase, located in a nanowell in the flow cell of a chip. This well holds thousands of oligonucleotides in a cluster, and accordingly, thousands of ligated library DNA strands bearing the complimentary adaptors can hybridise, forming the template for the construction of a cDNA strand (Figure 3.5).

This process generates thousands of forward and reverse copies of the original library DNA. Snipping off one of these copies leaves the remainder to be sequenced in the usual manner, i.e. with the four nucleotides each bound to its own specific fluorochrome. Thus, the progress of the sequencing can be followed by the signal from the fluorochrome received by a detector. The entire slide may contain up to 25 million clusters per flow cell, allowing the overall sequencing output to exceed a billion bp, resulting in a considerable signal to be analysed by the analyer's software. Read lengths for this method are often in the range of 60–150 bps, although some have read sections as short as 36 bps. The NovaSeqX analyser can read 1.6–52 billion reads per 15–48 hour run.

In 2008, the developers of the method published details of the process and its use in sequencing a female X chromosome with 278 million 30–35 bp reads, identifying over 92000 SNPs. They also sequenced the genome of a male from Nigeria, generating 4 billion 35 bp reads and identifying around 4 million SNPs.

The Practice

As with all technologies, Illumina sequencing was initially used as a research tool. One of the first studies used gel electrophoresis and other techniques to purify small RNAs, including miRNAs, from human pluripotent embryonic stem cells, and then determine their sequence. However, clinical studies followed, such as an analysis of tumour tissue from 28 cases of low-grade paediatric astrocytomas, which found a high frequency of 7q34 duplications. The method was used to catalogue the intestinal microbial gene population in faeces from 124 individuals of European

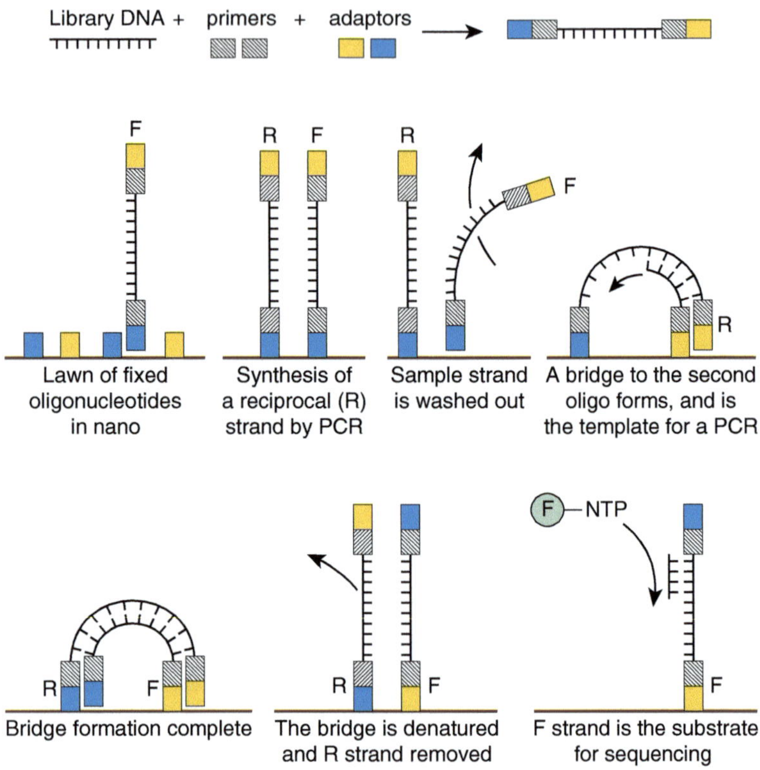

Figure 3.5 Essentials of Illumina sequencing. The process starts at the top left with the ligation of two adaptors to each end of a sample of library DNA with a read length of 30–40 bps. This DNA is then added to a chip containing thousands of nanowells, each containing thousands of copies of oligonucleotides complementary to the adaptors on the DNA (top panel, far left). The strand of DNA binding to the oligonucleotide is a template for the generation of a reciprocal strand attached to the nanowell, following which the sample DNA is washed out. This constructed stand forms a bridge to a second nearby oligonucleotide and serves as a template for amplification (far right). When separated, these two attached strands are then able to bind to additional oligonucleotides to form further bridges (lower panel, left to right), and the cycles continue. Once the amplification phase is complete, one of the two strands in removed, and the remainder forms a template for sequencing with NTPs linked to a fluorochrome.

origin, finding that >99% were bacterial, summing to 3.3 million genes from over 1000 species, with the most common being of the Bacteroidetes family.

By the early 2010s, Illumina sequencing was an established tool in diagnosis, as is that for diagnosing trisomy 21 (Down syndrome), which occurs in 1 in 800 live births, with invasive sampling of foetal material by amniocentesis being (at the time) the preferred diagnostic tool used in women at high-risk, though it carried a 1% risk of miscarriage. The NGS method had 100% sensitivity and 97.9% specificity for the trisomy, and so a predictive saving of 98% of the invasive procedures. Von Willebrand factor is the most common genetic hypocoagulation defect manifesting as a spectrum from very mild to life-threatening forms, making precise diagnosis crucial. The established method of gel electrophoresis may be superseded by Illumina-based molecular genetics, searching for substitutions, indels and deletions, and so define with considerably more accuracy the degree of pathology.

More recently, the method has been used to show that 23% of a cohort of 770 Italians had at least one alternative allele in the gene responsible for cystic fibrosis, to determine the precise pathological genotype of subjects with malaria (such as in *HBB*, coding for the haemoglobin β subunit), and to confirm visceral leishmaniasis in a case of paediatric acute lymphoblastic leukaemia. The Illumina method was amongst those that used a genome-wide association study (GWAS; Section 3.1.8) to investigate genes linked to variation in skin colour in Europeans, focusing on *ASIP* at 20q11.22, *SCL45A2* at 5p13.2, *IRF4* at 6p25.3, *HERC2/OCA2* at 15q13.1 and *MC1R* at 16q24.3.

Key Point 3.1 Pyrosequencing, Illumina and some other forms of sequencing may be described as 'sequencing by synthesis' because the sample sequence is essentially the template for the generation, by synthesis, of a complementary strand by a polymerase.

Self-check 3.2 Describe the major innovation underlying the Illumina platform.

3.1.5 SOLiD Technology

The Theory

The sequencing methods described in previous sections rely on PCR: an alternative approach is sequencing by oligonucleotide ligation and detection, known as SOLiD. As with all methodologies, its development as a tool in molecular pathology relied on previous, often unrelated, techniques. One of these involved attaching bacterial DNA to a microbead that acts as a solid phase, whilst numerous reports described an oligonucleotide ligation assay. By 2008, the method had been refined for routine use and brought to market by Applied Biosystems. The principle relies on the binding of fluorescently labelled dinucleotides to the sample DNA to be sequenced, broadly as follows.

The first step is to ligate an adaptor oligonucleotide to each end of the DNA fragments to be analysed. This preparation is then presented to an excess number of microbeads in an emulsion, each with sections of DNA complementary to one of the adaptors, allowing the two to hybridise. The next step is a standard PCR so that the bound sample DNA is amplified on its own bead, which is then bound to the solid phase of a glass slide, forming an array. Following the attachment of a primer to the adaptor, what follows next is a complex series of reactions with an 8-mer probe carrying one of four fluorochromes and two nucleotides that may hybridise to the sample strand, and if so, a ligase will join the probe to the primer.

The unbound part of the probe is removed, fluorescence is recorded and a new 8-mer probe is provided, which will ligate to the remaining section of the previous probe, ultimately generating a product with a read length of 35 bps, although this can be extended to 40+. The cycle is repeated, but at each cycle a fresh primer is provided whose length is reduced by one nucleotide, and the sum of different fluorochromes is analysed to provide the nucleotide sequence. As much as 1–3 Gbps can be analysed per run. The analysis demands complex software, and the entire process is summarised as 'sequencing by ligation'.

The Practice

One of the first major uses of this technology was in the sequencing the entire genome of an individual of African origin. Compared with reference genomes, 3.8 million SNPs, 5590–226 539 insertion/deletions, 91 inversions and 4 gene fusions were detected. A subsequent clinical use of SOLiD was in the molecular pathology of deep vein thrombosis, a condition where a great deal of pathophysiology is already known, although in many cases precise mechanisms are unclear. The analysis of 186 haemostatic/inflammatory genes confirmed various features, such as the role of fibrinogen, and identified 127 new variants in 18 relevant genes, data that may help in diagnosis and management. A subsequent report used the technique to determine the presence of aneuploidy in circulating cell-free foetal DNA, reducing the need for the dangerous procedure of amniocentesis.

3.1.6 Luminex Multi-analyte Profiling (xMAP) technology

The Theory

In common with other platforms, this method uses microspheres as a solid phase to which sample DNA is attached, one difference being that the microspheres are dyed with one of two spectrally distinct fluorochromes. The DNA sample is amplified using standard PCR with primers and a standard master mix of Taq polymerase and dNTPs. The PCR products are then mixed with the beads, perhaps in a 96-well microtitre plate. The microspheres bear specific oligonucleotide capture probes for direct hybridisation with the sample DNA, followed by further chain-building. Reporter markers for the final stages include streptavidin-phycoerythrin binding to biotinylated PCR products. Further refinements to the method enabled the detection of up to 500 different mutations in a single well and analysis of up to 384 samples per run. Amplicon sequences are likely in the range of 100–300 bps to minimise interference with the beads. The assay can also be amended to accommodate mRNA binding to a bead, which can be the template for a cDNA.

The Practice

Several iterations of Luminex gene assays have been developed. One such assay used a quantitative approach to detect the *JAKV617F* variant product of *JAK2* at 9p24.1, coding a tyrosine kinase important in myeloproliferative neoplasms, directly from blood spots. Another used it as a rapid screening platform able to detect 31 mutations in *CFTR* responsible for cystic fibrosis. Others developed a multiplex assay for detecting miRNAs in lung cancer, for the simultaneous detection of ten viral pathogens in faeces causing gastroenteritis, and a test for the simultaneous identification of 36 mutations in four leading genes in colorectal cancer: *KRAS, BRAF, NRAS* and *PIK3CA*.

> **Key Point 3.2** SOLiD and Luminex platforms offer novel solutions to problems with certain types of analysis.

3.1.7 RNA Sequencing

The Theory

Section 2.4.2 has described methods for obtaining RNA, almost always for the purpose of generating a cDNA from which the sequence of the RNA can be implied. At this stage, sequencing RNA directly was inefficient because only short read lengths (25–55bps) were possible, and these had high (4%) error rates. By the mid to late 2000s, technology (such as the Illumina, Affymetrix and 454 platforms) had advanced to a position where high-throughput transcriptome sequences became possible, leading to the development of RNA-Seq. In some cases, mRNA is fragmented into 200–500bp pieces before cDNA is generated. In one study, RNA-Seq detected 20% more genes than did a microarray. However, another study reported that only 40% of exonic variants identified by WGS (see below) were captured by RNA-Seq. The method has been particularly useful in further studies of alternative RNA splicing, whilst another development has been merging with micro-cell biology in investigating mRNAs in single cells (hence scRNA-Seq) and single-nucleus RNA sequencing (snRNA-Seq).

The Practice

One of the first clinical reports used 454 pyrosequencing technology and cDNA formed from polyadenylated mRNA in a study of the transcriptome of cells from malignant pleural mesothelioma. When compared to normal lung tissues and lung carcinoma tissues, several novel mutations were detected, pointing to candidate genes such as *ACTR1A* and *MXRA5*. The classic gene fusion is *BCR*::*ABL1* in chronic myeloid leukaemia, but fusions are rarely seen in breast cancer. However, the use of RNA-Seq and array CGH found several fusions in breast cancer cell lines, one of which, *VAPB*::*IKZF3*, was postulated to have a role in cellular growth and survival, and if so, could be of diagnostic or therapeutic value. Moxon and colleagues used RNA-Seq in a case-control study of ischaemic stroke to report altered expression of mRNAs, miRNAs and lncRNAs, concluding that miR-374 and miR-486 may have a role in the diagnosis of this condition.

3.1.8 Wider Applications of Next-generation Sequencing

The power, speed and reducing costs of NGS permit large analyses not previously possible.

Genome-wide Association Studies

First described in the early 2000s, an early definition of a genetic association was simply an approach for identifying genetic variation that is linked to variation in phenotypes. Effectively the younger sibling of CGH, as described in Section 2.3.1, GWAS compares the genetic components of sections of DNA from patients with a particular disease or syndrome with those from subjects free of such abnormalities. Originally examining a small number of subjects, NGS technology such as those of Affymetrix and Illumina allowed these numbers to be increased to the thousands, providing much-needed assurance of the validity of their findings.

A common method to visualise differences between groups is the Manhattan plot, which places the chromosome number or gene locus along the X-axis and the probability that a particular SNP is linked to the phenotype along the Y-axis (Figure 3.6).

An early study of 1614 individuals used Affymetrix and Illumina technology to genotype 15 840 SNPs and 328 satellite markers in a study of alcoholism, whilst a case-control study involving 547 and 548 subjects, respectively, used an Illumina Genotyping Bead Chip to examine over 300 000 SNPs, finding a highly significant link between *IL23R* at 1p31 and Crohn's disease. Affymetrix technology was used to examine 520 000 SNPs, revealing almost 90 000 differentials between 134 nicotine-dependent smokers and 320 controls, with subsequent analysis identifying candidate genes such as *CNTN6, PDE4D, SLC9A9* and *ACTN2*. The Illumina Hap300 platform was used to compare over 4000 breast cancer patients with over 17 000 controls, finding several new SNPs, including one at 2q35 linked to a 1.44-fold increased risk of oestrogen receptor-positive disease and another at 16q12 linked to a 1.64-fold increase. As discussed in Chapter 4, these and similar methods help form panels of a defined series of genetic markers to directly help the clinician not only in diagnosis but also in management.

More recently, a GWAS using Illumina technology of over 300 000 SNPs in 199 women with gestational diabetes and 199 free of this condition implicated SNPs in *TRPV3, PRMT6, HHEX/EXOC6, GLIS3* and *LPIN2* (on chromosomes 17, 1, 10, 9 and 18, respectively). The same technology was used to determine the risk of leprosy in over 7000 cases and over 7500 controls. Data from over 450 000 SNPs confirmed 15 previously reported risk loci and identified 3 novel loci, increasing the known leprosy risk loci of explained genetic heritability from 23.0 to 38.5%. Other examples of the use of GWAS are shown in Table 3.1.

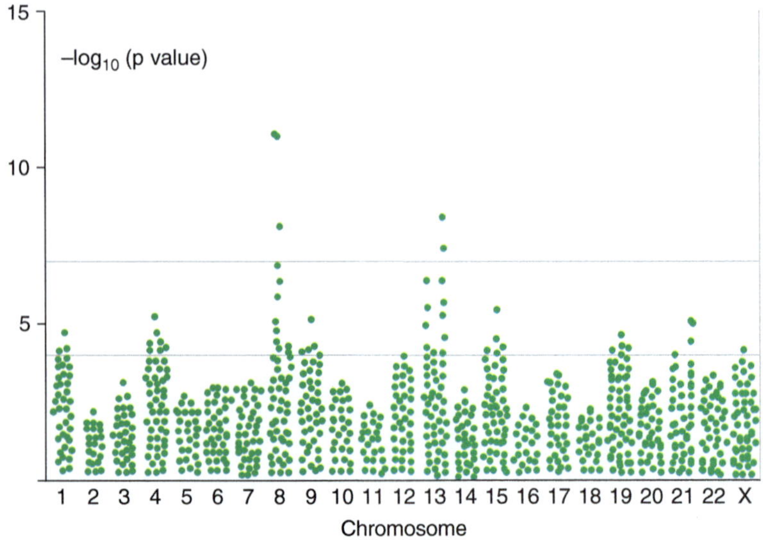

Figure 3.6 A Manhattan plot. In this very simplistic sketch, each dot represents a SNP. The vertical axis shows the inverse log of the *p*-value for the probability of the difference between two groups being significant. The horizontal lines are analyst-defined cut-off points, set at $p < 0.001$ (lower line) and $p < 0.000001$ (upper line), and are defined by the numerical power of the study and other factors. The most significant difference is a SNP in chromosome 8, whilst three other SNPs exceed the upper threshold of probability. SNPs between the two cut-off lines possibly demand additional attention. Note that the plot for chromosome 13 reveals two columns, potentially SNPs at different parts of the chromosome, but this may be an artefact.

Whole Genome Sequencing

This form of examination collects the sequence of all DNA, including both non-coding introns and coding exons. The former are important, as there are many examples of mutations in introns that influence the regulation of exons. As with many other technologies, whole genome sequencing (WGS) has its roots in microbiology – Sanger and colleagues published the entire genome of the bacteriophage virus φX174 in 1977, whilst that of the bacterium *Haemophilus influenzae* was published by Fleischmann et al. in 1995. In 1996, Goffeau and colleagues published the genome of the 16 chromosomes of the yeast *Saccharomyces cerevisiae*, whilst Hinds et al. probed the sequence in the entire human genome with resequencing microarrays in 2005, genotyping over 1.5 million

Table 3.1 Examples of the use of genome-wide association studies.

Analysis of 29,000 people reports 26 loci linked to epilepsy.
PHACTR1, at 6p24, is implicated in coronary artery disease, migraine headache, cervical artery dissection, fibromuscular dysplasia and hypertension.
A rare SNP 60 bp upstream of *ACE2* reduces the risk of infection by SARS-CoV-2 by around 40%.
Key features of over 12,000 SNPs can generate a polygenic score explaining 11–13% of the variance in educational attainment.
Body mass index is linked to 16 genes and to 500 distinct loci.
APOE4 causes widespread molecular and cellular alterations associated with Alzheimer disease.
Analysis of over 250,000 Europeans indicates that genetic input into height is characterised by a very large but finite number (thousands) of causal variants.
A meta-analysis of 7500 cases and 9000 reference samples found 41 distinct loci linked to type 1 diabetes.

SNPs. As discussed in Chapter 1, the Human Genome Project published the sequence of various chromosomes in the early 2000s.

The WGS literature is dominated by microbiology, such as examining antibiotic resistance and viral genome mutation. Recent clinical uses of the method include the identification of monozygotic twins, preimplantation testing for Merkel–Gruber syndrome, and phenotyping plasma cell-free DNA in preeclampsia, where differences in DNA methylation and nucleosome positioning were observed. A WGS study of 213 patients with age-related macular degeneration and 2704 controls confirmed links to four genes, but also reported new SNPs in *PARK7*, the lncRNA *AC103876.1*, and *TENM3*. These four reports all used Illumina technology. A GWAS meta-analysis of thyroid pathology in 2287 subjects found links to novel variants in *SYN2*, *PDE8B*, *NRG1* and a variant near *B4GALT/SLC25A52*, whilst another study of over 25000 subjects reported new loci at 17q21 and 6q22 with a strong effect on intra-cranial volume.

Self-check 3.3 A GWAS has been described as a genetic 'case-control' study. How can you explain this?

Whole Exome Sequencing

Since only around 1–2% of the genome codes for gene-active DNA (around 180 000 exons, of perhaps 30 Mb–6 Gb), whole exome sequencing (WES) is therefore considerably more efficient than WGS (around 90 Gb) in the pursuit of genetic aberrations that have a clinical dimension. This principle therefore relies on probes that can recognise intron/exon boundaries, and after having obtained fragments of exomes using exome capture methods, these can be sequenced using NGS methods. An alternative is to simply collect the transcriptome, which by definition is composed only of exon-derived material. To some extent, the basis of the technique has much in common with RNA-Seq described in Section 3.1.6.

Often cited as amongst the first clinical uses of WES in 2010, Ng and colleagues reported *DHODH* (at 16q22.2, coding for dihydroorotate dehydrogenase) to be the gene linked to the rare Mendelian disorder Miller syndrome, whilst Bowden et al. used Agilent and Illumina technology to link a variant of *ADIPOQ* (located at 3q27.3) with plasma levels of its product, adiponectin, a finding of potential pathological interest in diabetes, insulin resistance and obesity. More recently, the WES of 113 individuals from 19 families with adolescent idiopathic scoliosis revealed mutations in *BICD2* (at 9q22.31, coding for a molecule with a potential role in Golgi body physiology), which are already known to be linked to spinal muscle atrophy, and Neubauer and colleagues performed WES on 515 cases of sudden infant death syndrome, reporting potentially causative variants in 192 genes in 20% of the cases.

Table 3.2 summarises key features of leading NGS platforms.

> **Key Point 3.3** Driven by factors such as new techniques and automation, NGS was a major turning point in the widespread development and adoption of routine clinical molecular pathology.

3.1.9 Global and Other Perspectives

Sections 1.1.1 and 2.5.3 described the importance of the Human Genome Project, a turning point in the international recognition of the importance of molecular pathology. Since then, several other such global collaborations have further advanced our knowledge of molecular pathology.

Table 3.2 Key features of leading NGS platforms.

Pyrosequencing	Phosphate groups 'left over' from a nucleotide addition to an immobilised and developing DNA strand are substrates for a luciferase-based detection system.
Ion Torrent	As with pyrosequencing, this method exploits the proton released when a nucleotide is added to a growing chain and so effectively relies on an extremely sensitive pH meter.
Illumina sequencing	Library cDNA is ligated at both ends to adaptors that hybridise with reciprocal oligonucleotides in nanowells, forming a bridge for amplification and sequencing.
SOLiD technology	Library cDNA is ligated to oligonucleotides that hybridise to reciprocals on an emulsion microbead, followed by the binding of a dinucleotide part of an octamer linked to a fluorochrome.
Luminex xMAP technology	cDNA is amplified with a probe sequence that matches a reciprocal sequence on a bead with an internal fluorochrome, with subsequent steps adding a reporter marker to the developed chain.

The International HapMap Project

As the next step on from the Human Genome Project, this collaboration to map haplotypes saw scientists all over the world pooling resources to determine common patterns of DNA sequence variation. Mendelian genetics states that all genes have the potential to be expressed, or recombine, independently from each other, but if this is not the case and a set of genes (a haplotype) is not inherited with full independence, then linkage disequilibrium (LD) is a likely cause. One of the objectives of the HapMap project was to investigate patterns of LD in the genome.

One of the first major reports, published in 2005, was of the genotypes obtained from 269 individuals from diverse geographical regions (Europe, West Africa, East Asia). Using an Illumina BeadArray and other methods, phase 1 of the HapMap reported over a million SNPs, and much else. Subsequent publications provided details on 3.1 million SNPs and large inverse polymorphisms, highlighting potential uses in pharmacogenomics. Differences between geographic groups included positive selection for the *LARGE* and *DMD* genes in West Africa, probably linked to the consequences of Lassa fever virus infection, *SLC45A2* in skin pigmentation in Europe, and *EDAR* and *EDA2R* in the development of hair follicles in Asia. An important part of the Project is that all data were made freely available and in many cases are used as a reference genome for comparative studies.

Online Mendelian Inheritance in Man

A further free resource is a continuously updated catalogue of genes – that of the Online Mendelian Inheritance in Man (OMIM), which provides not only genetic information (size, location, form of inheritance, number of exons), but often also includes the function of the gene, its product, physiological and pathological aspects, and references to the relevant scientific literature.

The 1000 Genomes Project

This further international collaboration used GWAS methods to examine the genome of 1000 subjects from numerous global locations, seeking to define differences in structural variants, CNVs and SNPs, and accordingly had much in common with the HapMap and other projects. Active from 2008 to 2015, one of the last publications reported a theoretical typical genome, which differed from a reference genome at 4–5 million sites, with ~1000 large deletions, ~160 CNVs, ~1094 insertions and ~10 inversions, together affecting ~20 million bases of the nucleotide sequence. Most of these variants are rare – around 8 million have a frequency of >5% of the population, whilst 12 million have a frequency of 0.5–5%, and some 64 million are very rare, with a frequency of <0.5%. The scientific nature of the project is now maintained by the International Genome Sample Resource.

Reference Genomes

The sheer volume of the increasing number of genome sequences being published and freely accessible online led to their merging into a 'common' reference human genotype, against which an individual or groups of genomes can be compared. As the variety of published genomes increased, so did the number of sub-reference genomes, such as male/female and geographic region, which could be interpreted as race and/or ethnicity.

The leading curator of these data is the USA's National Library of Medicine, and within it the National Centre of Biotechnology Information. One key resource is RefSeq, which includes a diverse collection of genetic information, for both eukaryotic and prokaryotic organisms; as of the time of writing, GRCh38.p14 is the principal resource for the human genome.

There are many other resources freely available, such as ENCODE, an encyclopaedia of DNA elements, originally established as an extension of the Human Genome Project in 2003. One of the many products was an extensive map of DNA sections that are hypersensitive to DNAses, which may have a role in the functioning of regulatory elements such as enhancers and promotors.

At a national level, many countries have established collections of DNA and/or tissues from their own populations, often described as a Biobank. The UK Biobank was established in 2006, seeking to collect general health metrics and materials from 500 000 subjects aged 40–69. A case-control GWAS of over 24 000 heavy smokers and 24 000 never-smokers found links between four novel loci and lung function, and that these were linked to chronic obstructive pulmonary disease in the never-smokers. Other studies used an Affymetrix array in a GWAS to identify 15 new loci linked to coronary artery disease, which were used to develop a score to predict recurrent events and mortality.

Self-check 3.4 Which of the four aspects described in this section do you feel is the most important, and why?

3.2 Third-generation Sequencing

Three major new technologies developed in the 2000s were brought to the market in the 2010s. These technologies further reduced the cost of genetic analysis, partly by increasing the read length of the section of DNA strands being analysed. As described earlier, a disadvantage of certain NGS methods is that many can only generate shortish reads, but the key factor of Third Generation Sequencing (TGS) is that read lengths can be over 1000 bps and so may be described as long-read sequencing. A further common feature is that there are no pauses between read steps and the ability to sequence single DNA molecules free of the need for amplification and certain other processing steps, although certain pre-analytic preparation steps are necessary.

3.2.1 Nanopore Sequencing

The Theory

Although described as a clinically applicable TGS method in 2012, the derivation of the technology can be traced back to the 1980s. The principle relies on exquisitely sensitive and sophisticated software that detects changes in the electrical properties of electrolytes when different nucleotides of a single strand of DNA or RNA (including miRNA) pass through a micropore with a diameter of around 2 nm. This principle of objects interfering with the passage of an electrical charge is familiar to haematologists, as it is the basis of the numeration and size of red and white blood cells, and platelets, in the routine full blood count analysis. The nanopore nucleotide detection rate can exceed a thousand bases per second per pore, with the pore themselves formed from a bacterial cytolysin (such as α-haemolysin) or from a synthetic membrane (such as graphene or silicon nitride). In many cases, the nucleic acids to be sequenced require PCR amplification and other preparation before nanopore analysis (Figure 3.7).

The technology was brought to the wider clinical market by Oxford Nanopore Technologies, initially with the portable MinION analyser, which has 512 pores. It has proved popular amongst those in the field and with limited resources, an example being in identifying pathogens such as the Zika virus during localised epidemics. The GridION analyser is effectively five MinIONs scaled up to operate together, whilst the high-throughput analyser PromethION has 144 000 pores, such that the read length of a sample can exceed 2 million bps.

The Practice

The MinION analyser has been used to determine the sequence of 50 kbp sections of DNA of *CYPD6, HLA-A* and *HLA-B*, to clarify antibiotic resistance islands in *Salmonella typhi*, to sequence a complete influenza virus genome, and to detect aneuploidies such as Turner syndrome and trisomies of chromosomes 12 and 21. Others used this method to determine *TP53* deletions in chronic lymphocytic leukaemia, to map DNA

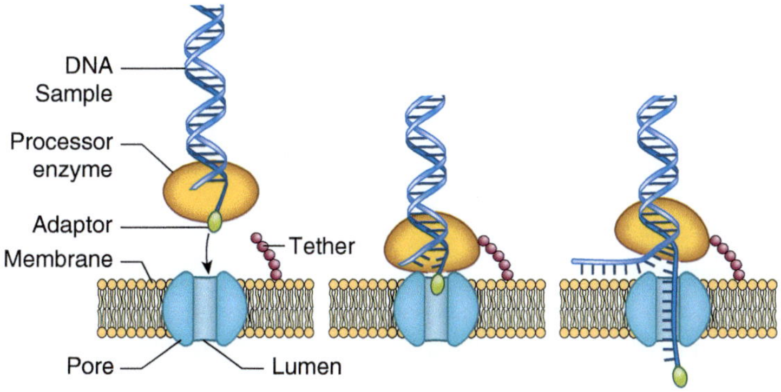

Figure 3.7 Basics of nanopore analysis. On the left, the DNA sample, complexed with a processor enzyme (a helicase) and an adaptor sequence, approaches a pore. As it is drawn in, the complex is held in place by a tether (to the right of the complex); the adaptor sequence locates with the entrance to the pore, and the helicase unzips the dsDNA. On the right, ssDNA passes through the pore, and the differences in the electrical charge linked to each nucleotide is detected and then recorded. The complimentary chain passes away from the pore, and once the ssDNA has been sequenced, the complementary strand and helicase move off the aperture and the pore is free for a further dsDNA complex.

methylation, in a routine setting for typing HLA class I molecules, and to further investigate the role of premature termination codons in *ABCA7* in 928 cases of early-onset Alzheimer disease.

The GridION analyser has been used to locate and define balanced and inversion translocations, and haplotypes around breakpoint regions, to report the genome of variants of SARS-CoV-2, the genome of a monkeypox virus and the genome of *M. tuberculosis*. These latter reports point to the value of the method for microbiology, a further example being the WGS of multi-drug resistance in a Shigella species, and so an examination of the genetic impact of anti-microbial resistance in this organism.

The power of the PromethION platform was shown by its 30x coverage of the human genome in a single flow cell during an investigation of tandem repeats in *ABCA7*, a 10 kbp gene associated with Alzheimer disease. Additional examples include sequencing 11 genomes (i.e. 33 billion bps) in 9 days, obtaining data relevant to pharmacogenomics on over 1000 genes (such as *CYP2C8, UGT1A1* and *VKORC1*). Another example is the characterisation of circular RNAs. Clinical use included long-read sequencing of cancer cell lines, the genome of SARS-CoV-2 and an investigation of genome AAGGG or ACAGG repeats in 216 patients with late-onset ataxia.

3.2.2 Pacific Biosystems

The Theory

With a history going back to (at least) 2006, this sequencing technique relies on a DNA polymerase fixed to the bottom of a nanowell (diameter 70 nm, depth 100 nm) which is illuminated from below by a detection systems called a zero-mode waveguide (ZMW), a platform launched in 2010. Sequencing begins with the preparation of a dsDNA with hairpin loops at each end, to which adaptors are added by ligation, in common with many other systems. The DNA sample to be analysed is placed in the well (which has some 150,000 ZMWs) and forms the template for the construction of a second strand, much like other methods that use sequencing by synthesis (Figure 3.8).

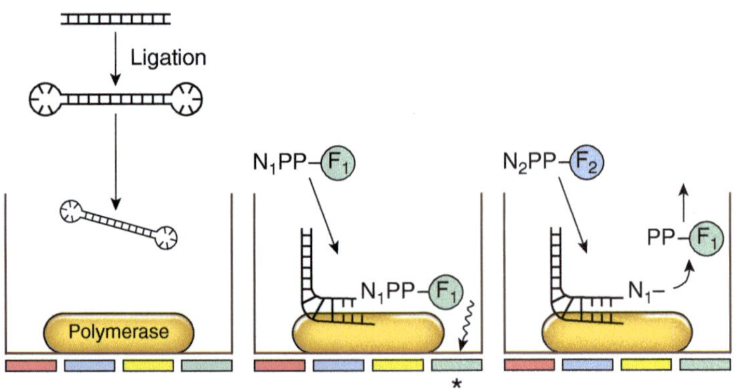

Figure 3.8 Basics of Pacific Biosystems sequencing. The process begins (top left) with hairpin sections ligated to the dsDNA to be sequenced. This is added to a nanowell with a DNA polymerase fixed to the bottom, below which are the ZMW detectors. Four forms of nucleotide pentaphosphate, linked to a fluorochrome (NPP-F) are added. Each time the polymerase adds a nucleotide the fluorochrome is detected by the ZMW imaging software (*). Upon completion on the polymerisation, the excess phosphate-fluorochrome (PP-F) diffuses away and the polymerase is ready for the next nucleotide.

The DNA polymerase unravels the dsDNA and builds a new chain. However, unlike other methods the nucleotide building blocks have a chain of five phosphates linked to a nucleotide-specific fluorochrome. The ZMW imaging system recognises the fluorochrome of each nucleotide as it is added to the developing strand, and so may be described as a single molecule real time (SMRT) method. With this system, long reads of 100 kbps are possible at a rate of 1–3 nucleotides per second, and although the random error rate may reach 10–15%, with multiple reads the true sequence will be formed. Detection of methyl- modification of bases is also possible.

The Pacific Biosystems (PacBio) Sequel System, developed in 2015, presented a million ZMWs generating 500 000 reads of an average length of 10 kbps, whilst the Sequel II, with 8 million SMRT wells, provided 4 million reads in a 30-hour run.

The Practice

Early uses of PacBio technology included sequencing the genome of a bacteriophage from *Listeria monocytogenes*, the sequence of the 4.4 Mbp genome of *Mycobacterium tuberculosis*, and the sequencing of 14 WHO gonococcal reference strains enabling the identification of those with low- to high-level resistance to cephalosporin. The method has also been used in the sequencing of the influenza virus, hepatitis B virus and hepatitis C virus, and of bacteria such as *Leishmania*, *E. coli, Klebsiella, H. pylori* and *Haemophilus* species, where a multiple drug-resistance gene island was identified.

Clinical studies have used PacBio technology to identify and define pathology such as *TCOF1* (at 5q32–33.1, coding a participant in ribosome formation) in Treacher–Collins syndrome, *PTPN11* (at 12q24.13, coding a tyrosine phosphatase) in male subfertility, the detection of tyrosine kinase inhibitor resistance mutations in the entire 1578 bp *BCR::ABL1* fusion, novel fusions in *RLN1* and *RLN2*, both coding for isotypes of relaxin, in prostate cancer, the former being under-represented in common cell lines, and of *IKBKG* (at Xq28, coding an inhibitor of an NFκB subunit), loss-of-function mutations in which lead to recurrent microbial infections, and a number of other features. Correct identification of mutations in this gene are important because of *IKBKGP1*, a nearby pseudogene.

An ultra-deep PacBio Sequel System was used to identify 1058 newly identified viral genomes in a sample of faeces, whilst a subset of 14 viruses could be used as colorectal cancer biomarkers. A further investigation of the enteric virome (as harvested from faeces) of over 22 000 phage genomes reported a high degree of methylation that may reflect virulence.

Self-check 3.5 How would you describe the advances made by third-generation sequencing over NGS?

3.2.3 DNA Nanoball Technology

The Theory

This technique has adapted a method used by various microorganisms (such as certain herpes and papillomaviruses) to replicate their DNA. In this reinvention for our own species, as with almost all other methods, step one is to prepare the sample DNA and render it down to fragments, in this case from 100 to 350 bps. Step two, also well known, involves using ligation to place adaptor molecules onto the ends of the DNA, and the sample is amplified by standard PCR. The process is also suitable for RNA processing.

The unique step with this platform is that further adaptors (splint oligos) are ligated onto the amplicons to convert the sample into an ssDNA circular molecule. A φ29 polymerase from a bacteriophage that infects *Bacillus subtilis* is then applied to the sample to affect a rolling circle replication, causing thousands of the products to self-associate into what is described as a nanoball, which has a diameter of some 230nm. The nanoball preparation is loaded into a flow cell on a microwell, and stepwise sequencing begins through a process known as combinatorial probe anchor synthesis. As with other platforms, the next step involves sequential presentation of nucleotides (linked to a particular fluorochrome) to the nanoball. These nucleotides bind to the sample strand, and the binding is monitored, with the data collected for analysis, i.e. a form of sequencing by synthesis.

The developers of the method, Complete Genomics, reported sequencing three human genomes, the technology since being acquired by the Beijing Genomics Institute, which in 2015 amended it to suit a dedicated analyser, the BGISEQ-500. A subsequent use was in the investigation of miRNAs. Figure 3.9 summarises the technology.

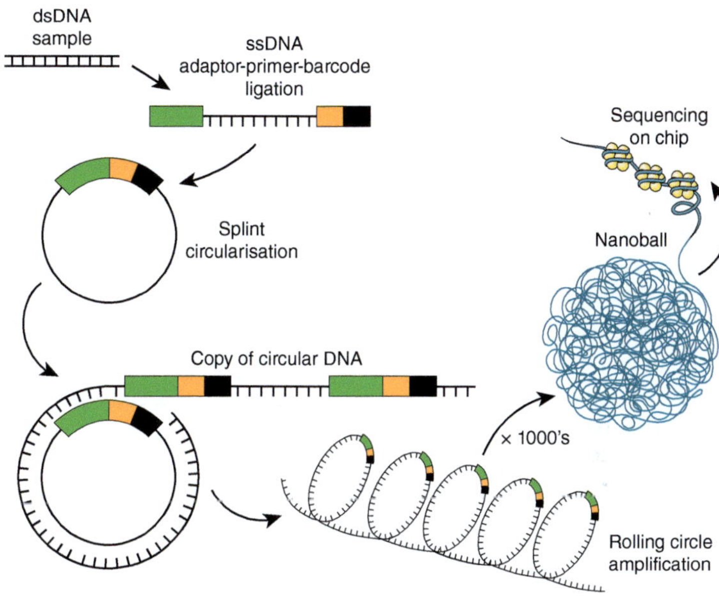

Figure 3.9 Nanoball technology. The DNA sample is fragmented and loose ends are blunted. The dsDNA is converted to ssDNA, and the ends are ligated to dedicated adaptors that allow splint circularisation. PCR amplification of the synthesised circular DNA proceeds and a critical mass is formed, i.e. a nanoball. This is then fed to a chip with nanowells, each of which accepts a nanoball. In the nanowell, NTPs linked to a fluorochrome build a strand by combinatorial probe anchor synthesis. The attachment of an NTP is detected by an imaging system, with the data being passed to an analytical programme.

The Practice

One of the first uses of the technology examined primary tumour tissues, metastatic lymph nodes and whole blood from a case of breast cancer to determine the heterogeneity of the genetic material. Although the proportion of SNPs, insertions, deletions and substitutions were similar between primary tumour and involved lymph node, mutations in *PDS5B, TGIF2* and *CCNB2* were unique to nodal tissues, pointing to genes linked to metastatic potential. Subsequent studies reported its use in detecting prenatal sex chromosome aneuploidies, autosomal dominant polycystic kidney disease, one of the first descriptions of the SARS-CoV-2 transcriptome, roles of miRNAs in erythropoiesis, and in determining bacterial genomes.

3.3 Bioinformatics

For the uninitiated, bioinformatics can appear an exceptionally complex field: what follows is the briefest of introductions. The earliest 'ladder' Sanger sequences, with each base instantly recognised, were translated with relative ease, although fully manual and time-consuming. However, as sequencing in the era of multiplexing and automation became more complex, previous methods for analysis became inadequate, calling for new, more bespoke methods, i.e. bioinformatics.

Once sequencing data was digitised, computer analysis became possible and, in fact, necessary due to the mass of information. Regardless of the platform, all digital material must be deciphered and converted into an understandable result, and the quality of the sequence reads assured. There is no shortage of software packages to assist the practitioner, some of which are specifically linked to a particular method, whilst others can be customised. One such family of software is sequence alignment, which seeks to find consensus sections of a number of separate fragmented

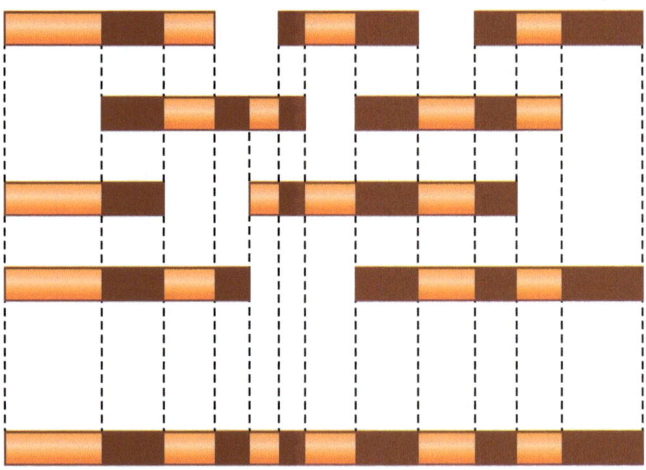

Figure 3.10 Sequence alignment. A simple view of the alignment of nine short sections of DNA of varying lengths (upper four rows), possibly derived from different analyses or platforms, which can be merged to form a consensus view (bottom row) highlighted by open and shaded sections. Consensus sections range from two to four per section, the latter giving more confidence. The sequence of the long consensus construct at the bottom may be checked against established sequences available from online databases.

DNA sections (Figure 3.10), so that when a gene is recognised, examples of which include ELAND, MAQ, BLAST, Burrows-Wheeler, MiniMap2 and Bowtie2.

For many years, the sequencing community has been placing its findings on freely available databases, and as such it can be accessed. Thus, sequences collected by a particular analysis can be checked to confirm any particular identity.

3.4 Chapter Summary and Conclusions

3.4.1 The Technology

Sections 3.1 and 3.2 have introduced a series of technological platforms for performing a variety of genetic analyses, and as such the question that naturally arises is which is best. The answer depends on what the practitioner is seeking to achieve, and in what arena – research, epidemiology, routine practice (diagnosis and management), the minimum mass of nucleic acid required, etc. Although there are common themes, certain features will dominate the choice from a number of criteria, which include ease of operation, error rate, read length, speed of action/turnaround time, etc. There are numerous published comparisons of two or three platforms, and some comparing up to five are shown in Table 3.3.

In many cases, suppliers can provide a complete package of analysers and reagent kits for both general and specific analyses, although the market has become so sophisticated and the reagents so standardised that practitioners are no longer obliged to use certain products with certain analysers.

Caveat Data shown in Table 3.3 is time-limited: it is very likely that certain of these metrics will have changed since they were published. The reader must refer to current information.

3.4.2 The Practice of Molecular Pathology

Compared to other pathology disciplines, many suggest that molecular pathology is of increased complexity and relies more and more heavily on technologically advanced equipment that is often robotic. Its recent

Table 3.3 Key features of major platforms.

Platform	Error rate[a]	Error rate[b]	Read length[b]	Read length[c]	Read length[d]	Max. run time[c]	Total run time[d]
Roche 454	0.72%	1.0%	500	–	–	–	–
Illumina	0.29%	0.58%	2 × 190	2 × 150	150	44	48
Ion Torrent	1.63%	1.78%	200	600	360	12	48
Complete Genomics/ BGI	2.32%	1.0%	2 × 35	400	100	40	36
PacBio	16.19%	13.0%	40000	25000	40278	30	30
Nanopore	–	38.2%	2000	4000	60869	72	–

Some data pooled from different analysers.
[a] Laehnemann D, Borkhardt A, et al. Briefs Bioinformat. 2016;17:154–179.
[b] Ambardar S, Gupta R, et al. Indian J Microbiol. 2016;56:394–404.
[c] Pervez MT, Hasnain MJU, et al. Biomed Res Int. 2022. doi:0.1155/2022/3457806.
[d] Meslier V, Quinquis B, et al. Sci Data. 2022;8:694 (which focuses on microbial metagenomics).

development has enabled a centralisation of specialised and reference labs, wherein the required degree of expertise can more effectively manage the workload. As indicated in Section 1.5.2, in the UK, this centralisation has resulted in the establishment of a small number of Genomic Laboratory Hubs, situated in areas of high population density and generally linked to a nearby university.

As a sub-discipline of pathology, molecular pathology serves clinicians, many of whom have medical qualifications, who seek support in managing their patients. This can be seen not only in terms of disease diagnosis and its management (with drugs, surgery, radiotherapy, etc.) but also by the age of the presenting patient. The next four chapters will explore how molecular pathology can provide this support in the major disease groups and conditions likely to develop in routine clinical practice.

Self-check 3.6 Confirm with yourself that you have completed the learning objectives at the beginning of this chapter.

Further Reading

Genomes Project Consortium; Auton A, Brooks LD, Durbin RM, et al. A global reference for human genetic variation. Nature. 2015;526:68–74.

Athanasopoulou K, Boti MA, Adamopoulos PG, et al. Third-generation sequencing: The spearhead towards the radical transformation of modern genomics. Life. 2021;12:30. doi:10.3390/life12010030.

Giani AM, Gallo GR, Gianfranceschi L, et al. Long walk to genomics: History and current approaches to genome sequencing and assembly. Comput Struct Biotechnol J. 2019;18:9–19.

Glotov OS, Chernov AN, Glotov AS. Human exome sequencing and prospects for predictive medicine: Analysis of international data and own experience. J Pers Med. 2023;13:1236.

Larson NB, Oldberg AL, Adjei AA, et al. A clinician's guide to bioinformatics for next generation sequencing. Thorac Oncol. 2023;18:143–157.

MacKenzie M, Argyropoulos C. An introduction to nanopore sequencing: past, present, and future considerations. Micromachines. 2023;14:459. doi:10.3390/mi14020459.

Muzzey D, Evans EA, Lieber C. Understanding the basics of NGS : From mechanism to variant calling. Curr Genet Med Rep. 2015;3:158–165.

Uhlen M, Stephen R, Quake SR. Sequential sequencing by synthesis and the next-generation sequencing revolution. Trends Biotechnol. 2023;41:1565–1572.

Websites

The NHS England site for clinical genetics: https://www.england.nhs.uk/publication/national-genomic-test-directories/

International Genome Sample Resource: https://www.internationalgenome.org/

Online Mendelian Inheritance in Man (OMIM): www.omim.org

Repository of various human genomes: https://www.ncbi.nlm.nih.gov/datasets/genome/?taxon=9606

ENCODE, an encyclopaedia of DNA elements: https://www.encodeproject.org/

4

Solid Tumour Cancer 1: Carcinoma

Having introduced the subject of molecular pathology and described key laboratory platforms in Chapters 1–3, we now move to the application of the techniques in the study of human disease. Table 1.1 of Chapter 1 shows the classification of diseases, whilst Table 1.2 points to the leading causes of death in England and Wales. We begin our exploration of human disease with cancer, as it is the most common cause of death worldwide, most probably also in Scotland and Northern Ireland, and the World Health Organization (WHO) reports that, globally, cancer caused 10 million deaths in 2020, which is almost one in six of all deaths.

However, there are many forms of cancer, and in this chapter, we will explore the most common form – carcinoma – leaving other forms for the chapters to come. We begin in Section 4.1 with a general introduction to cancer as a whole, which details the various forms of the disease, and followed by Section 4.2 with a discussion on the causes of cancer, an essential topic in view of the changes in cell biology linked to the development of malignancy. The remaining Sections 4.3–4.8 focus on the molecular pathology of leading carcinomas, broadly in order of their mortality rate within the population, starting with lung, followed by colorectal, prostate and breast cancer.

Molecular Pathology: A Primer for Laboratory Scientists, First Edition. Andrew Blann.

Learning Outcomes

After studying this chapter, you should be able to...

- Describe the process of carcinogenesis
- Explain the meaning and importance of tumour suppressor genes
- Understand the basics of the classification of cancer
- Provide a list of the most frequent and most lethal cancers
- Comment on the role of molecular pathology in lung and colorectal cancer
- Summarise those genes most relevant to prostate and breast cancer

4.1 Introduction to Cancer

4.1.1 General Comments

Basic Concepts

As with many aspects of science, cancers are named according to an established system, often with Greek or Latin roots. Cancer is derived from the Greek word *karkinos*, meaning a crab, as a cancer often has a central body with radiating 'legs', and we also have *ónkos*, which translates to mass or tumour, and so oncology, the study of cancer. A major feature of cancer is that it spreads from its original size (the body of a crab) and hence known as the primary tumour. The malignant cells that spread to other tissues and organs are therefore known as secondary (the legs of a crab) tumours. A primary tumour sending out secondary tumour is said to be metastatic, with the secondary tumours themselves being metastases. Certain organs are often the target of specific primary cancers, such as breast and prostate cancers sending secondaries to the bone.

Not All Tumours Are Cancers

There are several examples of non-physiological cell masses that are not malignant. A mass of lipids or fat is called a lipoma, which (generally) has no occult significance. A haemangioma is a knot of blood vessels, whilst a fibroma is a mass of fibrous connective tissue, which also generally has no severe pathological implications. But some masses can be pathological – an example being a granuloma, which is a mass of macrophages and neutrophils, brought together by a particular inflammatory event.

Not All Cancers Are Fatal

A further aspect is not merely the mortality of cancer but also its morbidity. The International Agency for Research on Cancer, a WHO-affiliated body, describes 19.2 million new cases of cancer in 2020, and the fact that these cases are linked to the 10 million deaths implies a great deal of disease that is not (yet) fatal. Indeed, some cancers are very slow-growing, and the patient may succumb to a different disease process. It follows that the overall burden of cancer (mortality + morbidity) and the impact it has on the friends and relatives of those with the diagnosis represent a considerable burden on society and the health care industry.

4.1.2 Classification of Cancer

Cancers may be classified in a number of ways, such as by the organ in which they appear: those of a 'hard' or 'solid' nature, typically a solid mass of abnormal cells, and those of a 'soft' nature, where there is no clear solid mass, with the cancer having a fluid composition. The former includes cancers of all major discrete body organs (lung, colon/rectum, breast, prostate), whilst the latter effectively refers only to cancers of the blood. Most will agree that the principal classification system for cancer is based on the underlying embryonic/cell biology nature of the transformed cells. This system is also present in the National Genomic Test Directory for England, although those of the other UK nations may differ. It follows that the reader must refer to their local directory. The most common forms of cancer (from the cell biology aspect) are as follows:

Carcinoma

These tumours are derived from epithelial tissues and are the most common form of cancer, such as lung cancer and colo-rectal cancer. An adenoma is a non-malignant tumour of a gland and can be endocrine or exocrine. However, should it become malignant (as many do), it is then classified as an adenocarcinoma and often secretes its product at an increased rate, such as mucus in colorectal cancer, or insulin in pancreatic cancer. Carcinomas may also be named according to the size of the malignant cell – large or small – or by the type of epithelium, such as squamous cell carcinoma or basal cell carcinoma.

Sarcoma

These tumours are derived from mesenchymal cells that give rise to connective tissues and may be further named by a particular tissue that has become malignant. Hence, a leiomyoma is a tumour of smooth muscle cells, whilst a rhabdomyoma is a tumour of striated muscle cells. An osteosarcoma is a cancer of the bone, and Ewing sarcoma is a tumour that forms in soft tissues or in bone. Further details are provided in Chapter 5.

The Central Nervous System

An important feature of carcinogenesis is the frequency of mitosis, at which point abnormalities may appear in the newly formed chromosomes that, if not repaired, may lead to cancer. Cells of the brain and peripheral nervous system (i.e. neurons) are terminally differentiated, that is, they do not undergo mitosis, and so it is most unlikely that true neurological cells will become malignant in this way. However, there are other cells that support the neuronal tissues, which do undergo mitosis, and so may become malignantly transformed. These non-neuronal glial cells have several subtypes such as Schwann cells, astrocytes and oligodendrocytes, and some specialised macrophages. Further details are provided in Chapter 5.

Paediatric Cancer

Cancers in young people are often placed in their own section as they are generally more aggressive than those in adults, often with features that are ‘embryonic’, i.e. they resemble rapidly growing tissues at an early stage of development in the uterus, and so should not be active after this stage of development. However, the anatomic features of many of these cancers bear much in common with adult diseases, such as leukaemia, kidney sarcoma and cancers of the central nervous system. Conversely, several ‘adult’ cancers are almost never seen in a paediatric setting, including the breast, prostate, lung and colorectal regions. Further details are provided in Chapter 5.

Blood Cancer

The type of cancer described earlier forms solid masses of abnormal tissue within its ‘host’ tissue and hence is called solid tumour cancer.

In contrast, blood cancers do not form a solid mass – the malignant cells are often dispersed as single cells, or clumps of cells, all over the body in the vascular system. However, although it could be argued that one type of blood cancer – the lymphoma – is indeed a solid mass of tissue within lymph nodes, the root malignant cell is a lymphocyte. It could equally well be argued that many of the blood cancer cells of leukaemia or myeloma form a semi-solid mass within bones. Furthermore, leukaemia may metastasise to the lymph nodes, and lymphoma may metastasise to the bone marrow. Nevertheless, all three types of blood cancers, and other allied blood-related neoplasms, are better explained when grouped together in Chapter 6.

> **Key Point 4.1** Cancer is the leading cause of death in England and Wales, and so probably also in the whole of the UK.

4.1.3 The Incidence and Mortality of Malignant Cancer

The frequency of deaths caused by specific cancers is shown in Table 4.1. However, any data set is only as good as the method for collecting the data, which in this case depends on the accuracy of the diagnosis of the cancer. The Office for National Statistics (ONS) data sets include the International Classification of Disease (ICD) classification of 'malignant neoplasms of ill-defined, secondary, and unspecified sites', which, in England and Wales in 2023, accounted for 8911 deaths, making it the 6th largest category (6.2% of cancer deaths) and thereby exceeding cancers of the pancreas, oesophagus, liver and others. Many are likely to reflect metastatic disease where the primary tumour cannot be located.

There is considerable data on global and national cancer rates. The WHO and UK government collect and publish data on the number of new cancer cases and cancer-related deaths reported annually. Although these two sets of data are not from the same time period, they do nonetheless give an indication of the relative mortality of each.

Overall, 327,174 new cases of cancer were reported, and with 148,818 deaths, this gives a crude 'mortality rate' of 45%, implying 55% of cases are either cured or the subject died of another cause. Data from Europe estimates 199,995 cancer cases in women in 2022, with 84,171 deaths,

Table 4.1 Frequency of leading malignant cancers in England and Wales in 2023.

Type of cancer	Number	%
All malignant neoplasms	152 418	100
Bronchus and lung	29 398	18.3
Colorectal	14 869	10.2
Blood	12 193	8.0
Prostate	11 072	7.3
Breast	9973	6.5
Pancreas	9151	6.0
Oesophagus	6775	4.4
Liver and intrahepatic bile ducts	5616	3.7
Bladder	5202	3.4
Brain, meninges and CNS	4041	2.6
Kidney	3680	2.4
Oral cavity and throat	3326	2.2
Ovary	3319	2.2
Stomach	3173	2.1
Uterus	2401	1.6
Melanoma	2401	1.6
Mesothelioma	2018	1.3

Figures for deaths do not sum to 100% as only selected categories are presented.
Source: Adapted from Nomis web / https://www.ons.gov.uk/peoplepopulationandcommunity/birthsdeathsandmarriages/deaths/methodologies/mortalitystatisticsinenglandandwalesqmi / Last accessed at 08/09/2024.

giving a mortality rate of 42.1%. Parallel data in men is 217,526 cases (8.8% higher than in women) with 96,123 deaths (14.2% higher) and an overall mortality rate of 44.2% (5% higher). Crucial to both diagnosis and treatment of cancer are imaging techniques (such as ultrasound, X-rays, MRI and CT scanning) and laboratory tests, with molecular genetics playing an important part.

Self-check 4.1 Blood cancer is the third most common cause of cancer-related death – what are its three major forms?

4.2 The Cellular Basis of Cancer

4.2.1 Carcinogenesis

By and large, all cells will perform their functions until they reach the end of their 'natural' life. The death of a cell may be brought about prematurely by factors such as an insurmountable infection, trauma or cancer. The transition from physiology to cancer pathology is described as a malignant transformation and may be caused by factors such as toxins or radiation. These factors are carcinogens, acting to induce genetic changes and so induce the process of carcinogenesis.

The consequences of carcinogenesis depend on the nature of the transformed cell. A key feature is that the cell loses its normal growth parameters and replicates inappropriately, forming a mass that will grow into a tumour, often described as a neoplasm (i.e. new growth), from which we get neoplastic transformation. In many cases, the basic physiological nature of the transformed cell continues when it becomes cancerous, so that a tumour whose origin was a cell of the endocrine system would result in a high level of the particular hormone. As described earlier, an example of this is a tumour of the β cell in the Islet of Langerhans (hence insulinoma) secreting high levels of insulin, with consequences for levels of blood glucose.

A key component of a growing tumour is that it will have an increasing demand for oxygen and nutrients, calling for an enhanced blood supply. The development of a vascular system is angiogenesis, which can have an entirely physiological component, but it can be induced by a growing malignancy and as such may be described as neoangiogenesis.

4.2.2 Genes, Proto-oncogenes and Oncogenes

A carcinogen, in its broadest sense, is likely to affect the changes of a normal gene to a proto-oncogene, and then a 'resting' proto-oncogene to a fully functioning oncogene that mediates the malignant transformation.

The leading process by which this occurs is errors in passing messages from the outside of the cell to the nucleus. These abnormal genes exert their effects on this messaging system via a number of diverse mechanisms:

1. Many cells rely on the steady supply of low levels of growth factors for their day-to-day survival. Increased plasma levels of these growth factors, such as platelet-derived growth factor, vascular endothelial growth factor (VEGF) and fibroblast growth factor, can push the cell into an accelerated growth phase, and so a potential tumour. These increased growth factor levels may themselves be driven by oncogenes acting on the promotor region of the particular gene of the growth factor in question.
2. These soluble growth factors interact with their target cell via a specific receptor, e.g. the VEGFR, on the outer surface of the cell membrane. The docking of these soluble ligands with their receptor sets in motion the stimulation of a complex series of molecules within the cytoplasm described as second messengers, and which pass the information to the nucleus. Over-expression and over-stimulation of these growth factor receptors by their ligands therefore lead to over-stimulation of the cell.
3. Oncogenes may code for mutant receptors that may be permanently switched on in the absence of their ligand binding, which has the same effect as mechanism 1, i.e., an over-stimulation of the second messenger pathways.
4. Similarly, abnormal genetic regulation of these intracellular second messengers may result in them permanently passing inappropriate instructions to the nucleus regardless of the state of activation of the ligand/receptor (points 1–4 are explained in the following section).
5. Once inside the nucleus, messengers are subject to additional regulation that controls the extent to which genes are moved to transcribe DNA into RNA. Failure of this regulation therefore leads to inappropriate gene activity with potential malignant consequences.

A separate aspect of carcinogenesis is apoptosis, which is a natural process for eliminating damaged or defective cells that are controlled by various genes and their products. Should there be loss-of-function mutations in these genes, apoptosis may not occur, and the damaged cells survive. Should these damaged cells have malignant potential, cancer may develop.

4.2.3 Genes and Cancer

Molecular genetics has identified hundreds of genes whose products have major roles in these processes, and all are subject to mutations that can lead to cancer. A selected group of these genes and their role in cancer are shown in Table 4.2, with a brief explanation of the roles of certain of these described in the section that follows.

Notably, several of these genes (the receptors in particular) are linked to tyrosine kinase (the importance of which will be described shortly), whilst others are tumour suppressors: molecules that act to prevent the formation of a cancer, as discussed in Section 1.1.2.

Tumour suppressor genes are crucial regulators, and dysfunction are implicated in many diverse cancers: *TP53* is mutated in 46% of cancers, *KRAS* in 32% and *EGFR* in 27%. Some of these abnormalities impact on the choices of chemotherapy. The same oncogene may also be found in many cases – in one study a pathological variant of *KRAS* is found in pancreatic adenocarcinoma (75% of cases), colorectal carcinoma (40%), colorectal adenoma (39%), bile duct cancer (31%), lung cancer (25%), liver, stomach and renal cancers (all 7%), but in one study, none of these were found in 20 melanomas.

Referencing Genes in This Book As we move to clinical molecular pathology, additional details of genes will be provided. Table 4.2 shows a selection of genes of interest to oncologists, their location in the genome, and a brief descriptor of their function. In the remainder of the text, as other genes are first mentioned, so too will be their location and a note of their role in cancer. It follows that a gene not followed by its location or function will already have been mentioned. The index provides a shortcut to the referrals of all genes.

Abnormal Cytoplasmic Functioning in Cancer

As introduced in the previous section, the cell receives signals from the extracellular environment via its cell membrane. An example of this is the effect of insulin in regulating the passage of glucose into the cell, the first step being the binding of the hormone on to its receptor, which is a

Table 4.2 Selected genes linked to cancer.

Gene (location)	Relevance
ABL1 (9q34.12)	A proto-oncogene that forms a neo-oncogene when fused with *BCL* (at 22q11.23) to form the Philadelphia chromosome and which causes chronic myeloid leukaemia.
ALK (2p23.2-23.1)	Anaplastic lymphoma kinase. This forms a fused oncogene when translocated next to the nucleophosmin gene.
BRCA1 (17q21.31)	This tumour suppressor gene (Breast Cancer type 1) codes a protein with a number of functions, including repair of damaged DNA.
CDKN2A (9p21.3)	Cyclin-dependent kinase inhibitor 2 codes for p16, a part-regulator of the cell cycle and so a tumour suppressor.
EGFR (7p11.2)	Codes for the epidermal growth factor receptor, which has tyrosine kinase activity. It is often over-expressed in various cancers such as of the breast and lung.
KRAS (12p12.1)	Codes for a second messenger K-Ras, part of the signal transduction pathway. Mutations are present in a large and varied number of malignancies of solid tumours and in blood cancers.
MYC (8q24.21)	A family of genes coding for transcription factors. *N-MYC* is linked to numerous cancers, such as neuroblastoma, *C-MYC* to lymphoma.
PDGFRA (4q12)	Codes for the alpha subunit of platelet-derived growth factor receptor, over-expression linked to various blood cancers.
PTEN (10q23.31)	Codes for tumour suppressor phosphatase and tensin homologue (PTEN), often linked to glioblastoma and prostate, breast and lung cancer.
RB1 (13q14.2)	Codes for the tumour suppressor retinoblastoma protein: loss-of-function mutations have been reported in many cancers.
SMARCA4 (19p13.2)	Codes for transcription factor BRG1, expressed in elevated levels in some breast and pancreatic cancers, melanomas and gliomas.
TP53 (17p13.1)	Codes for tumour suppressor p53: absence has been reported in all major cancers and is linked to poor overall survival.

component of the cell membrane. Similarly, in order for a cell to grow, it requires a signal such as that provided by epidermal growth factor (EGF), which must bind to its receptor in the cell membrane, i.e. the epidermal growth factor receptor (EGFR, coded for by *EGFR*), also known as Erb-1, and HER1. This interaction initiates a chain of biochemical reactions within the cytoplasm resulting in the activation of certain genes in the nucleus, the entire process being signal transduction.

Once a receptor such as the EGFR has bound to its ligand (EGF), an enzyme on the intracellular part of receptor mediates the transfer of a phosphate group from its donor substrate (e.g. ATP or GTP) to another molecule, which is therefore phosphorylated. Enzymes of this type are kinases, and in many cases, the receiving molecule is the amino acid tyrosine, hence being referred to as tyrosine kinase (TK), in addition to other kinases mediating the phosphorylation of serine/threonine. Once phosphorylated, the stimulus is cascaded down to one or more intra-cytoplasmic second messengers that include cyclic AMP, phosphatidyl inositol and calcium, whilst further downstream messengers include NFκB, and the enzymes Raf (coded for by *RAF1* at 3p25.2) and Ras (coded by *KRAS*). Following Ras down the signal transduction chain are other messengers such as a member of the mitogen-activated protein kinase (MAPK) family, which can act on a nuclear phosphoprotein transcription factor product of *MYC*. A very simplified cartoon of this process is shown in Figure 4.1.

Key Point 4.2 A common feature of cancer is the increased expression of growth factor receptors at the cell surface, which may be used in diagnosis.

It follows that kinases are important steps in the signal transduction pathway that leads to activation of second messengers. Accordingly, inhibition of the enzyme may be a therapeutic option in those with an over-active tyrosine kinase. This will be discussed in a clinical setting in sections to come.

Self-check 4.2 How many of the genes in Table 4.2 are tumour suppressor genes?

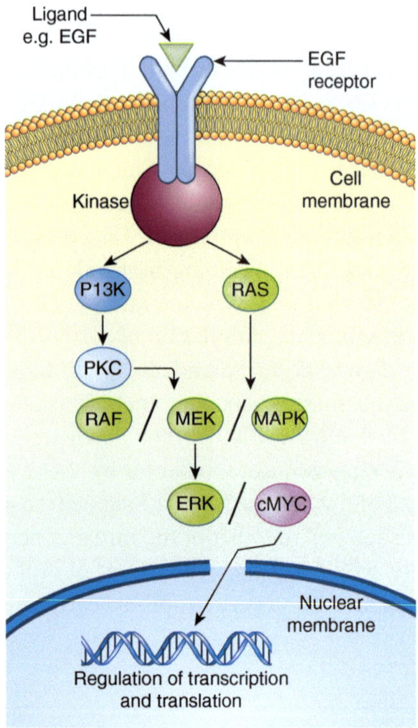

Figure 4.1 Simplified representation of a signal transduction pathway.

4.2.4 Developing Initiatives in Cancer Genetics

Non-coding RNAs

As introduced in Section 1.3.1, and mentioned in passing elsewhere, there is growing evidence of a role for ncRNAs in several diseases. The principal interests are in micro-RNAs (miRNAs) and long non-coding RNAs (lncRNAs), as they appear to have a role in certain cancers. A key feature is that many have the potential to bind to, and so potentially regulate, mRNA and possibly nucleic acids within the nucleus. This may lead to reprogramming metabolism in tumour cells and in acting as tumour suppressors/promotors (i.e. OncoMiRs, TSmiRNAs). There is also evidence that lncRNAs may fuse with mRNA or other lncRNAs, the full significance of which is unclear.

Examples from non-small cell lung cancer (NSCLC) include miR-128B, miR-545-3P and miR-494-3p targeting *EGFR*, miR-100-5p and miR200c targeting *ALK*, and miR-181a-5p, miR199b and miR-21 targeting *KRAS*. Examples in small cell lung cancer (SCLC) include miR-126 targeting *CrK* (at 17p13.3, coding an adaptor molecule), *VEGFA*, at 6p21.1, coding for VEGF A, and *SLC7A5*, at 16q24.2, coding an amino acid transporter, whilst miR-184 has been shown to target *c-Myc*. Similarly, data on lncRNAs also point to interactions such as LINC00680 as a sponge for miR-410-3p in enhancing the expression of *HMGB1*, at 13q12.3, with links to histones and nucleosomes. This may also promote adenocarcinoma growth via binding to *GATA6* (at 18q11.2, coding a transcription factor), thus negating the *GATA6* suppression of *SOX12*, at 20p13, which codes for another transcription factor. This pathway underlines the complex nature of these molecules.

Cell-free DNA

The presence of DNA and RNA in the plasma in health and disease has long been known, with one commentary estimating increased levels to be present in a third of patients with a malignant disease. Advances in molecular pathology demonstrated changes in circulating tumour DNA (ctDNA) and cell-free DNA (cfDNA), such as DNA satellites with LOH in melanoma, and of specific genes in gastrointestinal (*KRAS*) and breast cancer (*PIK3CA* at 3p26.32, coding for a subunit of phosphatidylinositol-4,5-bisphosphate 3-kinase (PI3K), which has a second messenger role).

Identification of patients with neuroendocrine prostate cancer may be assisted by the use of a targeted set combining genomic alterations (*TP53*, *RB1*, *CYLD* at 16p12.1, coding for a deubiquitinase, and *AR* at Xq12, coding the androgen receptor) and epigenomic alterations (hypomethylation and hypermethylation of 20 differential sites).

Meta-analyses have reported that the presence of any preoperative ctDNA to be a strong indicator of poor overall survival and recurrence-free survival in a form of lung cancer. In one study, blood from 41 patients with lung cancer was tested by Ion Torrent hotspot methodology, finding mutated *KRAS* in 44%, *ALK* in 37%, *TP53* in 32% and *PIK3CA* in 29%, with 6 patients having both *KRAS* and *TP53*, and 7 patients having both

KRAS and *PIK3CA*. Of these, the overall cell-free DNA burden and *TP53* mutation status both predict poor progression-free survival, the latter also predicting shorter overall survival.

As with ncRNAs, it may be predicted that assessment of cfDNA and ctDNA will become increasingly important and may also enter routine clinical practice.

Practice Point – Non-coding RNAs Despite the considerable interest in these molecules, as yet, ncRNAs do not have a place in the routine practice of molecular pathology. However, they are beginning to take a place as therapeutics – siRNAs have been used to reduce blood pressure in patients with hypertension by targeting *AGT*, at q42.2, coding for angiotensin, and another, designed to silence *ANGPTL3*, at 1p31.3, coding for an angiopoietin-like protein, is effective in reducing lipid levels. There are several other examples.

4.2.5 Summary of Section 4.2

The section has attempted to describe the leading aspects of the cell biology of cancer, many of which apply not only to carcinomas but also to sarcomas and other forms of solid organ cancers, and to soft and fluid malignancies such as those of the blood.

For the past three decades, Hanahan and colleagues have summarised key features of the cell biology of cancer and carcinogenesis, coining the phrase 'The hallmarks of cancer'. These point to numerous independent aspects of the events within the nucleus and cytoplasm, as cells transform to a malignant phenotype, many of which are directly attributable to molecular genetics and therefore molecular pathology, and are as follows:

- Evading growth suppressors, avoiding immune destruction and enabling replicative immortality
- Tumour-promoting inflammation, activating invasion and metastasis and inducing or accessing vasculature
- Genome instability and mutation, resisting cell death, deregulating cellular metabolism and sustaining proliferative signalling

In addition, a number of emerging hallmarks and enabling conditions include non-mutation epigenetic reprogramming, polymorphic microbiomes, senescent cells and unlocking phenotypic plasticity.

Having been introduced to the epidemiology and cell biology of cancer, we now move to an examination of the molecular pathology of specific malignancies, working in sequence down as shown in Table 4.1, i.e. the lung, colorectal, prostate, breast and pancreatic cancers, followed by those less frequently presenting.

4.3 Lung Cancer

4.3.1 Introduction

Unsurprisingly, a great deal is known about the epidemiology and aetiology of this disease, which kills almost twice as many people as the second most frequent malignancy, colorectal cancer. Over the past decades, rates in England and Wales have been falling, but a consistent feature being the increased death toll in men compared to women (8.1%). A potential (or perhaps likely) reason for this sex difference is the increased rate of tobacco smoking (the leading risk factor, responsible for over two-thirds of all lung cancer) and exposure to industrial toxins (e.g. inspired asbestos and heavy metals such as arsenic and cadmium) and other pollutants, and exposure to radon released from naturally occurring uranium (which brings a 14–29% increased risk). Non-smoking spouses of smokers have a 20–30% increased risk, whilst a positive family history of lung cancer brings an increased risk of 1.7-fold, rising to 2–4-fold in a first-degree relative, for this disease and so points to a heritable component, now fully established. In the USA, some 10–20% of lung cancers occur in never-smokers, also pointing to an intrinsic cause.

Diagnosis is made in secondary care. Once the patient has been referred by their signs and symptoms, they will undergo X-ray and more complex imaging (CT/MRI). Ideally, biopsies (ideally obtained at the same time as a bronchoscopy) of any abnormal growth will be taken and will determine the cellular nature of the pathology. This is based on the size of the malignant cells as large or small, with non-small cell lung carcinoma (NSCLC) as the most common (85%) and small-cell lung carcinoma (SCLC). A NSCLC can be further classified as adenocarcinoma (55% of cases), squamous cell carcinoma (34%) or 'other', often described as large cell carcinoma (11%). From the gathered tissue, molecular genetics can then determine mutations in several key genes, knowledge of which will be pertinent to certain treatments. Routine blood tests (haematology, biochemistry) are not especially helpful at this stage, unless the disease is well established and of high grade or stage.

4.3.2 Molecular Pathology

Chromosomes Maps

One of the first (1984) reports of the chromosomal location of a candidate gene used a human-mouse hybrid cell panel, placing an oncogene isolated from a lung carcinoma cell line to chromosome 12 (later found to be the location of proto-oncogene *KRAS*), whilst others used restriction fragment length polymorphisms to note deletions from 3p (now known to host several tumour suppressor genes) in several SCLC tissues. An ABI 3100 analyser, using capillary sequencing, was used to report a major susceptibility locus influencing lung cancer risk to a region on 6q23–25. This region was consequently reported to house *ARID1B* (at 6p25.3, coding part of the chromatin remodeling complex), *UTRN* (at 6q24.2, coding for utrophin, a component of the cytoskeleton) and *PHACTR2* (also at 6q24.2, coding a phosphatase and actin regulator), and so all potential candidate genes.

The Genes

Lung cancer was one of the first malignancies where molecular genetics were explored. Of these, the earliest reported an activated *rasK* (later renamed *KRAS*) transforming gene and amplification of oncogene *c-Myc* in lung cancer cell lines, subsequently reported in tumour tissues, but not in normal lung tissue. Amplification of *c-Myc* and *c-k-Ras-2* was later reported, and a family of *Myc* genes described, with *N-myc* amplified up to 170-fold in some SCLC cell lines. In addition to those described earlier, abnormalities were also found in *EGFR, PIK3CA*, *BRAF* and *ERBB2* (at 17q12, coding human epidermal growth factor receptor 2, HER2).

Other genes, several of which are proto-oncogenes, including *ALK, PTEN, ROS1* at 6q22.1, coding a receptor tyrosine kinase, *MET* at

7q31.2, coding a hepatocyte growth factor receptor, *cKIT, SRC* at 20q11.23, coding a non-receptor tyrosine kinase, and *RET* at 10q11.21, coding a receptor tyrosine kinase, were later added to the list, These have become established as potential markers of the disease, all with different specificities and sensitivities. Increased immunohistochemical expression of neurotrophic tyrosine kinase receptors 1 and 2 (NTRK) was demonstrated in lung cancer, to be followed by fusion of their genes (i.e. *NTRK1*, at 1q23.1, and *NTRK2* at 9q21.22) with other genes, thereby forming a neo-oncogene, whilst *NTRK3* at 15q25.3 was subsequently also implicated, as were rearrangements of the nucleotide sequences in all three. Further details of this important triad are presented in Section 4.9.1.

However, whilst the single gene approach has a place, a combination may be more informative. A multi-gene panel of *CDKN1A* at 6p21.2, coding a cyclin-dependent kinase inhibitor (and therefore a tumour suppressor), *E2F1* at 20q11.22, coding a transcription factor, *ERCC1*, at 19q13.32, *ERCC4*, at 16p13.12, *ERCC5*, at 13q33.1, (all three coding an excision repair enzyme subunit,) *GPX1* at 3p21.31, coding for glutathione peroxidase, *GSTP1* at 11q13.2, coding for glutathione transferase, *KEAP1* at 19p13.2, coding an inhibitor of transcription factors, *RB1*, *TP53* and *TP63* at 3q28, coding a transcription factor, and *XRCC1* at 19q13.31, coding a DNA damage repair protein, in bronchial epithelial cells can predict lung cancer with sensitivity and specificity that exceeds 90%.

A further study of squamous cell carcinoma that identified 444 prognostic genes reported a 10-gene panel, that was able to predict prognosis. This consists of *ITGA3* at 17q21.33, coding an integrin subunit, *VAV2* at 9q34.2, coding a nucleotide exchange factor, *FLNC* at 7q32.1, coding filamin C, *FLT4* at 5q35.3, coding a receptor tyrosine kinase, *HGF* at 7q21.11, coding hepatocyte growth factor, *MYL2* at 12q24.11, coding a myosin light chain, *ITGB1* at 10p11.22, coding an integrin subunit, *PDGFRA, CCND2* at 12p13.32, coding cyclin D2, and *PPP1CB* at 2p23.2, coding a phosphatase subunit.

Of particular interest are the cell surface proteins CD279, coded for by *PDCD1* at 2q37.3, and CD274, coded for by *CD274* at 9p24.1, also known as 'programmed cell death protein 1 (PD1)' and as 'programmed death-ligand 1 (PD-L1)', respectively. An association between PD-L1 and lung cancer was reported in 2014, its expression failing to link to the mutational status of *EGFR*, *KRAS* and *HER2*, or to the fusion status of

ALK-EML4 (the latter at 2p21), thus establishing it as an independent factor. Further studies indicated that the over-expression of PD-L1 is related to the epithelial-to-mesenchymal transition, a process linked to metastatic transformation.

> **Key Point 4.3** The precise identity of the mutated gene is important because in one study, median overall survival was 36.9 months in *EGFR*-mutated patients and 55.4 months in *ALK*-rearranged patients.

4.3.3 Which Genes to Test?

Table 4.3 shows the genes most commonly linked with NSCLC, headed by *EGFR* and *KRAS*, although the frequencies vary with geographical region (e.g. 10.4% in Italy, 16.6% in England and Wales and 50% in Japan), with smoking/never smoked and with type of tumour. For example, *EGFR* is mutated in around 50% of lung cancer tumours in lifetime non-smokers, around 10% in cases of NSCLC, whilst *KRAS* is mutated in around 25% of those with a NSCLC, and *ALK* is mutated in 5% of NSCLCs and in 10–15% of lung cancers in those who have never smoked. There is also variation in frequency according to the cellular basis of the tumour: in adenocarcinoma, the order of frequency is *KRAS*>*EGFR*>*NRAS*, whereas in a squamous cell disease, the order is *FGFR*>*EGFR*>*PIK3CA*. However, notably absent from this list is *TP53*, whose frequency has been reported to be as high as 41%.

In contrast to NSCLC, mutations in *EGFR* are rare (2.6–7.1%) in SCLC, although, where present, increased CNV has been linked to progressive disease. This variant of lung cancer has been linked to *ASCL1* (at 12q23.2), *NEUROD1* (2q21.3), *POU2F3* (11q23.3), all three coding for transcription factors and other checkpoint factors. These, in turn, may have a role in interferon signalling and in epithelial-to-mesenchymal transition.

A key document in the practice of molecular pathology in England is the National Genomic Test Directory, a guideline outlining possible procedures for investigating cancers (see Key Point 4.4). In NSCLC, several types of analyses have been described:

- A multi-panel NGS panel seeking to detect small variants, including *EGFR*, *ALK*, *BRAF*, *KRAS* and *MET* is suggested, whilst another

Table 4.3 The frequency of genes linked to non-small cell lung cancer.

Gene	Product	Frequency[a]
EGFR[b]	Epidermal growth factor receptor	10–35%
KRAS[b]	K-Ras, a GTPase second messenger	15–25%
FGFR1[c]	Fibroblast growth factor receptor	20%
PTEN[b]	Phosphatase and tensin homologue	4–8%
DDR2[b]	Discoidin domain-containing receptor 2	~4%
ALK[d]	Anaplastic lymphoma kinase	3–7%
HER2[b]	Human epidermal growth factor receptor 2	2–4%
MET[c]	Hepatocyte growth factor receptor	2–4%
NTRK1[e]	Neurotrophic growth factor receptor	3%
BRAF[b]	Serine/threonine-protein kinase B-Raf	1–3%
PIK3CA[b]	A subunit of phosphatidylinositol-4,5-bisphosphate 3-kinase	1–3%
AKT1[b]	A serine-threonine kinase	1%
MEK1[b]	Mitogen-activated protein kinase	1%
NRAS[b]	A second messenger GTPase	1%
RET[d]	A cell membrane receptor with tyrosine kinase activity	1%
ROS1[d]	A cell membrane receptor with tyrosine kinase activity	1%

[a] Frequency varies with population.
[b] Nature of abnormality: mutation.
[c] Nature of abnormality: amplification.
[d] Nature of abnormality: rearrangement.
[e] Nature of abnormality: fusion.
Source: Modified from Schabath, Cote. Cancer Epidemiol Biomarkers Prev. 2019;28: 1563–1579 and elsewhere.

panel focusing on structural variant detection includes *ROS1*, *RET*, *EML4::ALK* fusion, *NTRK1*, *NTRK2*, *NTRK3* and *MET*.

- The CNV in *MET* may also be sought by an NGS method or by FISH.
- Small variant detection may be used to examine hotspots in *EGRF* in both tumour tissues and circulating tumour DNA in serum or plasma, and for *ALK* hotspots.

- Structural variant rearrangements in *ROS1* and *RET* may be detected by FISH and/or RT-PCR, as may the structural variant of the fusion neogene of *EML4::ALK*.

For SCLC analysis, a multi-target NGS panel can be used to detect the CNV of *RB1*, whilst a second panel can detect structural variants in *NTRK1, NTRK2* and *NTRK3*. The latter is important because if a rearrangement is detected, patients are likely to be eligible for an *NTRK* inhibitor (see below).

In day-to-day practice, analysis is simplified by the availability of several commercial panels, including all genes in the panels described earlier, many of which are produced by the manufacturer of a particular analytical platform. The power of these panels resides in the particular form of multi-target NGS, some demanding dedicated analysers.

Self-check 4.3 Why is a panel of several genes valuable in testing for lung cancer?

4.3.4 Molecular Pathology Directing Treatment

In the UK, the National Institute for Health and Care Excellence (NICE) leads in the publication of guidance for the management of many conditions. NG122, 'Lung cancer, diagnosis and management', published on 26 July 2023 (but subject to updates), suggests options that include surgery, radiotherapy and chemotherapy. However, any treatment mode is dependent on any particular gene mutation present in an excised tumour, in certain forms of more advanced disease, such as with metastases and that which has proved resistant to primary chemotherapy. Molecular pathology has an important place in several instances.

Epidermal Growth Factor Receptor

As described earlier, alterations in the gene for this molecule are present in many cases of lung cancer (and in many other cancers), the key mutation being in the tyrosine kinase activity, hence *EGRF-TK*. The clinical importance of this mutation is that a change leading to over-activity of the enzyme is a key feature in carcinogenesis, and as such TK is a target for pharmacological intervention.

NICE NG122 refers to DG9 (updated in 2019) about guidance on *EGRF-TK* testing in adults with locally advanced or metastatic NSCLC. This document recommended tests and test strategies that include the Therascreen EGRF RGQ PCR kit (Qiagen), the Cobas EGFR mutation test (Roche) and Sanger sequencing of samples with >30% tumour cells. Although, at the time, there was insufficient evidence to make recommendations regarding high-resolution melt analysis, pyrosequencing with fragment length analysis, and NGS, all these platforms may well be suitable once more recent data is produced.

Patients with *EGRF-TK*-positive locally advanced or metastatic tumours are likely to be eligible for treatment with a TK inhibitor. Options include gefitinib (NICE TA192) and erlotinib (TA258). Later, clinical trials of other TK inhibitors were shown to be effective, including afatinib, dacomitinib and osimertinib, the latter conferring a significantly improved progression-free survival of 15.2 months compared to 9.6 months on standard care ($p < 0.0001$). The success of these drugs prompted their use in other gene mutations in NSCLC where TK is over-expressed (Table 4.4).

Several drugs are recommended in the presence of different genetic lesions, and also in other diseases, such as nintedanib in idiopathic

Table 4.4 Options for tyrosine kinase inhibitors in advanced NSCLC.

Genetic lesion present	Squamous NSCLC	Non-squamous NSCLC
No targetable mutation	None	Nintedanib
RET fusion	Selpercatinib	Nintedanib, selpercatinib
NTRK fusion	Larotrectinib, entrectinib	Nintedanib, larotrectinib
KRAS G12C[a]	Sotorasib	Nintedanib, sotorasib
*MET*ex14 skipping alteration[b]	Tepotinib	Nintedanib, tepotinib
ALK	Not described	Brigatinib, alcedinid, ceritinib, crizotinib
ROS-1	Not described	Nintedanib, entrectinib, rizotinib

[a] G12C glycine to cysteine at codon 12.
[b] Mesenchymal-to-epithelial transition, part of exon 14 is skipped leading to activation.

pulmonary fibrosis. These drugs may be used alone (rarely) or in combination (commonly) with other forms of chemotherapy such as carboplatin and paclitaxel. Note that all TK inhibitors carry the suffix *-nib* as the last three letters, specific for this class of targeted chemotherapy.

Programmed Death-Ligand 1

As described earlier, this cell-surface molecule is over-expressed in several malignancies, and accordingly is a further target for treatment, and targeted by several monoclonal antibodies to prevent it from binding to its ligand (i.e. PD-1) on other cells. These PD-L1 inhibitor antibodies include atezolizumab, avelumab, durvalumab, ipilimumab, pembrolizumab and nivolumab. NICE 122 has published a comprehensive list of instances where they may be used in squamous and non-squamous NSCLC, dependent on the extent of the expression of PD-L1. Options are extended to situations where there are no targetable mutations, or if there is a targetable mutation, such as a *RET* fusion, an *NTRK* fusion, *KRAS-G12C, METex14* skipping alteration, *EGRFex20* insertion, or *BRAF V600E*.

Key Point 4.4 Much of the material in the final part of this section refers to NICE and also to the National Genomic Test Directory for England. Whilst the former is UK-wide, the latter has been posted online by NHS England; the other UK nations of Scotland, Wales and Northern Ireland have their own guidelines, also available online. The guidelines for England are cited here because they are the most comprehensive, but the reader is advised to refer to the guidelines applicable to their location.

Self-check 4.4 What are the key differences in the mode of action of TK inhibitors and monoclonal antibodies?

4.4 Colorectal Cancer

4.4.1 Introduction

Colorectal cancer is the second most frequent cause of cancer deaths in England and Wales. Its incidence has been slowly rising, increasing by 11.6% from 2013 to 2023, compared to the overall increase in all malignant neoplasms of 5.1%. Modifiable risk factors include a sedentary lifestyle, obesity, poor diet (e.g. insufficient dietary fibre and enriched for red meat), smoking, excessive alcohol consumption, diabetes and insulin resistance, and the intestinal microbiome. Non-modifiable risk factors include race (Black>White), sex (frequency of death in males is 19.5% higher than in females, compared to 14.4% in all cancers and 8.4% in lung cancer), age (mean age adjusted mortality rate in years 30–49 = 30.8, 50–69 = 272.1, 70 and over = 1617.5), inflammatory bowel diseases (Crohn disease, ulcerative colitis) and abdominal irradiation. The finding that 10–20% of patients with colorectal cancer have a positive family history suggests a hereditable factor(s), the most common being Lynch syndrome.

There are no specific signs and symptoms, although constipation, abdominal discomfort, diarrhoea and mucus and/or blood in the faeces are semi-specific. If these symptoms are present, a faecal occult blood test and calprotectin test will be prescribed. The next step will be ultrasound, with colonoscopy or sigmoidoscopy, whilst blood tests for serum CA 19-9, mucin-1, carcinoembryonic antigen (CEA) and cytokeratin-1 will be performed.

Should the scoping find clear evidence of tumour, it will be biopsied for standard histopathology and molecular pathology. However, an additional finding may be numerous intestinal polyps (polyposis), a common finding in many middle-aged people and may range from hundreds to thousands. Although some consider these to be of little pathological significance, an alternative view being that a polyp is a benign tumour, and if secreting mucus, an adenoma. Over 40 years ago, a link between polyposis and colorectal cancer was hypothesised, with autosomal dominant familial adenomatous polyposis (FAP) being a further risk factor with a high malignant potential. This evolved with the establishment of the adenoma-carcinoma-metastasis model, likely to be present in 80% of cases unless addressed.

4.4.2 Molecular Pathology

The Role of Polyps

Data in favour of the hypothesis that polyps can develop into malignant tissue came from a number of sources, such as that non-malignant polyps express a similar degree of c-Myc protein as do malignant tissue. Adenoma cells from patients with FAP have been shown to have loss of, or rearrangements in, a section of chromosome 5, a locus where the gene adenomatous polyposis coli (*APC*, at 5q21-22) is found, now known to be a tumour suppressor. Since mutated *APC* is often present in normal tissue, it follows that a second event must be present to initiate the malignant transformation, a proposal reminiscent of Knudson's two-hit hypothesis.

Genes in Colorectal Cancer

As mentioned earlier, the challenge lies in understanding the precise aetiology of 'sporadic' colorectal cancer, i.e. one arising with no clear history, this being the case in ~85% of cases – did it evolve from undetected polyposis? and/or did it arise after the sequential action of a number of genes? As with lung cancer, one of the first genes to be implicated in this disease was *KRAS* (said to be present in 20–40% of cases), followed by reports of the amplification deregulation of *c-Myc* (60%), and inactivation or deletion of tumour suppressor *TP53* (75%). Findings in other genes followed, such as *EGFR* (60%), and a gene named deleted in colorectal cancer (*DCC*, at 18q21.2, 70%, coding a receptor for netrin), also known to be a tumour suppressor. It follows that, with these high frequencies, many cases are likely to have multiple gene lesions. Genes less frequently found include *MUTYH* at 1p34.1, coding for an enzyme required for base excising repair, *BRAF*, and *PTEN*. To this list may be added *NTRK1*, *NTRK2* and *NTRK3*, with around a quarter of cases seeing these genes fused with *KRAS, NRAS* or *BRAF*. Notably, many of these genes are also present in Table 4.3 on lung cancer.

A small (around 3.5%) number of colorectal cancers fall into the category of Hereditary Non-Polyposis Colorectal Cancers (HNPCRC, also known as Lynch syndrome). The aetiology of this autosomal dominant disease is linked to mutations in *MSH2* (at 2p21-p16.3), *MLH1*, *MSH6* (at 2p16.3), *PMS2* (7p22.1, all four coding for enzymes that repair mismatched DNA) and *EPCAM* (at 2p21, coding CD36, the epithelial cell adhesion molecule).

Key Point 4.5 The list of genes implicated in CRC bears considerable overlap with those linked to lung cancer.

4.4.3 Which Genes to Test?

The molecular genetics of colorectal cancer is, compared to lung cancer, poorly developed. NICE NG151 (updated December 2021) recommends testing for variants of *RAS* and *BRAF V660E* in all people with metastatic disease suitable for systemic anticancer treatment (i.e. chemotherapy). In addition, a supplement refers to the potential value of *NRAS*, *PIK3CA* status and the PD1/PD-L1 genes *PDCD1* and *CD274*.

NICE DG27 focuses on Lynch syndrome, recommending using immunohistochemistry for *MLH1*, *MSH2*, *MSH6* and *PMS2*, with further suggestions of testing for *BRAF V600E*, for microsatellite instability and for *MLH1* promoter hypermethylation. The National Genome Test Directory for England has several entries for this disease, almost all as described by NICE, as follows:

- A multi-target NGS panel for the small variant detection of mutations in *KRAS, NRAS, BRAF, MLH1, MSH2, MSH6, PMS2, POLD1* (at 19q13.3) and *POLE* (at 12q24.33, both coding for a subunit of a DNA polymerase)
- Another panel for NGS structural variant detection in *NTRK1, NTRK2* and *NTRK3*
- Analysis of hotspots in *KRAS* and *NRAS*
- Microsatellite instability analysis
- *MLH1* promoter hypermethylation
- Small variant detection in *DPYD* (see Box 4.1)

4.4.4 Molecular Pathology Directing Treatment

In the event of relevant mutations in *KRAS, NRAS* or *BRAF*, anti-EGRF therapy would be considered. Similarly, use of a tyrosine kinase inhibitor such as entrectinib or larotrectinib is relevant in the presence of an appropriate mutation, as may be coded by an abnormal variant of *NTRK*. Protein kinase inhibitors such sorafenib and vemurafenib may be used in certain *BRAF* mutations. Selected tyrosine kinase inhibitors (as may be used in lung cancer) are described in Table 4.4.

Box 4.1 Pharmacogenomics: Dihydropyrimidine dehydrogenase and thiopurine methyltransferase

The commonly used forms of cytotoxic chemotherapy for many cancers include 5-fluorouracil (5-FU), capecitabine and tegafur (which are both metabolised to 5-FU). These are detoxified principally by dihydropyrimidine dehydrogenase, levels being under the control of *DPYD* (located at 1p21.3) such that low activity leads to slow removal of the drugs. This therefore endows the drugs with a longer half-life, and so greater cytotoxic effect, which has potentially fatal consequences for the patient, such as liver and renal failure. Accordingly, the National Genome Test Directory for England and NICE make a point that any patient starting these drugs will need to have their *DPYD* status determined, a corollary being that the doses of the drug will be adjusted accordingly so that the toxic effects are minimised.

Similarly, thiopurine methyltransferase, coded for by *TPMT* at 6p22.3, adds a methyl- group to thiopurine drugs such as azathioprine, 6-mercaptopurine and 6-thioguanine. Variants of this gene produce different levels of the active enzyme and so influence levels of the antimetabolites, the high levels of which can also cause undesirable effects such as bone marrow suppression.

Self-check 4.5 What genes are implicated most strongly in both lung cancer and colorectal cancer?

4.5 Prostate Cancer

4.5.1 Introduction

Whilst the number of male lung/bronchus cancer deaths in England and Wales has fallen by 13.8% from 2013 to 2023, deaths due to prostate cancer have increased by 13.9%, with 11,072 deaths in 2023. Globally, it ranks 4th in incidence (1.47 million) and 8th in mortality (398,000). As with so many cancers, leading risk factors are age and obesity, although there is a long-established link with African descent, and family history implies a genetic component.

Signs and symptoms are unsurprisingly genitourinary. The GP may undertake some formal testing, such as digital rectal examination, where the practitioner estimates the size of the prostate with their fingers. Blood may be taken for the prostate-specific antigen (PSA, a 26 kDa peptidase coded for by *KLK3* at 19q13.33), and if increased, an urgent referral to secondary care will follow, and initial testing will be repeated. Prostate-specific membrane antigen (an enzyme coded for by *FOLH1* at 11p11.12) is highly expressed by prostate cancer cells, is linked to decreased survival and can be used in imaging to detect metastases. The title of NICE guideline NG131 is 'Prostate cancer: diagnosis and management.'

4.5.2 Molecular Pathology

Early studies reported increased levels of mRNA for *Ha-ras, myc, n-ras, Ki-ras, myb, fos, fms* and *sis* in prostate cancer cell lines, with immunocytochemistry showing increased expression of the p21 protein product (Ras) of the *ras* oncogene. Later data reported increased activity of *c-Myc*, with potential roles for *RB1* and *TP53*, and loss of heterogeneity on chromosomes 8p, 10q and 16q, with a potential role for tumour suppressor *CDH-1* at 16q22.1, coding for the adhesion molecule E-cadherin (CD234). Subsequent research reported the potential of *KAI1* (a tumour suppressor at 11p11.2, coding for CD82) and *PTEN*, to be followed by *TMPRSS2* (at 21q22.3, coding for a transmembrane protease). The latter links with carcinogenesis via a fusion with a very close transcription factor neighbour, notably *ERG*, at 21q22.2. A plausible explanation for this is that the fusion product interferes with normal androgen receptor signaling and therefore failure of prostate cells to differentiate correctly. AR-V7 is a splice variant of the androgen receptor (coded for by *AR* at Xq12), its mRNA being detectable in circulating tumour cells.

Other putative genetic markers include tumour suppressors *CDKN1A, CDKN1B* and *NKX3-1* at 8p21.1, coding a transcription factor, oncogenes *BCL2*, at 18q21.33, coding an apoptosis regulator, *EZH2*, at 7q36.1, coding a methyltransferase, and *ERBB2*, and members of the PI3K-Akt-mTOR signal transduction pathway.

In addition to many of the genes described in this section, *PMS2*, *PALB2* (at 16p12.2, coding for a mismatch repair endonuclease named Partner and Localiser of BRCA2), tumour suppressor *CHEK2* (at 22q12.1, coding a serine-threonine kinase) and *HOXB13* (at 17q21.32, coding for a transcription factor) may be important in hereditary prostate cancer. A pedigree study of high-risk men identified *LRBA* (at 4q31.3, coding an anchor protein) as a likely predisposing mutation for prostate cancer.

In contrast to the high frequency of gene lesions in lung/bronchus and colorectal cancer, the most frequent candidate gene in prostate cancer, *BRCA2*, at 13q13.1, coding a DNA damage repair protein, is present in only 8.7% of cases, followed by *CDK12* at 17q12, coding a cyclin-dependent kinase (6.3%), *ATM at* 11q22.3, coding a serine/threonine kinase (5.9%), *CHEK2* (1.2%) and *BRCA1* (1%). Much of this data is supported by a study from Pritchard and colleagues, showing that 11% of 692 patients with metastatic prostate cancer had inherited mutations in 16 DNA damage repair genes, the most frequent being *BRCA2* (5.3%), followed by *CHEK2* (1.9%), *ATM* (1.6%), *BRCA1* (0.9%), *GEN1* at 2p24.2, coding an endonuclease (0.5%) and by *PALB2* and *RAD51* at 15q15.1, whose product assists in repairing damaged DNA (both 0.4%). Further details of *BRCA1* and *BRCA2* are provided in Section 4.9.2.

4.5.3 Which Genes to Test?

The key document, NICE NG131, updated in December 2021, makes no mention of molecular genetics, although there is a growing literature base to be discussed later. Nevertheless, the National Genome Test Directory for England makes the following three points:

- A multi-target NGS panel that may be used to check for small variants in *BRCA1* and *BRCA2* in any prostate cancer, *ATM* and *CDK12* in metastatic castration-resistant cancer.
- A second NGS panel, for detecting structural variants in *TMPRSS2-ERG, NTRK1, NTRK2* and *NTRK3.*
- FISH for detection of a structural variant in *TMPRSS2-ERG*, should there be a doubt regarding the morphological basis of the aetiology of a carcinoma.

4.5.4 Molecular Pathology Directing Treatment

The 2020 European Society for Medical Oncology guideline points out that some 20% of metastatic prostate cancers have mutations in DNA damage and repair genes, and *BRCA2* is the most commonly altered.

Tumours with defects in mismatch repair genes such as *MSH2, MSH6, PMS2* and *MLH-1* (perhaps by PCR, or its product by immunohistochemistry) or microsatellite instability (by PCR) are likely to respond to the monoclonal antibody pembrolizumab, which targets PD-L1.

A study in the *New England Journal of Medicine* assessed *ATM, BRCA1, BRCA2, BRIP1* at 17q23.2, coding a helicase, *BARD1* at 2q35m, coding for a BRCA1 stabiliser, *CDK12, CHEK1* at 11q24.2 and *CHEK2* at 22q12.1, both coding a mitosis checkpoint kinase and so are tumour suppressors, *FANCL* at 2p16.1, coding for a ligase (see also Section 7.7.7 on *FANC* genes), *PALB2, PPP2R2A* at 8p21.2, coding a phosphatase, *RAD51B* at 14q24.1, *RAD51C* and *RAD51D* both at 17q22, and *RAD54L* at 1p34.1, all four being paralogues of *RAD51* that are required for repair of DNA breaks, in metastatic castration-resistant disease. Those with alterations in *ATM, BRCA1* or *BRCA2*, and those taking the drug olaparib (an inhibitor of poly-ADP-ribose polymerase [PARP]) had better progression-free survival or overall survival. Previous sections have pointed to those therapies such as tyrosine kinase inhibitors where the appropriate gene lesion is present.

Key Point 4.6 Compared to other cancers, there are relatively few firm examples of the importance of molecular pathology in prostate cancer.

4.6 Breast Cancer

4.6.1 Introduction

In parallel with prostate cancer and male cancer deaths, breast cancer is the second leading cause of female cancer deaths, the first being the bronchus/lung. From 2013 to 2023, it killed over 10,000 people in England and Wales annually, but in 2022, it killed 9934 women and in 2023, 9891 women (with 72 and 82 men, respectively). As regards this number of around 10,000 women annually, since the number of women dying of all causes has increased from 260,756 to 285,441 over this period, and those dying of any cancer from 68,243 to 70,775, it follows that the proportion of women dying of breast cancer is falling, from 3.88% of all female deaths in 2013 to 3.46% in 2023, and from 14.8% of all female cancer deaths in 2013 to 14.0% in 2023. In 2022, the WHO reported a global incidence of

2.3 million cases and almost 670,000 deaths, giving a crude mortality index of 29%, markedly better than that in all cancers of 42%.

Physical signs and symptoms are well-understood, whilst axilla lymphadenopathy may be present in advanced diseases. First steps in secondary care are X-ray and ultrasound, the latter to give enhanced views of potential lesions. However, a specialised form of X-ray, a mammogram, takes an image only of the breast, not of the entire chest. An important finding is ductal carcinoma *in situ* (DCIS) and may be regarded as a pre-cancerous state. Later, there may be CT/MRI, the latter helping in determining disease in the lymph nodes, primarily in the local axilla, and so an idea of stage. Serum tumour marker CA 15-3 (derived from the glycoprotein mucin-1 [MUC1], a product of *MUC1*, at 1q22) is often increased not only in breast cancer, but also in benign breast disease, which can help in differential diagnosis. NICE documents include NG101 on early and locally advanced breast cancer, CG81 on the diagnosis and management of advanced breast cancer and CG164 on familial breast cancer.

The composition of a lump is likely to be determined with a biopsy or fine needle aspiration for standard histology and immunohistochemistry for the expression of the oestrogen receptor and the progesterone receptor (ER and PR, respectively), and HER2, (closely related to HER1 already described). These three markers dominate diagnosis and management (NICE NG101); those cancers failing to express these markers are described as triple-negative, present in 10–20% of cases. The presence or absence of a further cell surface molecule, the androgen receptor, coded for by *AR* may be used to classify some tumours as quadruple-negative. Some guidelines suggest that immunocytochemical assessment of the expression of the nuclear marker Ki67 (coded for by *MKI67* at 10q26.2) may be valuable, it being associated with cell proliferation. In this respect, NICE DG4 recommends the use of the IHC4+C score, based on immunohistochemical assessment of the oestrogen and progesterone receptors, HER2 and Ki67, in guiding chemotherapy in early breast cancer.

4.6.2 Molecular Pathology

Formal and well-conducted studies on familial breast cancer emerged in the middle of the last century, implying autosomal dominant susceptibility genes: cancer of this form is now known to comprise 4–9% of all cases. Cytogenetic studies reported complete or partial monosomy of chromosomes 1, 6, 8, 11, 13, 16, 17, 22 and X, and trisomy of chromosomes 7 and 18. Soon after the development of molecular pathology, *H-ras* and *N-ras*, followed by *c-Myc* (in one study, amplified 2- to 15-fold in 32% of tumour tissue), *KRAS* and *TP53* were implicated, although *BRCA1* and *BRCA2* have come to dominate clinical practice. Section 4.8.5 discusses the further importance of these genes in ovarian cancer. Later, these were joined by oncogene *ERBB2* (previously known as c-erbB2), *PTEN*, *BCL-2, PALB2, CHEK2*, *ATM* and others. Notably (and, perhaps, curiously, as we have seen), many of these genes also have a role in prostate cancer. In 2015, a genome-wide association study (GWAS) of 60,000 cases and 60,000 controls reported 15 new susceptibility loci and identified likely target genes of *SETBP1* at 18q12.3, coding for a DNA-binding protein, *RNF115*, coding a ubiquitin ligase, and *PDZK1*, coding a protein with links to an ion exchange. Both of the latter genes are at 1q21.1.

4.6.3 Which Genes to Test?

The large number of candidate genes may be classified according to factors such as penetrance, which may be quantified by odds ratio and lifetime risk of breast cancer (Table 4.5). A study of 2595 women with metastatic breast cancer found a pathogenic variant in *BRCA2* to be the most common (at 2.9%), followed by *CHEK2* (2.2%), *BRCA1* (2.0%), *PALB*2 (1.1%), *ATM* (0.9%) and *TP53* (0.6%). Other potential candidate genes include *BARD1*, *RAD51B*, *RAD51C*, *RAD51D*, *BRIP1*, *PITPNB* at 22q12.1, coding a phosphatidylinositol transfer protein, *TTC28* also at 22q12.1, coding a molecule required for microtubule physiology, *MSH2*, *MSH6, PMS2, MAP3K1* at 5q11.2, coding a kinase involved in signal transduction, *XRCC2* at 7q36.1, and *XRCC3* at 14q32.33, both coding a DNA repair protein. These data have informed national guidelines.

NICE NG101 on early and locally advanced breast cancer: diagnosis and management (updated January 2024) recommends that patients under 50 years with triple-negative disease, including those with no family history of breast or ovarian cancer, be offered testing for *BRCA1* and *BRCA2*, but does not suggest further activity based on the result. However, this guideline does refer to NICE CG164 (updated November 2023), which focused on familial breast cancer: classification, care and management of cancer and related risks in people with a family history of breast cancer. The principal recommendations refer to pretest counselling with

Table 4.5 Leading candidate genes for breast cancer.

Gene	Penetrance	Odds ratio (95% CI)	Breast cancer risk (%)
BRCA1	High	8.1 (5.5–10.8)	52–79
BRCA2	High	5.9 (4.6–10.5)	46–79
PTEN	High	5.8 (2.4–14.0)	40–67
TP53	High	5.4 (2.8–10.4)	20–48
STK11	High	4.4 (0.7–29.6)	41–54
PALB2	High	4.1 (2.7–7.3)	34–59
CDH1	High	3.3 (1.7–7.3)	51–69
CHEK2	Moderate	2.7 (2.3–7.8)	23–25
ATM	Moderate	1.8 (1.3–2.6)	22–50

Source: Data pooled from Yadav S, Couch FJ. Amer Soc Clin Oncol Educ Book. 2019;39:61–74; Hu C, Hart SN, et al. NEJM. 2021;384:440–451; Kurian AW, Hughes E, et al. J Clin Oncol Precis Oncol. 2017;1:1–12. doi:10.1200/PO.16.00066; Lukasiewicz S, Czeczelewski M, et al. Cancers. 2021;13:4287; Petrova D, Cruz M, et al. Breast. 2022;61:66–76; Zhu JW, Charkhchi P, et al. Cancers. 2023;15:1917; and elsewhere.

mutation searching/screening for appropriate genes such as *BRCA1*, *BRCA2* and *TP53*.

Although NICE CG81 (updated August 2017) does not mention molecular genetics, DG34 on tumour profiling to guide adjuvant chemotherapy decisions in early breast cancer recommends a number of proprietary scoring systems. These are listed in the National Genomic Test Directory for England, which describes a number of multi-target expression arrays to be discussed later. The Directory also refers to the following:

- An NGS panel for the detection of structural variants in *NTRK1*, *NTRK2* and *NTRK3*
- Detection of a small variant in *PIK3CA* by NGS
- Detection of a structural variant in the hybrid *ETV6 – NTRK3* by FISH and/or RT-PCR (where there is a suspicion of a secretory carcinoma; *ETV6* at 12p13.2 codes for a transcription factor)
- Variants in *DPYD* for those likely to receive fluoropyrimidine treatment (Box 4.1)

Self-check 4.6 Which of the genes listed in Table 4.5 are tumour suppressors?

4.6.4 Molecular Pathology Directing Treatment

As described earlier, both NICE and the National Genomic Test Directory for England refer to proprietary panel assays, the science of which is derived from the large and complex analyses described in Chapter 3, such as microarray, GWAS, WES, WGS and other 'massive' NGS techniques. These panels are listed as follows:

- Oncotype DX, which includes 15 genes (derived from 250 candidate genes), focusing on a number of different aspects of carcinogenesis alongside five genes as reference controls (Table 4.6). Results from the analysis of these genes can form a mathematically complex score that is prognostic and predictive of chemotherapy benefit. Notably, *BRCA1, BRCA2, TP53* and other genes in Table 4.5 are not included in the panel.
- The EndoPredict panel, based on RNA, focuses on eight cancer-related genes and three reference genes for patients with oestrogen receptor-positive, HER2-negative breast cancer (Table 4.7). Despite the fact that only one gene is common between the two panels (*BIRC5*), concordance and correlation are moderate (75% and $r = 0.65$, respectively). This panel also excludes many 'high-profile' genes.
- The Prosigna PAM50 was derived from an extensive microarray panel from 189 breast tumour samples and 29 normal samples, reducing 161 candidate genes down to 50. The preparation of tissue for this panel calls for the extraction of RNA from which DNA is synthesised. As with other panels, these included not only genes coding for receptors (those for oestrogen, progesterone and fibroblast growth factor) and regulation of the cell cycle (cyclins B1 and E1) but also genes for structural proteins (actin-related protein, keratins), and others whose links to cancer at first sight seem unclear (biliverdin reductase A, phosphoglycerate dehydrogenase; Table 4.8). It also fails to include *BRCA1, BRCA2, TP53*, etc.

- Developed in the mid 2000s, the MammaPrint (Agendia) panel of 70 genes has evolved to assess the risk of distant recurrence of disease within 5 and 10 years, and whether patients with stage 1 or 2 disease would benefit from chemotherapy. The tumour is probed for genes spanning the 'Hallmarks of Cancer' grouping (Section 4.2.5), such as differentiation, local invasiveness and adaption to a secondary site. The panel places patients at low or high risk and also predicts the cellular nature of the tumour. In one study, patients defined as low risk by the panel had a 1.3% chance of tumour recurrence, whilst those classified as high risk had an 11.7% chance of recurrence. An 80-gene signature (BluePrint) classified tumours as basal, luminal or HER2 subtypes.

The development of these panels represents a turning point in the application of molecular pathology directly into the clinic where patient care is paramount. Their objective is to help guide chemotherapy, an example being that patients may be eligible for an NTRK inhibitor, such as entrectinib or larotrectinib, should they carry the appropriate

Table 4.6 The Oncotype DX panel.

Group	Genes (location, product)
Proliferation	*MKI67* (10q26.2, Ki67), *AURKA/STK15* (20q13.2 serine/threonine protein kinase 6), *BIRC5* (17q25.3, survivin), *CCNB1* (5q13.2, cyclin B1), *YBL2* (20q13.12, Myb-related protein B)
Invasion	*MMP11* (22q11.23, stromolysin 3), *CTSL2* (9p22.33, cathepsin L2)
Growth factor receptors	*GRB7* (17q12, growth factor receptor-bound protein 7), *ERBB2* (17q12, epidermal growth factor receptor HER2)
Sex hormone related	ESR1 (6p25.1-25.2, oestrogen receptor), *PGR* (11q22.1, progesterone receptor), *BCL* (18q21.33, B-cell lymphoma 2), *SCUBE2* (11p15.4, lipid-binding protein, potentially involved in angiogenesis)
Others	*GSTM1* (1p13.3, glutathione S-transferase Mu 1), *CD68* (17p13.1, CD68, marking monocyte lineage cells), *BAG1* (9p13.3, BAG co-chaperone 1)
Reference	*ACTB* (7p22.1, β-actin), *GAPDH* (12p13.31, glyceraldehyde 3-phosphate dehydrogenase), *RPLP0* (12q24.33, ribosomal phosphoprotein P0), *GUS* (7q11.21, β-glucuronidase), *TFRC* (3q29, transferrin receptor1)

Table 4.7 The EndoPredict panel.

Group	Genes (location, product)
Cancer related	*BIRC5* (survivin), *UBE2C* (20q13,12, a ubiquitin-conjugating enzyme), *DHCR7* (11q13.4, 7-dehydrocholesterol reductase), *BBP8* (18q11.2, retinoblastoma-binding protein 8), *IL6ST* (5q11.2, IL6 signal transducer), *AZGP1* (7q22.1, zinc-alpha-2-glycoprotein), *MGP* (12p12.3, matrix Gla protein), *STC2* (5q35.2, stanniocalcin-2)
Reference	*CALM2* (2p21, calmodulin), *OAZ1* (19p13.3, ornithine decarboxylase antizyme), *RPL37A* (2p35, 60S ribosomal protein L37a)

Table 4.8 Genes, and their products, forming the Prosigna panel.

ACTR3B (actin-related protein 3B), *ANLN* (anillin), *BAG1* (BAG family molecular chaperone regulator 1), *BCL2* (apoptosis regulator B-cell lymphoma 2), *BIRC5* (survivin), *BLVRA* (biliverdin reductase A), *CCNB1* (cyclin B1)
CCNE1 (cyclin E1), *CDCA1* (calcium-dependent domain-containing protein A1), *CDC6* (cell division control protein 6 homologue), *CDC20* (cell division cycle protein 20 homologue), *CDH3* (cadherin 3), *CENPF* (centromere protein J), *CEP55* (centrosomal protein of 55 kDa), *CXXC5* (CXXC-type zinc finger protein 5)
EGFR2 (epithelial growth factor receptor 2), *ERBB2* (receptor tyrosine-protein kinase erbB-2), *ESR1* (oestrogen receptor alpha), *EXO1* (exonuclease 1), *FGFR4* (fibroblast growth factor receptor 4), *FOXA1* (transcription factor forkhead box protein A1), *FOXCI* (transcription factor foxhead box C1)
GRB7 (growth factor receptor-bound protein-7), *GPR160* (G protein-coupled receptor 160), *KIF2C* (kinesin-like protein 2C), *KNTC2* (centromere protein kinetochore associated 2), *KRT5* (keratin type 5), *KRT14* (keratin type 14), *KRT17* (keratin type 17)
MAPT (microtubule-associated protein-Tau), *MDM2* (p53 activator mouse double minute 2 homologue), *MIA* (melanoma-derived growth regulatory protein), *MK167* (Ki67), *MELK* (maternal embryonic leucine zipper kinase), *MLPH* (melanophilin), *MMP11* (matrix metalloproteinase-11), *MYBL2* (Myb-related protein B), *MYC* (transcription factor basic helix-loop-helix)
NAT1 (N-acetyltransferase 1), *ORC6L* (origin recognition complex subunit 6), *PGR* (progesterone receptor), *PHGDH* (phosphoglycerate dehydrogenase), *PTTG1* (anaphase promoting complex substrate), *RRM2* (ribonucleoside-diphosphate reductase subunit M2)
SFRP1 (signal regulator secreted frizzle-related protein 1), *SLC39A6* (zinc transporter ZIP3), *TMEM45B* (lysosomal transmembrane protein 106B), *TYMS* (thymidylate synthase), *UBE2C* (ubiquitin-conjugated enzyme E2 C), *UBE2T* (ubiquitin-conjugating enzyme 2 variant T)

For reasons of clarity, the location of these genes not otherwise noted has been omitted.

pathogenic variant, as discussed in NICE technology appraisals TA644 and TA630, regardless of the organ/cellular basis of their malignancy.

4.7 Pancreatic Cancer

4.7.1 Introduction

Whilst the overall rate of deaths from all cancers in England and Wales from 2013 to 2023 has increased by 5.1%, deaths due to the pancreas have increased by 21.4%, having proportionately increased from 5.3% to 6.0% of all cancer deaths. The male-to-female ratio is approximately 1. Globally, in 2022, it had an incidence of 510,000 with a rank of 12th most frequent, but with 467,000 deaths it is ranked as 6th. Most risk factors are those already noted for other cancers: increasing age, smoking, alcohol, obesity, meat intake, physical inactivity, industrial toxins such as benzidine, and socioeconomic status, but an additional risk factor is increased salt intake.

Clinical risk factors include pancreatitis, diabetes/insulin resistance and hepatitis B infection, whilst signs and symptoms are similar to those of other digestive disease cancer – abdominal discomfort, nausea, vomiting, etc. Imaging can be helpful, as in demonstrating a mass, but biochemistry can provide clues as to diagnosis – carbohydrate antigen 19-9 (CA 19-9) is the most versatile marker, increased in over 80% of symptomatic patients, whilst many also use carcinoembryonic antigen, elevated in 30–60% of cases.

4.7.2 Molecular Pathology

Unsurprisingly, *K-ras* was one of the first genes to be linked to this disease, followed by *c-Myc, fos, H-ras* and *c-erb* (later named *ERBB2*), whilst cytogenetics was used to describe chromosome 5 allele loss (also lost in many other gastrointestinal tumours). Immunohistochemistry was used to show increased amounts of the protein product of *TP53*, whilst subsequent genetic analyses showed alterations in the gene itself. Other studies reported that the odds ratio (95% confidence interval) risk of pancreatic cancer if a first-degree relative had any cancer was 1.86 (1.42–2.44), rising to 5.25 (2.08–13.21) if that cancer was pancreatic. This familial data prompted further genetic analyses, which found links with *ATM, BRCA1* and *BRCA2*.

The advent of NGS enabled highly powered and exquisitely sensitive analyses that brought four genes to the fore – *KRAS, TP53, SMAD4* (at 18q21.1, coding a transcription factor) and *CDK2NA*, with reported incidences of mutations of 53–81%, 27–47%, 50–85% and 71–87%, respectively. Data that followed implicated genes such as *NTRK1, NTRK* fusions and others. More recent meta-analyses have focused on *ATM, BRCA1, BRCA2, PALB2, CDKN2A, EPCAM, MLH1, MSH2, MSH6, PMS2, STK11* (at 19p13.3, coding a serine/threonine kinase) and *TP53*, the first five also being linked to familial disease. Notably, all these genes have already been described, indicating a high degree of non-specificity.

4.7.3 Which Genes to Test?

NICE NG85 on pancreatic cancer in adults: diagnosis and management, recommends offering surveillance to those with hereditary pancreatitis and a *PRSS1* (at 7q34, coding for a trypsinogen) mutation, with a *BRCA1, BRCA2, PALB2* or *CDKN2A* mutation, and one or more first-degree relatives with pancreatic cancer, and people with Peutz-Jeghers syndrome (as this is linked to pancreatic cancer). NICE also recommends considering surveillance of pancreatic cancer in people with Lynch syndrome (with a *MLH1, MSH2, MSH6* or *PMS2* mutation), and any first-degree relative with pancreatic cancer. The National Genomic Test Directory for England has four components in this section:

- A multi-target NGS panel for the detection of small variants of *BRCA1* and *BRCA2*
- A multi-target NGS panel to detect structural variants in *NTRK1, NTRK2* and *NTRK3*
- Testing for microsatellite instability
- Testing for DPYD (in view of the possibility of fluoropyrimidine treatment: Box 4.1)

4.7.4 Molecular Pathology Directing Treatment

As described in the previous sections (breast, ovarian, prostate), PARP inhibitors such as olaparib may be appropriate for those with mutant *BRCA1* or *BRCA2*. Similarly, patients with a *NTRK1, NTRK2* or *NTRK3* mutation may be eligible for an NTRK inhibitor, such as entrectinib or larotrectinib.

> **Key Point 4.7** With a frequency-to-mortality ratio of over 0.9, pancreatic cancer is particularly lethal, and as such, investigations will be extremely urgent.

4.8 Other Carcinomas

The sections covered earlier have focused on the most common and the most pathologically dangerous cancers. We now move to those of a lesser frequency.

4.8.1 Oesophageal Cancer

Oesophageal cancer is the 7th most common cause of cancer death in England and Wales, and ranks 11th in incidence globally, but 7th in mortality. NICE has not published a specific guideline, but has merged oesophageal cancer with gastric cancer as NG12, whilst NG231 focuses on Barrett's oesophagus and stage 1 oesophageal carcinoma: neither document refers to molecular genetics in oesophageal disease. The National Genomic Test Directory for England describes testing for microsatellite instability and for *DPYD* variants.

4.8.2 Liver Cancer

Although ranking 8th in mortality in England and Wales, liver cancer (the most common being hepatocellular carcinoma) ranks 6th in incidence and 3rd in mortality in a global setting. NICE has published a series of technology appraisals of several drugs aimed at receptors for VEGF, which function via tyrosine kinase. However, molecular genetics is not mentioned, and the method for determining the presence or absence of the VEGF receptor is not described.

The National Genomic Test Directory for England refers to a multi-target NGS panel to investigate possible structural variants in the three *NTRK* isoforms, and, as in almost all solid cancers, for *DPYD* variants. Should a pathological *NTRK* variant be present, an agent such as entrectinib or larotrectinib may be recommended. Cholangiocarcinoma may also be tested for by the *NTRK* variants, as well as *FGFR2* (at 10q26.13, coding for fibroblast growth factor 2) and microsatellite instability.

4.8.3 Bladder Cancer

This neoplasm ranks 9th in mortality in England and Wales, 9th in incidence and 13th in mortality globally. There are numerous reports in the literature regarding the role of several genes, such as *FGFR3* (at 4p16.3, coding fibroblast growth factor 3), *KRAS, TP53, CCNE1, CDKN2A, TERT* (at 5p15.33, coding for telomerase reverse transcriptase) and *PLEKHS1* (at 10q25.3, coding a molecule that can bind phosphatidylinositol lipids) and evidence of alterations in chromosomes 9 and 22.

NICE NG2 recommends offering patients the UroVysion, a FISH test for detecting loss of the 9p21 locus (the location of *CDKN2A*) and aneuploidy in chromosomes 3, 7 and 17 in urine. The National Genomic Test Directory suggests three multi-target NGS panels for patients where platinum-based chemotherapy is appropriate:

- A panel for small variant detection in *FGFR2* and *FGFR3*
- A panel for CNV detection at exon levels in *FGFR2* and *FGFR3*
- A panel for structural variant detection in *NTRK1, NTRK2, NTRK3, FGFR2* and *FGFR3*

4.8.4 Renal Cancer

Renal cancer is the 11th most frequent cause of cancer death in England and Wales, and globally it ranks 14th in incidence and 16th in mortality. One of the first reports of a genetic lesion in hereditary renal cell carcinoma (RCC) was translocation t(3;8)(p14.2;q24.13), bringing *c-Myc* adjacent to a section of 3p14 previously known to be linked to certain cancers. This gene, alongside *c-fos, KRAS, PTEN, TP53, SMARCA4, NRF2* (also known as *NFE2L2*, at 2p31.2, coding a transcription factor), *MDM2, MDM4, PIK3CA, PBRM1* and *ERBB2*, was later linked to primary RCC. Analysis for the genetics of von Hipple-Lindau disease (VHL, caused by germline mutation and/or deletion of the tumour suppressor *VHL* at 3p25.3) is justified, as one of its clinical consequences is RCC. The role of lesions in chromosome 11 and Wilms' tumour is discussed in Chapter 5.

NICE does not have a dedicated guideline for renal cancer, whilst the National Genomic Test Directory has several sections.

- A multi-target NGS panel for small variant detection in *SDHA, SDHB, SDHC, SDHD, FH, VHL, ELOC (TCEB-1), TSC1, TSC2, MET* and *BRAF*
- A muti-target NGS panel for structural variants in *NTRK1, NTRK2, NTRK3, TFE3* and *TFEB*
- FISH and/or PCR for *TFEB* and *TFE3* rearrangement
- FISH for structural variant detection and/or CNVs, in chromosomes 3, 7, 8 and 17

4.8.5 Ovarian Cancer

Accounting for 2.2% of all cancer deaths in England and Wales in 2023, ovarian cancer ranks 18th in incidence and 14th in mortality globally. Whilst familial ovarian cancer was described in 1970, it had become linked to breast cancer by 1983, with the description of the activation of *KRAS, C-MYC* and *TP53* in the years that followed. Cytogenetic studies reported the involvement of chromosomes 1, 3, 6, 7, 10 and 12, with partial deletions of 3p, 6p, 8p and 10q, and loss of heterozygosity of 6q, 11 and 17. However, the study of this disease was linked to that of breast cancer, such as HER2/neu, and of a gene(s) predisposing to breast and ovarian cancer located at 17q12-21, which in 1995 was shown to be the site where *BRCA1* is located. Soon after, *BRCA2* was also linked to ovarian cancer, mirroring data in breast cancer, whilst a much later discovery was a role for *SMARCA4* (located at 19p13.2 and coding a transcription factor) in the form of small cell ovarian carcinoma.

NICE CG122 on the recognition and initial management of ovarian cancer points to imaging and the measurement of serum CA125, alpha-fetoprotein and human chorionic gonadotropin. The document also states that the point at which to use genomic biomarker-based therapy is solid tumour treatment is uncertain. However, it also refers to technology appraisal TA598, which recommends the use of PARP inhibitor olaparib for the maintenance treatment of *BRCA* mutation-positive advanced high-grade epithelial ovarian cancer, and to TA644 and TA630, which discuss entrectinib and larotrectinib for treating *NTRK* fusion-positive solid tumours.

The National Genomic Test Directory describes germline and tumour whole genome screening for histologically proven high-grade disease, and three entries:

- A multi-target NGS panel for detecting small variants of *BRCA1, BRCA2* and *SMARCA4*
- A multi-target panel for structural variant detection of *NTRK1, NTRK2* and *NTRK3*
- Homologous recombination deficiency status for *BRCA1* and/or *BRCA2*

> **Key Point 4.8** *BRCA1* and *BRCA2* are implicated in numerous forms of cancer but are most closely linked to breast and ovarian cancer: a malignancy in one is a strong risk factor for a malignancy in the other, leading to the combined syndrome known as 'Hereditary Breast/ Ovarian Cancer'.

4.8.6 Rare Carcinomas

In this final section, we move to carcinomas with a low frequency (<2.25% of all cancer deaths, <3500 deaths annually) in the general population: some are also described in Chapter 7. A small number of malignancies may be assessed by relatively non-specific analyses, such as microsatellite instability for small bowel cancer, and proto-oncogene *KIT* at 4q12, coding for a receptor tyrosine kinase, for thymic carcinoma. Where there is a cancer of unknown primary origin or a solid tumour not listed elsewhere, then WGS of germline and tumour analysis is appropriate, with a multi-target NGS panel for structural variants in *NTRK1, NTRK2* and *NTRK3*. Post-transplant patients in whom passenger lymphocyte syndrome is suspected may be tested for short tandem repeats and for sex chromosomes by FISH. The remaining solid tumour cancers are discussed in the following sections:

Gastric Cancer

NICE NG83 focuses on oesophago-gastric cancer but does not directly refer to molecular genetics. However, the guideline does refer to treatment with fluoropyrimidine, in which cases *DPYD* assessment would be called for. Investigating a gastrointestinal stromal tumour involves a multi-target NGS panel for small variants in *KIT* and *PDGFRA*, and another panel for structural variant detection in *NTRK1, NTRK2* and *NTRK3*, with testing for microsatellite instability.

Head and Neck Cancer

These cancers may be tested for by panels focusing on *CDNK2A*, *EGFR* and *TP53*, and on *RET, NTRK1, NTRK2* and *NTRK3*, although there may be a place for FISH for *RET* and *TP53*. A secretory carcinoma of the salivary gland is likely to be tested for *EVT6*::*NTRK3* fusion by FISH and/or RT-PCR, and structural variant detection in *NTRK1, NTRK2* and *NTRK3*. Structural variant detection for *MYB*::*NFIB* by FISH/RT-PCR and for *NTRK1, NTRK2* and *NTRK3* in adenoid cystic carcinoma is probable. Mucoepidermoid carcinoma, a salivary gland neoplasm, may be investigated with a multi-target NGS panel for structural variants in *MAML2* (at 11q21, coding a transcription factor), *NTRK1, NTRK2* and *NTRK3*, and for *MAML2* rearrangement by FISH.

Mesothelioma

CNV in *CDKN2A* by FISH may be useful for diagnosing mesothelioma (which killed 2018 people in England and Wales in 2023, 1.3% of all cancer deaths, 82% being men), alongside multi-target NGS panels for structural variants in *NTRK1, NTRK2* and *NTRK3*, and for CNVs in *CDKN2A*.

Endometrial Cancer

Cancer of the endometrium was linked to 1864 deaths (1.2%) in England and Wales in 2023. NICE DG42 recommends that those diagnosed with this condition should be tested for pathology linked to Lynch syndrome, i.e. loss of mismatch repair mechanisms (see Section 4.4). The National Genomic Test Directory for England refers to a multi-target NGS panel for structural variants in *NTRK1*, *NTRK2* and *NTRK3*, for *MLH1* promoter hypermethylation, a multi-target NGS panel for small variants in *MLH1, MSH2, MSH6* and *PMS2*, and a multi-target NGS panel for small variant detection in *POLE*.

Thyroid Cancer

Although thyroid cancer is the cause of a relatively small number (401) of deaths in 2023, the National Genomic Test Directory recognises several types, such as papillary, follicular and medullary carcinoma, and poorly differentiated anaplastic cancer, Hurthle cell cancer and non-invasive follicular neoplasm. Practitioners have a choice of nine multi-target NGS panels and some FISH analyses for all of the genes described in previous sections, i.e. *KRAS, NRAS, HRAS, RET, NTRK1, NTRK2, NTRK3, BRAF* and *TP53*. However, analysis of a promotor hotspot in *TERT*, may be relevant in papillary carcinomas. NICE's NG230 on thyroid cancer does not directly refer to molecular genetics, but there are technology appraisals of drugs aimed at *RET* alteration and other genetic lesions.

Adrenal Cancers

There are two major tumours of this organ, which together killed 134 people in England and Wales in 2023. A phaeochromocytoma may be investigated by an NGS panel looking for a small variant in *RET* or by a NGS panel detecting structural variants in *NTRK1, NTRK2* and *NTRK3*. Testing for carcinomas of the cortex includes analysis for CNVs and small variants in *TP53*, a panel for *NTRK1, NTRK2* and *NTRK3*, and FISH for del(17p), and *TP53* copy number. As with any cancer, *DYPD* hotpot testing is needed if they are to undergo fluoropyrimidine treatment.

Self-check 4.7 Test yourself on this chapter by reviewing the learning objectives.

4.9 Chapter Summary

The leading cause of death in England and Wales is cancer, and the major cellular form of cancer is carcinoma – a tumour derived from epithelial cells. The genetic basis of cancer falls to aberrant expression of cell surface receptors, and both cytoplasmic and nuclear second messenger and transcription factors. These can be manifested as an over-activity, such as over-expression and activation of tyrosine kinase receptors, and as an underactivity, where loss of tumour suppressor gene activity permits a potentially malignant cell to thrive.

The genetic basis of cancer in general (and carcinomas in particular) is heterologous: the same gene lesion can be present in several different neoplasms, whilst cancer in a particular organ may be caused by a number of mutated genes, of which two bear further scrutiny.

4.9.1 NTRK

The astute reader is likely to have noted frequent referrals to three variants of *NTRK*, a family of cell-surface receptors (hence R), whose ligand is one of a family of neurotrophin growth factors (hence N and T) and who have the phosphorylating enzyme tyrosine kinase (hence K) in their intracellular domain. Although located at difference sites in the genome (*NTRK1* at 1q23.1, *NTRK2* at 9q21.33, and *NTRK3* at 15q25.3, coding for tropomyosin receptor kinases A, B and C, abbreviating to TrkA, TrkB and TrkC, respectively), their intracellular activation pathways have much in common, in addition to the roles with second messenger metabolism such as Ras/MAPK.

These genes/receptors are linked to numerous cancers, as we have discussed throughout this chapter. For example, a mutation in any of the three *NTRK* genes (and their fusions with other genes) has been reported in cancer of the lung, colon/rectum, prostate, breast and many other organs, demonstrating the non-specific nature of the lesion. Accordingly, the genes coding these receptors are the target for a class of inhibitors that includes nintedanib, entrectinib and larotrectinib, which are recommended for these particular malignancies.

4.9.2 BRCA

The history of BRCA (**BR**east **CA**ncer) is now some 30 years old, with key research reports published in the mid 1990s. Subsequent work located *BRCA1* at 17q21.31 and *BRCA2* at 13q13.1, with the product of both genes found to be part of a complex that repairs damaged

DNA. It follows that loss of function of these genes results in failure to correct repairs to DNA, thus permitting an aberration to remain, and consequently are classified as tumour suppressor genes.

These genes are the second most cited in this chapter, and consequently also bear further scrutiny. Section 4.6.3 and Table 4.5 have already highlighted the epidemiology of *BRCA1* and *BRCA2* with respect to odds ratio for breast cancer. Pathogenic variants of these genes are present in 4–13.3% (*BRCA1*) and 0.6–8% (*BRCA2*) of women with unselected ovarian cancer and confer a 44% and 17% respective lifetime risk of this malignancy. In a genetic audit in 17 different global populations, the median (interquartile range) of frequencies were 9.8% (7.4–14.2) and 5.4% (1.7–6.5), respectively. In another study, the mean frequencies of germline mutation were 9.5% and 5.1% for *BRCA1* and *BRCA2*, respectively, but in somatic gene changes, data were 3.5% and 1.9%, respectively, indicating a large difference in somatic/germline carcinogenesis, pointing to other features linked to the malignancy.

With a long period in the scientific eye, many studies have been published, and a meta-analysis of either of the genes with a positive association in other cancers is shown in Table 4.9. The analysis found no statistically significant links between either of the genes and cancers of the brain, buccal cavity, pharynx and oral cavity, tongue, larynx and throat, oesophagus,

Table 4.9 Links between *BRCA1* and *BRCA2* and cancer.

Cancer	*BRCA1*	*BRCA2*
Liver	1.30 (0.11–15.01)	3.59 (1.04–12.34)
Gall bladder and ducts	1.93 (0.50–7.42)	4.70 (1.42–15.53)
Pancreas	3.30 (1.97–4.65)	3.41 (2.36–4.92)
Uterus	1.53 (1.01–2.31)	1.48 (1.24–1.78)
Endometrial	5.00 (3.06–8.16)	6.20 (2.33–16.52)
Bone	0.05 (0.00–0.74)	6.76 (1.09–42.07)

Data are relative risk with a 95% confidence interval. Those where the 95% confidence interval does not include 1 are statistically significant. *Source:* From Lee YC. Medicine. 2021;57:905.

stomach, small intestines, colon, rectum, colorectal, cervix, urogenital system or bladder, prostate, testis, kidney, thyroid or connective tissue.

Interestingly, the odds ratio (and its 95% CI) for bone cancer and *BRCA1* is markedly less than 1, which implies that the presence of an abnormality in the gene protects from the malignancy. There are also instances where only one of the genes is linked to a cancer, and some where both genes are linked. However, the lack of significance in this meta-analysis for prostate cancer is contradicted by data in Section 4.5.2 and from elsewhere, indicating a link with this malignancy, so caution in interpretation and practice must be exercised.

4.9.3 Panels of Genes

The discussion of breast cancer in Tables 4.6, 4.7 and 4.8 has emphasised the power of NGS and TGS in their ability to detect a large number of genetic abnormalities. This multiple-gene approach has been used in a population screening approach to molecular pathology. One such study probed not only for 25 genes of interest in cancer (Table 4.10) but also for genes in cardiovascular disease, these being *APOB* and *LDLR*. Amongst

Table 4.10 Links between gene mutations and certain carcinomas.

Gene	Linked cancer
APC, BMPR1A, MUTYH, THL1, PTEN, SMAD4, STK11	Colon polyps, colon cancer
ATM	Breast, prostate and pancreatic cancer
BRCA1, BRCA2	Breast, ovarian and pancreatic cancer
BRIP1, PALB2	Breast and ovarian cancer
CDH1	Breast and stomach cancer
CHEK2	Breast and colon cancer
EPCAM, MLH1, MSH2, MSH6, PSM2	Colon, endometrial and ovarian cancer
HOXB13	Prostate cancer
RAD51C, RAD51D	Ovarian cancer
TP53	Breast and many other cancers

Source: Rao ND et al., 2023 / Springer Nature / CC BY 4.0.

the cancer genes, the most commonly detected mutations were in *BRCA2* (23%), *CHEK2* (12%), *BRCA1* (8%), *ATM*, *RAD51C*, *PALB2* (7%), *MSH6*, *PMS2*, *BRIP1*, *MLH2* (4%), *HOXB13*, *APC* and *RAD51D* (3%).

4.9.4 Bridge to Chapter 5

Whilst the greater part of cancer is the carcinoma, there is considerable mortality and morbidity in other cancers, i.e. the sarcomas, neurological cancers and melanoma. These will be considered in Chapter 5, which will also address paediatric cancer.

Further reading

Bosse Y, Amos CI. A decade of GWAS results in lung cancer. Cancer Epidemiol Biomarkers Prev. 2018;27:363–379.

Hanahan D. Hallmarks of cancer: New dimension. Cancer Discov. 2022;12:31–46.

Shi Z, Lopez J, Kalliney W, et al. Development and evaluation of actseq: A targeted next-generation sequencing panel for clinical oncology use. PLoS One. 2022;17:e0266914. doi: 10.1371/journal.pone.0266914.

Taylor A, Brady AF, Frayling IM, et al. Consensus for genes to be included on cancer panel tests offered by UK genetics services: Guidelines of the UK cancer genetics group. J Med Genet. 2018;55:372–377.

Websites

ONS website for deaths by specific causes: https://www.nomisweb.co.uk/

WHO site on global frequencies of cancers: https://gco.iarc.fr/

European Union website on cancer: https://ecis.jrc.ec.europa.eu/

The NHS England site for clinical genetics: https://www.england.nhs.uk/publication/national-genomic-test-directories/

5

Solid Organ Cancers Part 2: Sarcoma, Neurological, Paediatric, Dermal and Others

Having examined the molecular pathology of the carcinomas in Chapter 4, we now move to the remaining forms of solid organ cancers, which may be classified according to the basic underlying nature of the cell biology of the tumour. As described in Chapter 4, sarcomas are tumours arising from connective tissues, themselves derived from the mesoderm, and will be discussed in Section 5.1. The ectoderm gives rise to all nervous tissues, these being the central nervous system (CNS) of the brain, spinal cord and related tissues, the object of Section 5.2.

The UK's Office of National Statistics (ONS) tells us that there is a proportional excess of total cancer deaths in those aged 1–4 years, possibly because many such malignancies are particularly aggressive, and so, accordingly, warrants a separate discussion in Section 5.3. A further aspect is that many cancers are found in the young and adults, leading to a question as to placement in a particular section, an example being Ewing sarcoma. Although present in the adult, the majority of these cases are in the young, as will be discussed in the section on paediatrics. Similarly, paediatric blood cancer will be part of Chapter 6. Cancers of the skin are discussed in Chapter 5.4, and the chapter concludes with Section 5.5, bringing together other cancers not described elsewhere.

Molecular Pathology: A Primer for Laboratory Scientists, First Edition. Andrew Blann.

Learning Outcomes

After studying this chapter, you should be able to...

- Describe the leading forms of sarcoma
- Summarise the various forms of cancer of the CNS
- Comment upon paediatric cancers
- Outline the key features of melanoma

5.1 Sarcomas

These tumours cannot be classified as easily as those of solid organs as the nature of the malignancy often relates to its cellular basis, which can be present in many different parts of the body. Consequently, most of this section will be driven not so much by organs but by other aspects of anatomy. Although it does not specify particular cancers, excepting the uterus and Kaposi sarcoma, the ONS reports 312 deaths in England and Wales due to malignant neoplasms of bone and articular cartilage, which is likely to include many of the cancers in this section.

The examination of these tumours in this chapter follows the system of the National Genomic Test Directory for England. In almost all sarcomas, there are places not only for WGS tumour/germline analysis, and an NGS panel for *NTRK1, NTRK2* and *NTRK3* (as in chondroblastoma), but also for testing a malignancy-specific gene (as in *NR4A3* at 9q31.1, coding a nuclear receptor) and extra-skeletal myxoid chondrosarcoma. Notably, as elsewhere in molecular pathology, the *NTRK* family is a feature of many multi-target NGS panels, and for clarity this is often omitted in the text to come. This to some degree parallels the discussion in Section 4.9.1, where we noted the high prevalence of this gene family in carcinomas.

5.1.1 Tumours Based on Bone

Of the three most common forms of bone cancer, two are shown below, comprising 36% and 30%, respectively, of tumours of this nature. The third, an Ewing sarcoma, forming 16% of bone cancers, is predominantly a paediatric or young adult cancer, and is discussed in Section 5.3.3. The cell biology of tumours in the bone needs to be defined as it may be the first manifestation of a metastatic deposit (e.g. a secondary from a primary breast or prostate cancer) or a primary malignancy.

Osteosarcoma

Perhaps the most well-known bone malignancy, this condition has a bi-modal presentation in children and adolescents and in the over-60s, and most with this disease are aged over 25. Inactivation in tumour suppressors *ATRX, RB1* and *TP53* is often found, as is over-activation of *MYC* and *MDM2*, and accordingly, determination of the copy number variation (CNV) of the latter by FISH or an NGS panel is appropriate.

Chondrosarcoma

The leading genes of interest in this malignancy are *IDH1* and *IDH2* at 2q34 and 15q26.1 (coding for subunits of isocitrate dehydrogenase), small variants in which can be detected by a multi-target NGS panel. A variant – the mesenchymal chondrosarcoma – may be investigated by RT-PCR for a deletion causing the fusion of *HEY1-NCOA2* at 8q21.13 and 8q13.3, coding a nuclear protein, and a nuclear receptor co-activator, respectively.

Other rare bone-related cancers are summarised in Table 5.1; many investigations have been made to determine differentials between closely related forms, and so they may be included or excluded from a diagnosis. Close attention to this table also shows the varied and non-specific distribution of numerous genes across different neoplasms, adding to the complexity of the final diagnosis. These include genes such as *USP6* at 17p13.2, coding a ubiquitin-specific peptidase, and *GNAS* at 20q13.32, coding a G-protein subunit.

5.1.2 Tumours Based on Muscle and Soft Tissues

Tumours of muscle are classified as smooth muscle (leiomyosarcoma) or striated muscle (rhabdomyosarcoma). Tumours of cardiac muscle are exceptionally rare. One study reported 70 forms of soft tissue sarcoma, the WHO estimating an incidence of 50 per million, so it would undoubtedly be described as 'rare', and a candidate for Chapter 7 were it not a malignancy.

Table 5.1 Genetics of selected bone-related sarcomas.

Sarcoma	Gene	Method
Extra-skeletal myxoid chondrosarcoma	*NR4A3*	FISH and/or RT-PCR
Fibrous dysplasia/myxoma	*GNAS*	Multi-Target NGS panel
Giant cell tumour of bone	*H3-3B*	Simple PCR for hotspot
Bone-forming soft tissue tumour differential	*MDM2*	FISH for CNV
	USP6	FISH for rearrangement
Cartilage-forming bone tumour differential	*H3-3A, H3-3B, IDH1* and *IDH4*	Multi-target NGS panel
	HEY1-NCOA2 fusion	RT-PCR for structural variant detection
Bone-forming differential and osteoclast-rich bone tumour differential	*H3-3A, H3-3B*	Multi-target NGS panel, single gene sequencing and hotspot
Round cell sarcoma of bone differential	*EWSR1*	FISH/simple targeted mutation testing
	WT1	Rearrangement by FISH/RT-PCR
Vascular tumour of bone differential	*MYC*	CNV by FISH and NGS panel
	SERPINE1-FOSB	FISH/RT-PCR
Spindle cell tumour of bone differential	*ALK, BCOR, CIC, EWSR1, FUS*	Multi-target NGS panel
	EWSR1, FUS, MDM2	Rearrangements or CNV by FISH/RT-PCR

For reasons of clarity, the location of these genes not otherwise noted has been omitted. CNV, copy number variation; FISH, fluorescence in situ hybridisation; RT-PCR, real-time polymerase chain reaction. Text modified from the National Genomic Test Directory for England: https://www.england.nhs.uk/publication/national-genomic-test-directories/.

> **Key Point 5.1 Kaposi sarcoma:** There is a degree of uncertainty as to whether this tumour, perhaps the most well-known disease of this group, is actually a sarcoma at all. If the presentation is dermal, it may rise from a transformed endothelial cell or fibroblast and has also been reported as developing in a lymph node. But whatever its aetiology, in 2023, it was linked to 4 deaths in England and Wales and to 16,169 deaths globally.

Uterine Cancer

One of the few organs where there is a clearly demarked sarcoma is the uterus, although a very common benign tumour is a leiomyoma – a fibroid – effectively treated by myomectomy. Globally, in 2022, there were over 42,000 known cases of cancer of the uterus, with almost 98,000 deaths. In the UK, these figures were 10,000 and 2700, respectively. In England and Wales in 2023, neoplasm of the uterus was linked to 2401 deaths (for comparison – cervix: 761 deaths, ovary: 3319 deaths).

Early reports in the literature describe rearrangements in 1p32, 3p24 and 10q22; later studies point to changes in *c-ERBB2*, *PLK4* (at 4q28.2, coding a serine–threonine protein kinase), *SPARC* (at 5q33.1, coding for osteonectin), *SMOC2* (at 6q27, coding a calcium-binding protein), *SATB2* at 2q33.1, coding a DNA-binding protein, *BRCA2* and *FOXP3* (at Xp11.23, coding a transcription regulator), with several *NTRK1* and *NTRK3* fusions, and possible roles for miRNAs.

The National Genomic Test Directory for England suggests a number of options for testing:

- A multi-target NGS panel for structural variants in *EPC1::PHF1* (at 10p11.22 and at 6p21.32, respectively, both coding for a transcription regulator), *JAZF1::PHF1* (the former at 7p15.2-15.1, coding a transcription repressor), *JAZF1::SUZ12* (the latter at 17q11.2, possibly coding a chromatin silencer), *NUTM2B::YWHAE* (at 10q22.3, coding a molecule of unclear function, and at 17p13.3, coding a signal transduction regulator, respectively), *NTRK1*, *NTRK2* and *NTRK3*.
- FISH for structural variant detection in *EPC1::PFH1*, *JAZF1::PHF1*, *JAZF1::SUZ12*.
- FISH for simple targeted mutation testing for structural variant detection in *NUTM2B::YWHAE*.
- WGS screening for all variant types.

Endometrial Stromal Sarcoma

Closely related to uterine cancer, molecular investigation of this variant includes all of those for uterine cancer, but also FISH/simple targeted mutation testing for *ZC3H7::BCOR*. The former, at 22q13.2, codes a molecule of unclear significance, whilst the latter at Xp11.4 codes a co-repressor for *BCL6*, itself a gene whose product is involved in lymphoma development (Chapter 6).

Alveolar Rhabdomyosarcoma

This principal sarcoma of the lung is linked to *FOXO1* at 13q14.11 (coding a transcription factor), and its fusions with *PAX3* and *PAX7* (at 2q36.1 and 1p36.13, respectively, also both coding for transcription factors), structural rearrangements of which can be detected by FISH and simple targeted mutation RT-PCR, and an NGS for all three genes. These may be useful in the differential diagnosis of embryonal rhabdomyosarcoma.

Alveolar Soft Part Sarcoma

This slow-growing tumour was so named in the early 1950s as the malignant cells resembled those of the alveolar rhabdomyosarcoma. However, it can develop in a variety of anatomical sites including the female genital tract, the kidney, abdomen, spine, pelvis, thigh, buccal cavity and eye orbit, the most common metastases being of the lung, brain and bone. Although heterologous in anatomy, a common genetic lesion is t(X;17)(p11.2;q25) forming an *ASPSCR1::TFE3* fusion (at 17q25.3, coding a molecule that interacts with *TFE3*, and at Xp11.23, coding for transcription factor TFE3). Accordingly, FISH/RT-PCR for detection of structural variants in *TFE3*, which is also a part of an NGS panel, may be used.

Angiomatoid Fibrous Histiocytoma

Despite the anatomical and histological pointers to a modified leukocyte (a histiocyte), blood vessels and fibrous tissue, this malignancy arises in soft tissues and is nevertheless classified as a sarcoma as opposed to a haematological malignancy. It is linked to translocations between *ATF1* at 12q13.12 (coding a transcription factor), and *FUS* and *EWSR1* (at 16p11.2 and 22q12.2, respectively, both coding an RNA-binding protein). *EWSR1* is also important in Ewing sarcoma (fully discussed in Section 5.3). Accordingly, rearrangements in *FUS* and *EWSR1* may be sought by FISH and/or RT-PCR, and a multi-target NGS that includes these genes.

Angiosarcoma

This group includes tumours of blood vessels and lymphatics, where the malignant cell is a transformed endothelial cell. There are several forms: an epithelioid haemangioendothelioma is characterised by the *WWTR1::CAMTA1* fusion (at 3q25.1, coding for a transcription co-activator, and 1p36.31-36.23, coding for a calmodulin-binding transcription activator, respectively) and a pseudomyogenic haemangioendothelioma, characterised by a *SERPINE1::FOSB* fusion (at 7q22.1, coding for plasminogen activator inhibitor 1, and 19q13.32, coding for a transcription factor, respectively), both structural variants being detected by RT-PCR. A radiation-induced angiosarcoma may be investigated by looking for CNVs of *MYC*, which may also be part of a multi-target NGS panel.

Other Soft Tissue Sarcomas

One of the problems with accurate diagnosis of many soft tissue sarcomas is the heterogeneity of their histology. However, several can be differentiated from each other by any particular genetic lesion, common methods being FISH, RT-PCR of individual genes, CNV and a multi-target NGS panel. For example, clear cell sarcoma of soft tissue is defined histologically and can be tested for by a structural variant of *EWSR1*, located at 22q22.2 and coding an RNA-binding protein, by FISH/RT-PCR, and as above, and an NGS panel that includes *EWSR1*. Those with multiple gene links include:

- Round cell sarcomas (and those not otherwise specified) with *BCOR, CIC, DDIT3, EWSR1, FOXO1, WT1, BCOR::CCNB3, HEY1::NCOA2, PAX3::FOXO1* and *PAX7::FOXO1*.
- Vascular sarcomas by *MYC*, *SERPINE1::FOSB* and *WWTR1::CAMTA1*.
- Spindle cell sarcoma by *ALK, BCOR, CIC, EWSR1, FUS, MDM2, SS18, USP6, COL1A1::PDGFB, TPM3::ALK, TPM4::ALK* and *ETV6::NTRK3*.
- Myxoid soft tissue by *BCOR, DDIT3, FUS, NR4A3, TGFBR3::OGA* and *BCOR::CCNB3*.
- Adipocytic sarcomas by *DDIT3* and *MDM2*.
- Epithelioid soft tissue sarcomas by *EWSR1, FUS, SS18, TFE3* and *WWTR1::CAMTA1*.

Other tissue sarcomas linked to one or two genes are shown in Table 5.2. In contrast to the bone-related neoplasms (Table 5.1), there are few examples of genes shared between different tumours.

Self-check 5.1 What genes are more commonly examined in the investigation of a sarcoma?

5.2 Tumours of the Central Nervous System

This organ of the brain and spinal cord can be seen as consisting of two types of cells: neurons and all other cells that are supportive. One of the more authoritative publications is that of the WHO (Louis et al. 2021), which describes tumours of these non-neuronal cells as gliomas, choroid plexus tumours, embryonal tumours, pineal tumours, cranial and paraspinal nerve tumours and meningiomas. The system also includes mesenchymal tumours (i.e. the sarcomas described in Section 5.1), melanocytic tumours (Section 5.4), haematolymphoid tumours (Chapter 6) and others.

However, as this volume focuses on scientific (as opposed to clinical) aspects, what follows is summarised from the National Genomic Test Directory for England, whose approach is different. The Directory also brings together neurological tumours into a number of groups, these being glial cells, which include astrocytes, ependymal cells, oligodendrocytes and microglial cells (which are modified macrophages, and have little, if any, neoplastic footprint in this setting), although some gliomas cannot easily be classified. Schwann cells, also known as satellite cells, are present in the peripheral nervous system but are nonetheless relevant to this section. In many cases, tumours of these cells are described as paediatric/infantile, and where appropriate will be discussed in Section 5.3.

5.2.1 Gliomas

Astrocyte Tumours

Astrocytomas are themselves a heterologous group, and so many genes are relevant and include methylation analysis of multiple CpG

Table 5.2 Genetics of selected soft-tissue sarcomas.

Sarcoma	Gene	Method
Myoepithelial tumours of soft tissue	*EWSR1*	Rearrangement by FISH or RT-PCR, or as part of a multi-target NGS panel
Myxoid/round cell liposarcoma	*DDIT3, MDM2*	FISH or RT-PCR
Myxoinflammatory fibroblastic sarcoma	*TGFBR3::OGA*	FISH or RT-PCR or as part of a multi-target NGS panel
	MDM2	CNV detection to genome-wide resolution or as part of a multi-target NGS panel
High-grade neuroepithelial tumour–BCOR group	*BCOR*, *BCOR-CCNB3*	Multi-target NGS panel for structural variant, or RT-PCR
Inflammatory myofibroblastic tumour	*TPM4::ALK*	Multi-target NGS panel for structural variant, or RT-PCR
Low-grade fibromyxoid sarcoma	*FUS*	Rearrangement by FISH/RT-PCR
Well differentiated/dedifferentiated liposarcoma	*MDM2, DDIT3*	CNV detection, multi-target NGS panel, RT-PCR, FISH
Desmoid-type fibromatosis	*APC, CTNNB1*	Multi-target NGS panel for small and structural variants
Desmoplastic small round cell tumour	*WT1*	Multi-target NGS panel for structural variant, or rearrangement by FISH/RT-PCR
Nodular fasciitis	*USP6*	Multi-target NGS panel for structural variant, or rearrangement by FISH/RT-PCR
Phosphaturic mesenchymal tumour	*FN1*	Multi-target NGS panel for structural variant or rearrangement by FISH
Sclerosing epithelioid fibrosarcoma	*FUS*	Multi-target NGS panel for rearrangement by FISH/RT-PCR
Synovial sarcoma	*SS18*	Multi-target NGS panel for structural variant or rearrangement by FISH

Adapted from https://www.england.nhs.uk/publication/national-genomic-test-directories/.

sites, a multi-target NGS panel for small variants of *IDH1, IDH2, ATRX, H3-3A* (at 1q42.12) and *H3-C2* (at 6p22.2, both coding for histones), *BRAF* and the *TERT* promotor. Other panels include *EGFR, CDKN2A* and *CDKN2B*, and of *EGRvIII* fusion with *NTRK1, NTRK2* and *NTRK3*. CNVs of *CDNK2A, CDKN2B* and *EGFR* by FISH may also be sought, and also a 1p19q co-deletion, an abnormality is present in 50% of astrocytomas. Hotspot analysis of *IDH1* and *IDH2* may also be appropriate.

A pilocytic astrocytoma may be investigated by FISH and/or RT-PCR for fusions between *BRAF* and *KIAA1549* (at 7q34, with a possible role in cilium function), *AGK* (at 7q34, coding for acylglycerol kinase), *AKAP9* (at 7q21.2, coding a part-regulator of protein kinase A), *CCDC6* (at 10q21.2, coding a molecule with a potential role in papillary thyroid cancer), *FAM118B* (at 11q24.2, coding a component of a nuclear Cajal body), *FXR1* (at 3q26.33, coding an RNA-binding protein), *GNAI1* (at 19p13.3, coding a subunit of a G protein) and *MACF1* (at 1p34.3, coding a molecule with a role in the cytoskeleton).

Other analyses include *BRAF* rearrangement by FISH, copy number of *KIAA1549* and *CDKN2A* by FISH or a multi-target NGS panel, methylation analysis of multiple CpG sites and a multi-target NGS panel for all these genes to detect any structural variants.

Ependymal Cell Tumours

Ependymomas develop from a layer of epithelial cells lining ventricles and the central canal of the spinal cord, these being the generators of cerebrospinal fluid. FISH and/or RT-PCR may be used to probe for rearrangements in *YAP1* (at 11q22.1, coding a transcription co-regulator) and *RELA* (at 11q13.1, coding a subunit of NFκB), and both may be part of multi-target NGS panel. Methylation analysis of multiple CpG sites is a further option.

Oligodendrocyte Tumours

A principal analysis of these malignancies is a multi-target NGS panel for small variants in *IDH1, IDH2, ATRX, H3-3A, H3-C2, BRAF* and in the *TERT* promoter. The 1p19q co-deletion may be sought by FISH/RT-PCR, as may hotspots in *IDH1* and *IDH2* by RT-PCR, and the methylation status of multiple CpG sites.

Gliomas of Uncertain Specificity

Should a glioma fail to fit easily into one of the types described earlier, options are available. Investigation of a diffuse midline glioma is likely to include a multi-target NGS panel for small variants in *IDH1, IDH2, ATRX*, the *TERT* promoter, and *H3.1* and *H3-3A* (for a point mutation in histones), hotspot analysis of *IDH1* and *IDH2*, 1p19q co-deletion by FISH/RT-PCR and for the methylation status of CpG sites.

Glioblastomas may be investigated using multi-target NGS panels for small variants in *IDH1, IDH2, ATRX, H3-3A, H3-C2, BRAF* and in the *TERT* promoter, another with *EGFRvIII* and the three NTRK variants, and for CNVs of *PDGFRA, MYC, PTEN* and *EGFR*, all of which may also be detected by FISH/RT-PCR. The 1p19q co-deletion may also be detected, as could *EGFRIII* by RT-PCR, and methylation of the promoter region of *MGMT* (at 10q26.3, coding for a DNA methyltransferase) determined.

Analysis for a high-grade glioma and a non-midline glioma calls for a multi-target NGS panel for small variants of *IDH1, IDH2, ATRX*, the *TERT* promoter, *H3-3A* and *H3-C2*, for *MGMT* promoter hypermethylation and the methylation status of multiple CpG sites, for the 1p19q co-deletion by FISH/RT-PCR, and its copy number by an NGS panel and for hotspots in *IDH1* and in *IDH2*. Investigation of a low-grade glioma is the same, but *ATRX* is omitted. In cases of low-grade glioma/glioneuronal tumours, *BRAF* variants may be sought. In addition to those mentioned earlier, an adult glioma may also be investigated with *TP53, VHL* (at 3p25.3, coding von Hippel–Lindau tumour suppressor), *H3-C2, H3-C3* (at 6p22.2) and *H3-C14* (at 1q21.2, all three coding a histone).

5.2.2 Meningiomas

These malignancies, arising from the three meningeal membranes, may be investigated with a multi-target NGS panel for the *TERT* promoter and for the methylation status of CpG sites.

5.2.3 Medulloblastomas

A large number of genes have been linked to this neoplasm and include *ALK, BCOR, CTNNB1* (at 3p22.1, coding for β-catenin), *DDX3X* (at Xp11.4, coding an RNA helicase), *EZH2* (at 7q36.1, coding a methyltransferase),

MYCN, PTCH1 (9q22.32, coding a tumour suppressor), *PTCH2* (at 1p34.1, also coding a tumour suppressor), *PTEN SMARCA4, SMO* (at 7q32.1, coding a G protein–coupled receptor), *SUFU* (at 10q24.32, coding a molecule that regulates a signaling pathway), *TERT* promoter, TP53 and *YAP1*. Together, these can form a multi-target NGS panel, whilst determination by FISH of the copy number of almost all can be informative. FISH can also be used to investigate rearrangements in *ALK, BCOR* and *MYC*, and, in conjunction with RT-PCR, the fusions *PVT1::MYC* (the former oncogene at 8q24.21, coding an miRNA), *YAP::C11orf9* and *YAP1::TFE3*. A wider range of candidate genes may be fruitful: a 23-gene signature derived from 142 patients achieved a classification accuracy of 95.2%.

5.2.4 Pituitary Tumours

The National Genomic Test Directory for England places cancer of this organ as neurological: others may classify it elsewhere, principally endocrine. Nevertheless, likely analyses include *MGMT* (at 10q26.3, coding a methyltransferase) and a multi-target NGS panel or FISH for CNV in *EGFR*.

Key Point 5.2 In almost all cases of adult neurological tumours, an NGS panel would include *NTRK1, NTRK2* and *NKRT3*, WGS germline, and tumour and DNA methylation studies.

Self-check 5.2 What is the cell biology of the various adult tumours of the CNS?

5.3 Paediatric Cancers

5.3.1 Introduction to Paediatrics

Although what we may call 'adult' cancers described in the sections earlier are exceptionally rare in the neonate, they may develop as the infant matures. Nevertheless, there are many non-malignant abnormalities that are evident soon after birth (i.e. are congenital), and whilst often well-defined clinically, there is still a place for genetic analysis, which is likely to add to the general pool of information about the particular disease, and is likely to be required if the patient is to be enrolled into any clinical trials. Indeed, Chapter 7 is dedicated to rare, inherited and metabolic diseases, many of which may be found in a paediatric setting, and accordingly there will be some cross-over with the text in the present chapter. A testament to the complexity of paediatric disease is the establishment, in many large towns and cities all around the world, of dedicated children's hospitals, one of the leading such institutions being London's Great Ormond Street Hospital.

The Neonatal and Infant Period

Often defined as up to 28 days, this period often sees clinical signs such as failure to thrive, features such as poor motor development, and repeated infections. An example of the latter is chronic granulomatous disease, where the X-linked form is linked to mutations in genes such as *CYBB* at Xp21.1-11.4, coding for NADPH oxidase, resulting in the inability to kill numerous species of bacteria. Other autosomal recessive forms can occur due to mutations in *CYBA, NCF1, NCF2, CYBC1* or *NCF4*. Examples of unequivocal birth defects (such as spina bifida) include those of the brain, skull and spinal column, and many are linked to gene abnormalities, others to infections, where congenital syphilis comes to mind. Further examples include some cases of anencephaly, which may be linked to, and probably caused by, defects in *ALX1* at 12q21.31, coding for a protein with a role in cartilage metabolism. Other cases may be linked to mutations in *NUAK2* at 1q32.1, coding for a kinase. Certain whole chromosomal abnormalities, such as Down syndrome, are also relatively easy to diagnose at birth or shortly thereafter and can be confirmed in the laboratory by cytogenetics. Cystic fibrosis may be suspected in a neonate if there are large, greasy faeces, and confirmed or excluded by analysis for ΔF508, a mutation of the causative mutation in *CFTR* – further details are present in Chapter 7.

As the neonate moves to the infant stage, many of the same abnormalities may be noted, but others include failure to communicate normally, problems with the senses, lack of muscle tone and poor mental development. Should the infant be tired and lethargic, perhaps jaundiced, then a haemoglobinopathy such as sickle cell disease (often caused by a mutation in *HBB* at 11p15.4, coding for β-globin) may be present. As the infant skull develops, other abnormalities may be noted, suggestive of craniofacial features of conditions such as Williams syndrome, caused by a micro-deletion in 7q11.12 (detectable with a FISH probe for this region), with the loss of numerous genes.

Childhood

Once more, many of the clinical features described in the previous subsection apply as the infant enters childhood, often when starting school, perhaps at ages 4–5. Diagnoses such as Fragile X syndrome may be made at this stage if the child has impaired intellectual capacity compared to its peers. Data from the ONS shows a peak in deaths due to cancer in this age group, mostly of the blood (predominantly leukaemia) prompting focus on these conditions. The incidence of these diseases fall at ages 10–14 but rises in the teenage years and beyond.

Whilst the legal ages of the child/teenager/adult transition can be defined, many paediatric cancers are also present in young adults, often defined as up to age 25, an example being Ewing sarcoma, described in Section 5.3.3. A further example of extended age of presentation is a primitive neuroectodermal tumour, which arises from the embryonic neural crest and does not classify easily with other tumours, although some describe it as a neuroblastoma, and by others as a medulloblastoma. It may be present in the lung, stomach, heart and kidney, whilst potentially valuable gene analysis includes *TP53*, *SHH* at 7q36.3, coding a signalling molecule, amplification of *C19MC* cluster at 19q13.42, and small variants in *CDKN2A* may be sought with an NGS panel.

Epidemiology

Globally, cancer is one of the leading causes of paediatric death, with 140–150 cases per million, 60% being solid tumours. In England in 2017, there were 1459 cancer diagnoses in those aged 0–14 years, the leading cases being leukaemia (31%), tumours of the CNS and allied tissues (25%) and lymphoma (10%), with soft tissue sarcoma, neuroblastoma and other peripheral nervous cell tumours and renal tumours at 6% each. A follow-up of 309,473 neonates for a mean of 5.4 years from Denmark reported 332 cases of cancer, of which 124 were haematological (37%), 80 were of solid organ (24%) and 44 (13%) were CNS tumours. However, this data is incomplete as 84 cases (25%) had other and unspecified malignant neoplasms. In 2021, Public Health

Table 5.3 Childhood and teenage cancers.

Cancer	Age 0–4 years	Age 5–9 years	Age 10–14 years	Age 15–19 years
Leukaemia	38	37	23	17
Lymphoma	7	20	25	40
CNS tumours	16	35	25	10
Sympathetic nervous system tumours	15	5	3	2
Retinoblastoma	5	1	0	0
Soft tissue sarcomas	10	11	14	10
Epithelial tumours and melanoma	1	3	9	21

Data is provided in %. *Source:* Adapted from Steliarova-Foucher et al. (2017).

England reported that overall, 5-year survival to be ~84%, varying from ~70% for neuroblastoma, via ~88% for leukaemia, to ~99% for retinoblastoma. Table 5.3 shows the frequencies of different groups of cancers up to the age of 19.

Wider View of Paediatric Molecular Pathology

The conditions described earlier are relatively easy to categorise, but in many instances, genetic diseases of the young can be complex, and several genes may be implicated in a certain condition. Accordingly, multiple analyses may be called for in order to define the precise genetic lesion likely to be responsible for the clinical syndrome in question.

As with adult molecular pathology, many conditions are investigated with WGS for germline and tumour abnormalities, DNA methylation of CpG sites, and the *NTRK* family, genes that code for a receptor whose intracellular domain is linked to a tyrosine kinase. The importance of this enzyme in carcinogenesis is explained in Chapter 3.

> **Key Point 5.3** There are distinct differences in epidemiology, genomic complexity, cell of origin, mechanism of mutations and pattern of metastases between adult and paediatric cancers.

5.3.2 Blood Cancer

Epidemiology

As evidenced by the data earlier, this group comprises the most common forms of paediatric cancers. The Danish data described earlier reports 77% of blood cancers being leukaemia (the dominant group at 85% being lymphoid, with myeloid at 7%), with 23% being lymphoma (the dominant group being non-Hodgkin lymphoma). Parallel data from England reported registrations of leukaemia at 83% of blood cancers, lymphoma at 17%, with acute lymphoblastic leukaemia (ALL) being the most frequent, and of these, the culprit lesion is most often (85% of the time) a malignant B lymphocyte. The epidemiology of paediatric and adult ALL and acute myeloid leukaemia (AML) differ markedly. In the USA, the rate of paediatric ALL far exceeds that of AML, and that of adult ALL, whilst adult AML exceeds that of paediatric disease (Table 5.4).

Table 5.4 Prevalence of acute lymphoblastic and myeloid leukaemia.

Age (years)	ALL	AML
<1	20	14
1–4	78	10
5–9	35	6
10–14	22	8
15–19	18	10
20–24	8	12
25–29	6	13
30–34	7	14
435–39	7	15
40–44	8	18
45–49	8	24
50–54	8	32
55–59	10	46
60–64	12	70
70+	14	>110

Data is rate per million.

The leading risk factors for paediatric ALL include paternal age, paternal smoking, domestic painting 1–3 months before conception and exposure to pesticides in men, alongside fertility treatment, high consumption of coffee, pesticides and breastfeeding for less than 3 months As men don't tend to do a great deal of breast feeding, with high birthweight in the neonate, low doses of ionising radiation in early childhood and certain viral infections in the patient. Figure 5.1 illustrates the high white blood cell count in ALL.

In 2021, Public Health England reported an overall 5-year survival of all paediatric leukaemia at ~88%, with that for lymphoma being ~93%. Therefore, whilst many of these cancers are relatively easy and quick to diagnose, many have a good prognosis, although, of course, there are fatalities. Table 5.5 shows data from the ONS on death due to blood cancer in England and Wales in 2023, almost all of which are leukaemias, reflecting the frequency of diagnoses. Close attention to these frequencies demonstrates a peak in all blood cancer deaths at age 1–4, a trough at years 5–9, following which levels rise in sequence to give a 'J'-shaped frequency curve that continues into the elderly. There is also a peak/trough in myeloid leukaemias, but death due to lymphoid leukaemia increases at each stage, as did all cancers.

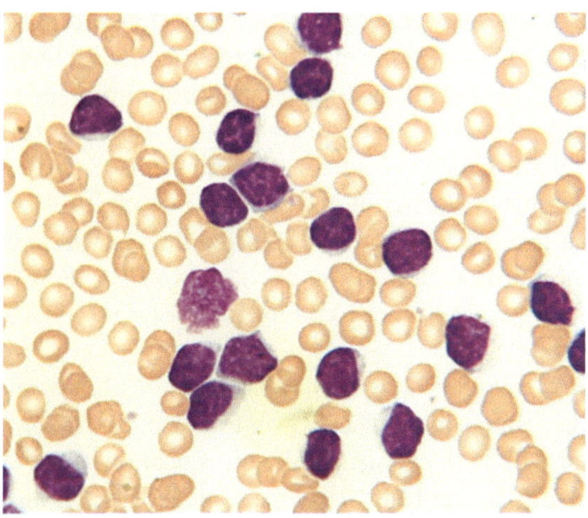

Figure 5.1 An acute lymphoblastic leukaemia. Blood film: blasts are stained purple.

Table 5.5 Deaths linked to all cancers and selected blood cancers in those aged up to 19.

Cancer	Age <1	Age 1–4	Age 5–9	Age 10–14	Age 15–19
All cancers	11	52	63	77	99
All blood cancers	1	19	13	23	24
Hodgkin lymphoma	0	0	0	0	1
Non-Hodgkin lymphoma	0	0	1	7	6
Multiple myeloma	0	0	0	0	0
All leukaemia	1	19	12	16	17
– Lymphoid	0	6	8	12	12
– Myeloid	1	11	3	3	5
– Monocytic	0	0	0	0	0

Adapted from https://www.nomisweb.co.uk/.

Molecular Pathology

As with many diseases, genetic analyses accelerated during the 1980s, with robust data being produced in the 1990s. Amongst these were links between *NF1* (at 17q11.2, coding for neurofibromin) and other RAS pathway mutations and juvenile myelomonocytic leukaemia, between *KMT2A* (at 11q23.3, coding lysine methyltransferase 2A), *TAL1* (at 1p33, coding an erythroid transcription factor), *LYL1* (at 19p13.13, coding a transcription factor), *TLX1* (at 10q24.31, coding a DNA regulator) and *MYC* and the translocation fusion oncogenes *TCF3::-PBX1* (at 19p13.3 and 1q23.3, respectively, both coding a transcription factor), and of *BCR::ABL1* and ALL.

Hyperploidy >50 chromosomes in paediatric ALL was shown in one study to exceed that in adult disease over fourfold, whilst another pointed to 22 cases of ALL with changes in the translocation *TEL::AML1* (physiologically at 12p13.2 and 21q22.12, respectively, both coding transcription factors), but only two adult cases, suggesting a strong effect of age on the presentation of the disease. Since renamed as *ETV6* and *RUNX1*, by the mid-2010s, this translocation had emerged as the most common genetic abnormality in paediatric B-cell ALL (26% of cases) and *TAL1* in T-cell ALL (47% of cases). In the present decade, *TCF3::PBX1, HLF::TCF3* (the former at 17q22, coding a transcription factor) and *KMT2A::AFF1* (the latter at 4q21.3–22.1, coding a transcription factor) were added to the list of genetic pathogens.

Key Point 5.4 The frequency of ALL in children far exceeds that in adults, whilst deaths peak at age 10–14, fall thereafter, and only exceed this figure at age 60 and beyond.

The molecular genetics of paediatric AML, as in adult AML, focuses on abnormalities such as t(8;21)(q22;q22) forming *RUNX1::RUNX1T1* (the latter at 8q21.3, coding a transcription factor) and inv(6)(p13q22), or t(16;16)(p13;q22). The pathophysiological link is that these aberrations disrupt the role of the core binding factor (coded for by *CBFB* at 16q22.1), which has a role in the regulation of haemopoiesis. Others include a rearrangement in *NUP98* at 11p15, coding for nucleoporin, a component of the nuclear pore complex, and *KMT2A*, located at 11q23, and mutated in around 18% of cases, more so in those aged under 2, where the frequency may be 50%, although this gene is mainly associated with mixed lineage leukaemia (Section 6.2.3).

Almost all cases of acute promyelocytic leukaemia may be identified by t(15;17)(q24;21), leading to the *PML::RARA* lesion. The former, a tumour suppressor at 15q24,1, codes for a protein with a role in nuclear body scaffolding, whilst the latter, at 17q21.2, codes for retinoic acid receptor alpha, present on the nuclear membrane, and which points to treatment with all-trans-retinoic acid (curiously, also treatment for acne), often alongside arsenic.

The National Genomic Test Directory for England list of potential gene targets is extensive, and most will be discussed in Chapter 6 on blood cancer. In almost all instances, where paediatric patients are referred to, WGS germline and tumour analysis is mentioned. Specific referrals to paediatric haematological tumours are paediatric-type follicular lymphoma, with a multi-target NGS panel focusing on *MAP2K1* (at 15q22.31, coding a signal transduction kinase), and juvenile myelomonocytic leukaemia, whose entry is extensive:

- A multi-target NGS panel for small variants in *PTPN11, KRAS, NRAS, NF1* and *CBL* (at 11q23.3, coding a ubiquitin-protein ligase).
- FISH for copy number of chromosome 8, for 7/7q, and for 5/5q.
- A multi-target NGS panel for CNVs in −7/del7q, −5/del5q and trisomy 8.
- CNV detection for genome-wide resolution by karyotype or equivalent.
- A multi-target NGS panel for small variants in *KRAS, NRAS, TP53, JAK2, CALR, MPL, ASXL1, CBL, CSF3R, CUX1, DNMT3A, EZH2, IDH1, IDH2, IKZF1, KIT, NFE2, SF3B1, SH2B3, SRSF2, TET2, U2AF1, HRAS, RUNX1, SETBP1, ZRSR2, BCOR, PTPN11, FLT3, NF1* and *NPM1*.
- The last bullet point of 31 genes underlines not only the complex nature of this disease, and so the difficulty in determining the correct approach to diagnosis and so potential treatments.

Self-check 5.3 Which type of blood cancer often seen in an adult is rarely seen in a child?

5.3.3 Neurological Tumours

As with adult/paediatric leukaemias, there is considerable duplication with Section 5.2).

These malignancies are second in frequency: their genetic basis can be complex, requiring numerous analyses to confirm/refute potential differential diagnoses. Ideally, tumours would be classified by histology, but this alone is insufficient. Where no further morphological classification of a paediatric brain tumour is possible, molecular genetics are called for. In these cases, a multi-target NGS panel for small variants in *IDH1, IDH2, BRAF, H3C2, H3C3, H3C14, H3-3A, H3-3B, TERT* promoter, *TP53, VHL, YAP1, SMARCA4, SMARCB1* (at 22q11.23, coding a chromatin regulator), *DDX3X* and *CTNNB1* may be required. In addition, there may be a place for a second panel looking for structural variants in *CIC* (at 19q13.2, coding a transcriptional repressor), *BCOR, EWSR1, MN1* (at 22q12.1, coding a transcription regulator), *RELA, YAP1, NTRK1, NTRK2* and *NTRK3*.

There are many examples of fused genes worthy of attention, which include *BCOR::CCNB3* (the latter at Xp11.22, coding a component of the cyclin group of regulators), *C11orf95::RELA* (the former also known as *ZFTA*, located at 11q13.1, coding a regulator of transcription), *YAP1::C11orf95, YAP1::TFE3*, all of which may be sought by an NGS panel, whilst estimates of CNVs in *C19MC, SMARCA4, SMARCB1* and *YAP1* are likely to be needed. However, many of these genes may be analysed individually, often by FISH.

Embryonal Tumours

Several types of cancer arise from primordial germ cells, those that give rise to gametes, the dominant clinical presentation being the gonads. However, in 3% of cases, presentation may be elsewhere, such as (in order) the mediastinum, retroperitoneum, cranium and sacrococcygeal region. Germ cell tumours are moderately metastatic, the most common sites being the lung, liver, bone and thyroid.

Genes of value in a differential diagnosis of an embryonal tumour include an NGS panel for small variants in *CDKN2A, DICER1, RB1, YAP1, SMARCA4, SMARCB1, MSH6, PMS2, PTEN* and *ATRX*, and another panel seeking structural variants in *BCOR, BRAF, MYC, MN1, RELA, YAP1, NTRK1, NTRK2* and *NTRK3*. As mentioned earlier (and elsewhere), fusions between many of these genes will be sought and include *BRAF::KIAA1549, BCOR::CCNB3, BRAF::AGK, BRAF::AKAP9, BRAF::CCDC6, BRAF::FAM118B, BRAF::FXR1, BRAF::GNAI1, BRAF::MACF1, C11orf95::RELA, TTYH1::C19MC, YAP1::C11orf95* and*YAP1::TFE3*. In many cases, CNV may be sought by FISH or an NGS panel.

A variant of this group is embryonal tumour with multi-layered rosettes. The leading genetic feature of this relatively rare aggressive tumour, named by its histological appearance, is *C19MC*, a miRNA cluster of 46 genes on chromosome 19, a second being *TTYH1*, at 19q13.42, coding for an anion channel. Analyses include methylation status of multiple GpC sites, CNV status, structural variation by FISH and RT-PCR and a multi-target NGS panel.

Note the considerable duplication of the genes in the previous paragraphs, which underlines the difficulties with the heterologous nature of the tumours. This theme is repeated in sections to come.

Neuroblastomas

This malignancy is one of the most common extracranial paediatric tumours, often of the adrenal medulla and spinal sympathetic ganglia, and accounts for 15% of deaths, with 70% exhibiting metastases. Clinical care is frustrated by the heterologous nature of its presentation and cell biology. The most common genetic abnormalities are in amplification/gain in function of *MYC-N* (present in 20–30% of cases), *AURKA* (at 20q13.2, coding a serine/threonine kinase) *ALK* and in *PHOX2B* (at 4p13, coding a transcription factor), with small variants and CNVs detectable by FISH, RT-PCR and a multi-target NGS panel, which may include *ATRX, FGFR1, NRAS* and *TP53*. Analyses of this form are important in suggesting treatment, as with the *AURKA* inhibitor alisertib and the *ALK* inhibitor crizotinib. FISH may also be used to determine CNV and rearrangement in the *TERT* promotor, and for abnormalities in 17q, 11q and 1p36.

Astrocytomas

The leading investigation of an anaplastic astrocytoma is with a multi-target NGS panel for the detection of small variants in *BRAF* and for the *BRAF*::-*KIAA1549* fusion by FISH/RT-PCR. A diffuse astrocytoma is likely to be examined with a multi-target NGS panel for the detection of small variants in *H3C2, H3C14, IDH1* and *IDH2*, and for hotspots in the latter two. CNV, rearrangements and structural variants in *MYC* may also be sought using FISH and as part of an NGS panel.

Ependymomas

As with adult cases, analysis focuses not only on DNA methylation status and *YAPI*, but also on *n-MYC* CNV, *YAPI*::*C11orf95*, *YAPI*::*TFE3* and *C11orf95*::*RELA* by FISH, RT-PCR and a multi-target NGS panel.

Glioblastomas

An *IDH1/IDH2* wildtype glioblastoma is likely to be investigated with a multi-target NGS panel with *RB1* and *NET1*, at 10p15.1, coding a molecule that activates and regulates the expression of rhodopsin (Figure 5.2).

Gliomas

The National Genomic Test Directory for England describes a number of gliomas, and in all cases, the 1p19q co-deletion is named. The unqualified entry of glioma highlights a multi-panel NGS panel with *H3-3B*,

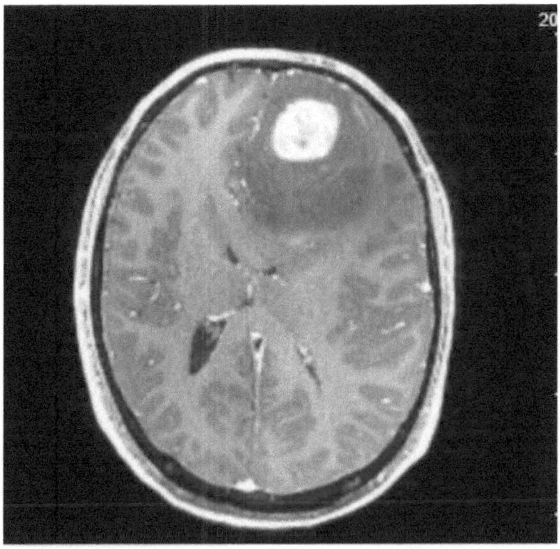

Figure 5.2 A glioblastoma. © *Credit:* Washington University School of Medicine. https://www.eurekalert.org/multimedia/897249.

H3C2, H3C3, H3C14, IDH1, IDH2, TP53 and *VHL*, seeking small variants, with hotspot analysis of *IDH1* and *IDH2* by FISH/RT-PCR. A diffuse midline glioma may be investigated with two multi-target NGS panels – one for a small variant in *H3-3A, NF1* and the *TERT* promoter, and a second for CNVs in *MYC* and *n-MYC*, with the copy number of the latter two genes, and *MYC* rearrangement, being sought by FISH. The Directory also has entries for the grade of a glioma: for high-grade gliomas, tumour suppressor *CDKN2B* at 9p21.3 (adjacent to *CDKN2A*), coding a cyclin-dependent kinase inhibitor, is named, whilst for low-grade gliomas, *TSC1* at 9q34.13, coding hamartin, and *TSC2* at 16p13.3, coding tuberin, the two molecules forming a complex that regulates certain second messenger pathways such as mTORCH1, and *IDH1, IDH2* and *FGFR1*, are named.

As in other sections (5.2) and subsections (5.2.1), there is marked duplication of genetic testing in different types of glioma, underlining difficulty in forming a firm diagnosis. Differential analysis of glial and glioneuronal tumours includes a multi-target NGS panel seeking small variants of *SMARCA4, SMARCB1, DDX3X, CTNNB1, CDKN2A, TP53, RAF1, PIK3CA, FGFR1, FGFR4, KIT, BRAF, H3-3A, ALK, ATRX, NRAS* and *PHOX2B*. A further panel focuses on structural gene fusion variants in *BRAF::KIAA1549*, *BRAF::AGK*, *BRAF::AKAP9*, *BRAF::CCDC6*, *BRAF::FAM118B*, *BRAF::FXR1*, *BRAF::GNAI1*, *BRAF::MACF1*, other rearrangements of *BRAF*, *ALK*, *C11orf95::RELA*, *EWSR1*, *FGFR1*, *RAF1::NF1A*, *RAF1::SRGAP3*, *RELA*, *YAP1*, *YAP1::C11orf95*, *YAP1::TFE3*, *EGFRvIII*, *NTRK1*, *NTRK2* and *NTRK3*. FISH and RT-PCR may be used not only for the detection of individual or fused genes but also for the 1p19q co-deletion.

Other Nervous System Tumours

Neurofibromatosis type 1 is a multisystem disorder that can manifest itself in various organs and tissues, such as those of the skin, referred to as neurofibromas, gliomas and meningiomas. It is caused by loss of function of tumour suppressor *NG1* (as mentioned earlier) at 17q11.2, coding for neurofibrillin 1. Neurofibromatosis type 2 is linked to *NF2* at 22q12.2, coding for a tumour suppressor described as Merlin, which is characterised by intracranial and/or spinal schwannomas, meningiomas and ependymomas. Schwannomas may also be caused by a mutation in *SMARCB1*.

> **Self-check 5.4** Which organs are most commonly the sites of an embryonal tumour, and why is this so?

5.3.4 Solid Organ Cancers

As with adult solid organ cancers, this group is defined anatomically.

Osteosarcoma

As discussed in Section 5.1.1, this malignancy has a bi-model distribution and is reported as accounting for 80% of bone cancers in children and adolescents. Around 15–20% of patients present with lung metastases and 40% develop subsequent metastases. The leading genetic alterations include *RB1* (present in up to 70% of cases), *AP-1* (~50%), *TP53* (35%), $p16^{INK4A}/p14^{ARF}$ (~10%) and *MDM2* (~10%). Deletions or loss-of-function mutations in *CDKN2A* and amplification in *CDK4* (at 12q14.1, coding cyclin-dependent kinase 4) can be sought by FISH and RT-PCR.

Ewing Sarcoma

Around a fifth of paediatric cancers are sarcomas. By far, the leading cancer in this group is Ewing sarcoma, presenting at a rate of 7.5 per million, and is the second most common primary malignant bone tumour in children after osteosarcoma. As with certain other malignancies, it can indeed be present in young adults, but the majority of cases are in the young, with metastases present in 20–25% at presentation. The leading candidate gene, *EWSR1* (at 22q12.2), codes for a protein that binds RNA, with a t(11;22)(q24;q12) placing it alongside a transcription factor coded for by *FLI1* (at 11q24.3) to generate a fusion protein.

Several forms of Ewing sarcoma are recognised by the National Genomic Test Directory for England, and in all, a WGS germline and tumour analysis and testing for the NTRK family is recommended. Other analyses include:

- A multi-target NGS panel for a CNS Ewing sarcoma family tumour with an alteration in *CIC*, alongside *NTRK1, NTRK2* and *NTRK3*.
- A multi-target NGS panel for structural variants in *BCOR* and *BCOR::CCNB3*, with RT-PCR for the latter for an Ewing-like sarcoma.
- A multi-target NGS panel for structural variants in *EWSR1*, with rearrangements detected by FISH or RT-PCR in cases of Ewing sarcoma of the bone.

Other Sarcomas

These include infantile fibrosarcoma, where key investigations include the *ETV6-NTRK3* fusion by FISH, RT-PCR or as part of an NGS panel, primitive mesenchymal myxoid tumour of infancy, focusing on *BCOR* and *BCOR::CCNB3* by RT-PCR or a multi-target NGS panel, and an undifferentiated round cell sarcoma of infancy, where *CIC* structural variant assessment by an NGS panel is relevant.

Wilms Tumour

This renal cancer may be described as a nephroblastoma, but the neoplastically transformed root leading to malignant cells (fibroblast, epithelial, endothelial and mesangial?) is difficult to identify. The disease is present at a rate of 1 per 10,000 live births and has a mean age at diagnosis of 43–48 months, and 95% of cases are aged under 10 years (Figure 5.3).

First described over a century ago, with a familial component recognised in the 1970s, the principal candidate gene, *WT1* at 11p.13, coding a transcription factor, was identified in the 1990s and subsequently shown to be a tumour suppressor gene. This was followed by a second, *WT2*, a locus located nearby at 11p14.1–15.5 and home to several genes, whilst changes to *WTX* (at Xq11, also known as *FAM123B*, whose product upregulates transcription activation by the Wilms tumour protein), *CTNNB1* and *TP53*, have been reported. However, the National Genomic Test Directory for England points to a multi-target NGS panel for structural variants in *NTRK1, NTRK2* and *NTRK3*, and for WGS germline and tumour analysis. However, the Directory also refers to a panel and single gene sequencing for *DROSHA*, located at 5p13.3 and coding a nuclear ribonuclease of the same name that part-regulates miRNA processing.

Retinoblastoma

As discussed in Chapters 1 and 2, this disease has a special place in molecular pathology (Figure 5.4). Knudson observed that hereditary retinoblastoma (and therefore germline, with the likelihood of genes on both chromosomes being mutated, hence biallelic) developed earlier,

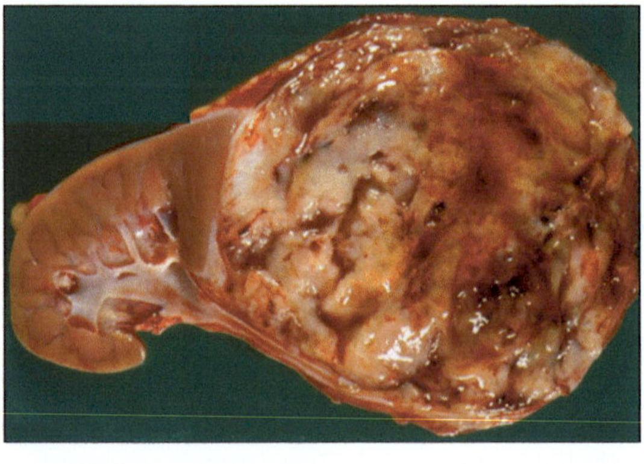

Figure 5.3 A Wilms tumour. https://pediatricimaging.org/diseases/wilms-tumor/.

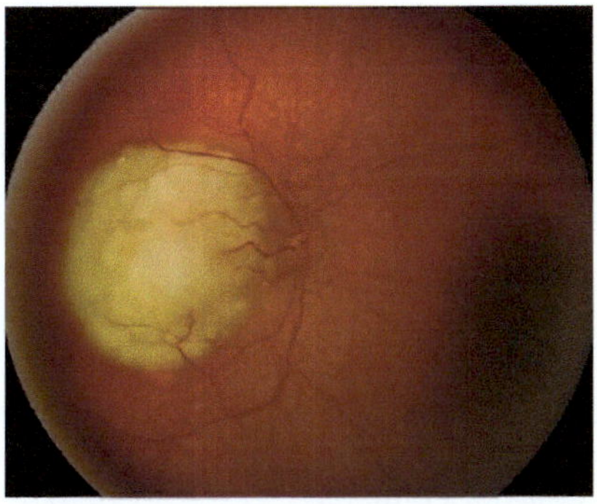

Figure 5.4 A retinoblastoma. Note the blood vessels over the tumour demonstrating tumour-promoted angiogenesis. © *Credit:* Kaliki S, Shields CL. Retinoblastoma: Achieving new standards with methods of chemotherapy. Indian J Ophthalmol. 2015;63:103–109 (Open Access).

and often in both eyes, than in uniallelic sporadic disease, which appeared later and was never observed in both eyes. From this, he built on the two-hit hypothesis of Ashley and others, presenting a focused genetic view of carcinogenesis.

First described in transformed cell line, the likely candidate gene, *RB1*, was traced to 13q14 in 1982. Its product, pRB, is part of the IBK4A–Cyclin D1–pRB–E2F second messenger pathway. Data such as the deletion of the relevant part of 13q14 in the disease, and loss-of-function mutations in the gene itself, later showed it to be a tumour suppressor gene. Subsequent work showed a role for the amplification of *n-MYC* in retinoblastoma, as it is in many other cancers.

The National Genomic Test Directory for England points to a multi-target NGS panel for small variants in *RB1* and another for structural variants in *NTRK1, NTRK2* and *NTRK3.* Other referrals are to WGS germline and tumour analysis and for the methylation status of multiple CpG sites.

Self-check 5.5 Name the leading genes in an Ewing sarcoma, a Wilms tumour and a retinoblastoma.

5.4 Dermal Cancers

Although the skin may not be seen by some as a solid organ, it nevertheless has an important immunological function and includes a number of cells that may transform into a cancer.

5.4.1 Squamous Cell Carcinoma and Basal Cell Carcinoma

These two cancers are the most common skin malignancies, and aetiology is varied. However, the key point is that they rarely metastasise and respond well to surgery and follow-up. Molecular pathology is very rarely called for, standard histopathology alone being sufficient.

5.4.2 Melanoma

Epidemiology

Despite a relatively high incidence of almost 16,000 annual cases of melanoma, it is linked to relatively few deaths (a little over 2000), giving a crude mortality rate of around 12.5%. These data can be contrasted with mesothelioma, with figures of some 2400 and 1900, respectively, so giving a rate of around 80%. There are numerous reasons for this difference, one of which is the likelihood of a quick and early diagnosis, whilst another is the success (or not) of treatment for the disease (a rapid surgical excision). Melanoma of the skin (Figure 5.5), with a high public health awareness, is more likely to be diagnosed early, whilst mesothelioma within the chest will be diagnosed late, partly explaining

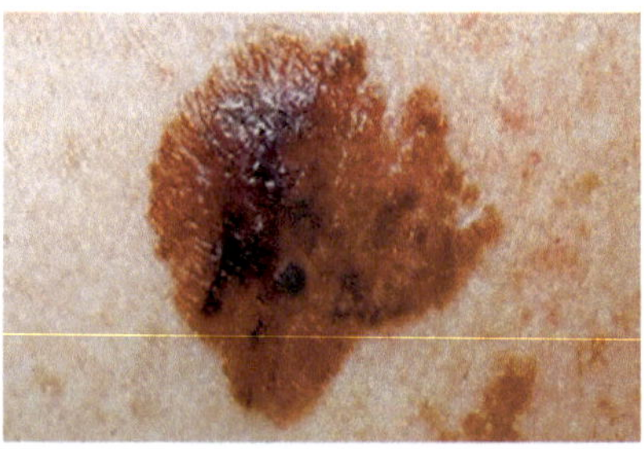

Figure 5.5 A melanoma. https://www.nhs.uk/conditions/melanoma-skin-cancer/symptoms/.

the difference in new cases/death ratio. A very rare (around 5 per million per year) uveal form mostly involves the iris (unsurprisingly, this being the coloured part of the eye), and metastatic potential is high.

Molecular Pathology

The National Genomic Test Directory for England lists several forms of melanoma. Adult melanoma may be investigated by a multi-target panel for small variants in *BRAF, KIT* and *NRAS*, for simple targeted mutation testing for *BRAF* and for structural variant detection in *NTRK1, NTRK2* and *NTRK3*.

A microarray may be used to look for CNVs in proto-oncogene *MYB* (at 6q23.3, coding a transcription factor), *RREB1* (at 6p24.3, coding a transcription element), *CCND1* (at 11q13.3, coding cyclin D1), *MYC* and *CDKN2A*. FISH can be used for CNV estimation in each of these genes in adult and conjunctival melanoma.

Multiplex ligation-dependent probe amplification or an equivalent may be used for CNVs in chromosome 1p, and chromosomes 3, 6 and 8, and small variant hotspot detection in *BRAF* in uveal melanoma. The use of two panels may be appropriate: multi-target NGS for small variant detection in *BRAF, NRAS* and *NF1*, and for structural variant detection in *NTRK1, NTRK2* and *NTRK3*. The latter three genes may also be useful in exploring a spitzoid tumour.

5.4.3 Dermatofibrosarcoma Protuberans

As implied by its name, this tumour is also based in the skin, and although there are several subtypes, ring chromosomes 17 and 22 may be found, and t17;22(q21.3;q13.1) is common. This brings together *COL1A1* at 17q21.33, coding the alpha chain of collagen type 1, and *PDGFB* at 22q13.1, coding subunit B of the platelet-derived growth factor, to form the fusion gene *COL1A1::PDGFB*. This structural variant can be detected by FISH and as part of a multi-target NGS panel.

> **Key Point 5.5** Melanoma is probably the most easily recognisable cancer, and as such generally brings a relatively good prognosis if acted upon urgently.

5.5 Other Cancers

There are, of course, quite probably hundreds of other cancers that are present very rarely, some that are inherited are described in Chapter 7, many of which we do not have the space to discuss in this volume. However, it has been estimated that rare cancers represent 22% of all cancers in Europe and are the cause of 30% of the deaths. A leading group of rare cancers is of the head and neck, which has a crude incidence of 188 per million per year. The system of classifying cancers is further flawed in that it focuses on those involving only one organ, of which one stands as an example.

5.5.1 Multiple Endocrine Neoplasia

The condition, abbreviating to MEN, has three major groups. Type 1, with a prevalence of around 1 in 35000, consists of a tumour expressing itself in the parathyroid (found in >90% of cases), pancreas (>80% of cases, often a gastrinoma and insulinoma) and pituitary (30–50% of cases), but there may also be lipomas and angiofibromas. The leading causative gene is *MEN1* at 11q13, coding for a tumour suppressor named Menin.

Type 2A, present at around 1 in 40,000, includes parathyroid, thyroid and adrenal tumours, the leading gene being the proto-oncogene *RET* at 10q11.21, and coding for a receptor tyrosine kinase. *RET* is implicated not only in type 2B, present in one in a million, which also involves the thyroid and adrenals, but also in a Marfanoid body habitus and a mucosal neuroma. Molecular pathology defining *RET* in type-2 MEN is important as its product is sensitivity to non-specific tyrosine kinase inhibitors such as cabozantinib and vandetanib: other drugs of this class are used in many other cancers, which include sorafenib and sunitinib in neoplasms, such as non-small cell lung cancer and breast cancer (Chapter 4, Sections 4.3 and 4.6 and elsewhere).

Other forms of MEN include cases with follicular medullary thyroid cancer alone, and MEN4, those with a phenotype of type 1 MEN, but whose driver is a mutation in *CDKN1B* at 12p13.1, coding a molecule that inhibits a cyclin-dependent kinase that participates in regulating the cell cycle.

5.5.2 Inherited Cancers

Thanks to NGS and third-generation screening, WGS, WES and others, the number of cancers with a heritable component is growing. Drawing to the end of this chapter, a review of these (and other inherited cancers not previously described) includes

- Epithelial ovarian cancer (without breast cancer), where testing for *BRCA1, BRCA2, BRIP1, MLH1, MSH2, MSH6, PALB2, RAD51C* and *RAD51D* are called for.
- Where there is inherited breast cancer and ovarian cancer, the focus is on *BRCA1, BRCA2, PALB2, RAD51C, RAD51D, ATM* and *CHEK2.*
- Hereditary nonpolyposis colorectal cancer (Lynch syndrome), calling for analysis of *MLH1, MSH6, POLE, POLD1* and *EPCAM.*
- Inherited polyposis and early-onset colorectal cancer, focusing on *MUTYH* and *APC.*
- Investigation of Peutz–Jeghers syndrome includes *STK11.*
- *PTEN*-hamartoma tumour syndrome.
- Suspicion of a Nevoid basal cell carcinoma syndrome or Gorlin syndrome requires *PTCH1* (also linked to oesophageal and bladder cancer) and *SUFU* testing.
- *CDH1*-related cancer syndrome (including lobular breast cancer and renal cancer: *CDH1* codes a cadherin).
- Li Fraumeni syndrome in a large number of cancers, the candidate genes being *TP53* and *POT1* (at 7q31.33, whose product has a role in telomere protection).
- Genes relevant to inherited renal cancer include *VHL* (as described in Section 4.8.4) and tumour suppressor *FLCN* at 17p11.2, coding for folliculin.
- Von Hippel–Lindau syndrome, caused by tumour suppressor *VHL*, and implicated in a wide range of cancers.
- *BAP1*, at 3p21.1, codes for a molecule with deubiquitinating activity, deficiency in which is linked not only to mesothelioma or melanocytic tumours, but also to a tumour predisposition syndrome.
- Inherited predisposition to a gastrointestinal stromal tumour (GIST), possibly with a family history of phaeochromocytoma, may be caused by mutations in *SDHA, SDHC* and *SDHD.*

- *DICER1*-related cancer predisposition may be suspected if there is a diagnosis of any of a large group of cancers that include cystic nephroma and pituitary blastoma.
- Tumours such as hereditary leiomyomatosis and renal cell cancer may be the result of mutation in *FH*, at 1q43, coding for fumarate hydratase.
- Inherited prostate cancer calls for testing for mutations in *BRCA1, BRCA2, MLH1, MSH2, MSH6, ATM, PALB2* and *CHEK2* (many of which are also part of the breast and ovarian cancer panels).

5.6 Summary

The previous chapter focused on carcinomas, which make up the majority of cancers. The present chapter focuses on cancer of other cells and tissues, i.e. the sarcomas, those of the CNS and those of the skin. A second aspect is a summary of the molecular pathology of paediatric cancer, of various cell types and tissues, and which in most cases mirror those of adult cancer, a major difference being that of ALL, which is more prevalent in the young.

In the next chapter, we move to the final major group of malignancies – those of the blood.

Self-check 5.6 Test yourself in a recall of this chapter by reviewing the learning objectives.

Bibliography

Chaiyawat P, Sangkhathat S, Chiangjong W, et al. Targeting pediatric solid tumors in the new era of RNA therapeutics. Crit Rev Oncol Hematol. 2024;200:104406.

Ekpa QL, Akahara PC, Anderson AM, et al. A review of acute lymphocytic leukemia (ALL) in the pediatric population: Evaluating current trends and changes in guidelines in the past decade. Cureus. 2023;15:e49930. doi:10.7759/cureus.49930.

Glembocki AI, Somers GR. Prognostic and predictive biomarkers in paediatric solid tumours. Pathology. 2024;56:283–296.

Louis DN, Perry A, Wesseling P, et al. The 2021 WHO classification of tumors of the central nervous system: A summary. Neuro Oncol. 2021; 23:1231–1251.

Munier FL. Special issue cancers. Retinoblastoma: Current challenges and promising new approaches. Cancers. 2023;15(8):2293. doi:10.3390/cancers15082293.

Sánchez-Molina S, Figuerola-Bou E, Sánchez-Margalet V, et al. Ewing sarcoma meets epigenetics, immunology and nanomedicine: Moving forward into novel therapeutic strategies. Cancers (Basel). 2022;14(21): 5473. doi:10.3390/cancers14215473.

Steliarova-Foucher E, Colombet M, Ries, LAG, et al. International incidence of childhood cancer, 2001–10: A population-based registry study. Lancet Oncol. 2017;18:719–731.

Urlić I, Jovičić MŠ, Ostojić K, et al. Cellular and genetic background of osteosarcoma. Curr Issues Mol Biol. 2023;45:4344–4358.

Websites

The European Cancer information System: https://ecis.jrc.ec.europa.eu/

Website of the American Academy of Neurological and Orthopaedic Surgeons: https://web.archive.org/web/20120430183809/http://www.aanos.org/epidemiologyboneca.html

UK Office for National Statistics, Cancer Registry: https://www.ons.gov.uk/peoplepopulationandcommunity/healthandsocialcare/conditionsanddiseases/bulletins/cancerregistrationstatisticsengland/2017

NHS England National Disease Registration Service: https://digital.nhs.uk/ndrs/data/data-outputs/ctya-uk-cancer-statistics-report-2021/summary

6

Blood Cancer

We complete our exploration of the molecular pathology of cancer by focusing on this important organ, bringing oxygen to the tissues and removing the waste products of metabolism, maintaining haemostasis, and protecting from pathogenic micro-organisms. Despite the clear 'organ' status of blood, the World Health Organization (WHO) presents the major forms of blood cancer as leukaemia, lymphoma and myeloma separately. However, if the data is merged, blood cancer ranks fifth in incidence (after lung, female breast, colorectum and prostate) but fourth in mortality (after lung, colorectum and liver cancer). But this data is an underestimate as there are other blood cancers of relatively low frequency that demand attention, and which include myelodysplasia (MDS), myelofibrosis, essential thrombocythaemia (ET), polycythaemia and others.

In England and Wales, in 2023, there were 12,193 deaths due to malignant neoplasms of lymphoid, haematopoietic and related tissues, i.e. blood cancer, which ranks third, after bronchus and lung (27,856), and colorectal (15,527), and so exceeding that of the prostate (11,072) and the breast (9,973 deaths). Within blood cancer, the most frequent cause of death is lymphoma (4,845 deaths, 39.2%, Section 6.1), followed by leukaemia (4,368 deaths, 35.8%, Section 6.2) and multiple myeloma and related neoplasia (2,879 deaths, 23.6%, Section 6.3). The remaining rare causes of blood cancer deaths, principally myeloproliferative neoplasms (MPNs), are discussed in Section 6.4.

Leading publications pertinent to this chapter include the fifth edition of the WHO classification of haematolymphoid tumours, one focusing on lymphoid neoplasms (Alaggio et al. 2022), whilst another

Molecular Pathology: A Primer for Laboratory Scientists, First Edition. Andrew Blann.

summarises myeloid and histiocytic/dendritic neoplasms (Khoury et al. 2022). Also of interest are the reports of the International Consensus Classification of myeloid neoplasms and acute leukaemias (Arber et al. 2022) and acute lymphoblastic leukaemia (ALL)/lymphoma (Duffield et al. 2023). These reports are very likely to be updated. Although the simple categorical classification of lymphoma, leukaemia and myeloma has much to offer, it is imperfect. An established development of leukaemia is its metastasising to lymph nodes (and the liver, spleen and elsewhere). It has long been noted that some lymphomas can transform, or perhaps break out, from their lymph node origins, with neoplastic cells appearing in the blood, much as a leukaemia, and also metastasising to the bone marrow. Further analysis of this phenomenon, tied into molecular genetics, prompted the need for a hybrid classification of lymphoblastic/lymphoma, linked to certain abnormal genes.

Learning Outcomes

After studying this chapter, you should confidently be able to...

- Describe major forms of blood cancer
- Recall the key features of Hodgkin and non-Hodgkin lymphoma
- List the leading types of leukaemia
- Describe the potential development of myeloma
- Understand the complex nature of myeloproliferative neoplasms
- Outline the importance of the 2022 WHO guidelines

6.1 Lymphoma

These cancers of lymph nodes can be classified by the nature of the malignant lymphocyte (B-cell, T-cell or NK cell) or by the demonstration of a characteristic Reed–Sternberg cell (generally described as a transformed B lymphocyte), whose presence defines a Hodgkin lymphoma, and whose absence defines a non-Hodgkin lymphoma (NHL). The cancer may also be defined by morphology, immunophenotype, the site of its host lymph node (such as gastric), whether it displays an 'early' or 'mature' phenotype, and the distribution of the malignant cells within the structure of the lymph node. Diagnosis relies on clinical signs and symptoms (notably lymphadenopathy) with imaging, and then a fine-needle aspirate or biopsy of the suspect node, which will go to the histology laboratory for standard tinctorial processing and immunocytochemistry. However, some tissue will inevitably be passed to the molecular pathology service. Involvement of the bone marrow, if present, generally develops in the later stages, so that (unlike leukaemias) anaemia and thrombocytopaenia are not usually part of the diagnosis.

The literature and practice of this cancer are dominated by two forms – the long-established (from 1832) Hodgkin lymphoma, and NHL, referred to separately by the WHO and other bodies. The former was linked to 290 deaths in England and Wales in 2023 (6% of lymphoma deaths), the latter 4,555 deaths. Globally, these mortality data are 8.3% and 92.7%, respectively. The WHO also publishes data on specific diseases in specific countries. In 2022, it reported an incidence of Hodgkin lymphoma in the UK of 2,270 persons, with 310 deaths, and so a crude mortality index of 14%. This is in contrast to data on NHL with an incidence of 18,317, with 5,981 deaths, giving a mortality index of 35%.

Key Point 6.1 Lymphoma has the highest rate of mortality of all blood cancers, possibly because of the difficulty with diagnosis, which is often late when the disease can be well established.

6.1.1 Hodgkin Lymphoma

Epidemiology and Subtypes

Globally, this disease ranks 26th in incidence (82,429 cases) and 28th in deaths (22,733), so giving a crude mortality index of 27.6%, suggesting almost three-quarters of cases will die of another cause, or will be cured of their disease. It has an incidence of around 30 per million per year in most Western countries, five-year survival being around 90%. Most cases (95%) are described as classical Hodgkin lymphoma, the remainder as nodular lymphocyte predominant Hodgkin lymphoma, the former having four recognised subgroups – nodular sclerosis, mixed cellularity, lymphocyte-depleted and lymphocyte-rich Hodgkin lymphoma. A genetic link was supported by the report that a first degree relative of a patient with Hodgkin lymphoma has a 3.1-fold risk of developing the disease.

Molecular Pathology

Genetic work from the early 1980s focused on a leading causative agent, the Epstein–Barr virus (EBV), but soon after, genes such as *N-ras* and *c-myc* were identified and deletion in chromosome 7 was reported. Rearrangements in immunoglobulin genes and in T-cell receptor genes were subsequently reported, pointing to either form of lymphocyte as the seed of the malignancy. Cytogenetic analysis of 49 Hodgkin lymphoma samples has revealed abnormalities in 11q23, 14q32, 6q11-21, 8q22-24 and 11q13, alongside instances of t(11;14)(q23;q32) and t(14;18)(q32;q21), the latter pointing to a role for *BCL2*. Notably, chromosome 14 is the site of the immunoglobulin heavy chain locus, genetically described as *IGH*, and is a common component of many lymphomas.

Genome-wide association studies have identified 67 SNPs from 42 loci in Hodgkin lymphoma. A meta-analysis reported links between the disease and *IL10* at 1q32,1, coding IL10, *IL6* at 7p15.3, coding for IL6, *GSTP1* (also linked to lung cancer), *GSTT1* at 22q11.23, coding glutathione-S-transferase theta, *XPG/ERCC5* (the former at 13q33, coding a molecule predisposing to xeroderma pigmentosum), *MTHFR* at 1p36.22, coding methylenetetrahydrofolate reductase, *COX2* (on the mitochondrial genome, coding cytochrome c oxidase II) and *UGT1A1* at 2p37.1, coding a glucuronosyltransferase.

Malignant Hodgkin lymphoma cells may succeed in evading the immune system by altered expression of *PDL1/2*, by a permanently active *JAK/STAT* pathway (as transcription transducer and activator *STAT6*, at 12q13.3, is frequently mutated) and by the avoidance of apoptosis through the PI3K–AKT–mTOR axis. In common with many other cancers, the presence of various species of miRNAs and lncRNAs has been reported, some with diagnostic and potential therapeutic roles. A further development is the potential to determine circulating cell-free Hodgkin lymphoma DNA, noting that falling levels after treatment correlated strongly with prognosis.

Genomic Testing

The National Genomic Test Directory for England has little specific to Hodgkin lymphoma to offer. A non-specific entry on lymphoma describes assessing WGS of germline and tumour tissues, but for paediatric patients only. Where there is a suspected lymphoma, investigations include karyotyping, rearrangement of immunoglobulin heavy and light chain genes, and those genes coding for the T-cell receptor. Thus, in the absence of specific Hodgkin lymphoma comments, practitioners are likely to consider investigations as NHL, if possible.

> **Key Point 6.2** Diagnosis of Hodgkin lymphoma requires the presence of Reed–Sternberg cells.

6.1.2 Non-Hodgkin Lymphoma

Epidemiology and Subtypes

Failure to identify a Reed–Steinberg cell defines this malignancy, of which there are many forms, based on features such as the intra-node anatomy and phenotype of the cell. The WHO's International Classification of Disease recognises a number of groups: follicular lymphoma (linked to 303 deaths in England and Wales in 2023), non-follicular lymphoma (1454 deaths), and mature and other specified types of T/NK cell lymphoma (311 deaths). However, the largest group is 'other and unspecified types of NHL', linked to 2,487 deaths, underlining how challenging the recognition and diagnosis of this disease can be. This data, of course, tells us nothing about the cure rate in each particular group, but overall, NHL has a five-year survival of around 72%.

Despite the above, a more authoritative document may be the WHO's classification of haematolymphoid tumours mentioned in the introduction, which recognises a large number of variants. Notably, this publication fails to mention NHL, but does refer to Hodgkin lymphoma, and so, for the sake of this chapter, all non-Hodgkin lymphoma entries in the document are taken to be NHLs.

Precursor B-cell Neoplasms

As discussed in the introduction, recognising the close association between certain B-cell leukaemias and B-cell lymphomas, in that there are numerous instances of CD markers and genetic changes in common, this section discusses a series of gene abnormalities each associated with a different form of the malignancy, many being described in the literature as B-lymphoblastic leukaemia/lymphoma. These include those with high hyperploidy (between 50 and 66 chromosomes), hypoploidy (24–45 chromosomes), intrachromosomal amplification of chromosome 21,

BCR::ABL1 (i.e. t(9;22)(q34;q11)), a form with *BCR::ABL-1*-like features, rearrangement in a *KMT2A* rearrangement (i.e. t(v;11q23.3)), an *ETV6-RUNX1* fusion (i.e. t(12;21)(p13.2;q22.1)), those with *ETV6::RUNX1*-like features, a *TCF3::PBX1* fusion (i.e. t(1;19)(q23;p13.3)), a *TCF3::HLF* fusion, (i.e. t(17;19)(q22;p13)), where *HLF* at 17q22 codes for a transcription regulator, those cases of leukaemia/lymphoma not otherwise specified, and those with other defined genetic abnormalities.

Mature B-cell Neoplasms

This large section has numerous subsections, many themselves with several variants. Some have only a single entry, these being lymphoplasmacytic lymphoma, primary cutaneous follicle centre lymphoma and transformations of indolent B-cell lymphomas. A well-known member of this group is Burkitt lymphoma, of historical, clinical and scientific interest.

First identified as a jaw tumour by the surgeon after whom it was named in 1958, this disease was shown to be a malignant lymphoma in 1961. It was subsequently linked to EBV, and so was one of the first descriptions of a human oncovirus, whilst later immunocytochemistry showed it to be a B lymphocyte disease. A notable feature of the aetiology of one form of this disease is the hypothesis that it develops as a consequence of an immunosuppression, which may be due to malaria (as many cases are present in areas of Africa where the pathogen is present, a model reminiscent of Knudsen's dual-hit theory), in a post-transplant setting or in an HIV infection. The disease may also develop in the absence of a micro-pathogen infection or immunosuppression, often described as 'sporadic'. Most cases of the lymphoma are the product of abnormalities in *c-MYC* at 8q24, generally involving its translocation adjacent to a gene for an immunoglobulin chain, i.e. t(8;10)(q24;q32) to a heavy chain gene, t(2;8)(p12;q24) to a kappa chain gene and t(8;22)(q24;q11) next to a lambda chain gene. The translocation t(8;14)(q24;q32) is also often present. Some cases have been linked to *TP53* and to *RB1*.

Several subsections of NHL are complex – some consider the anatomical position of the leading lymph node (spleen and abdomen), others of the site of the malignant cells within the lymph node. These include:

- Splenic B-cell lymphomas: including some marginal zone lymphomas, diffuse red pulp small B-cell lymphomas and B-cell lymphomas with prominent nuclei.

- Marginal zone lymphomas (MZL): including extranodal MZL of mucosa-associated lymphoid tissue (MALT), primary cutaneous MZL, nodal MZL lymphomas and paediatric MZL.
- Follicular lymphomas: (where the malignant cells are found within the follicles of the germinal centre) including in situ follicular B-cell neoplasms, follicular lymphomas, paediatric-type follicular lymphomas and duodenal-type follicular lymphoma.
- Mantle cell lymphomas: (where the malignant cell lies around the outer region of the lymph node – the mantle) include in-situ mantle cell neoplasms, mantle cell lymphomas and leukaemic non-nodal mantle cell lymphomas.
- Large B-cell lymphomas (LBCL): This extensive family includes many varied forms of the disease. These are diffuse LBCL not otherwise specified, T-cell/histiocyte-rich LBCL, diffuse LBCL/high-grade B-cell lymphoma with *MYC* and *BCL2*, high-grade B cell lymphoma with *MYC* and *BCL2* and/or rearrangement in *BCL6* (at 3q27.3, coding a transcription factor), *ALK*-positive LBCL, *IRF4*-rearranged LBCL (coding for an interferon regulator), high-grade B-cell lymphoma with 11q aberrations, lymphomatoid granulomatosis, EBV-positive diffuse LBCL, diffuse LBCL associated with chronic inflammation, fibrin-associated LBCL, fluid overload–associated LBCL, primary LBCL of immune-privileged sites, primary cutaneous diffuse LBCL, intravascular LBCL, primary mediastinal LBCL, mediastinal grey zone lymphoma and high grade lymphoma not otherwise specified.
- Lymphoplasmacytic lymphoma: This group may be classified as those involving IgM, i.e. Waldenstrom macroglobulinaemia (WM, which in 2023 killed 101 people in England and Wales) and those involving IgG or IgA monoclonal proteins. This is pertinent to the discussion of myeloma in Section 6.3.
- Kaposi sarcoma-related herpesvirus (KSHV, otherwise known as HHV8) including primary effusion lymphoma, KSHV/HHV8-positive diffuse LBCL and KSHV/HHV8-positive germinotropic lymphoproliferative disorder (LPD).
- Lymphoid proliferations and lymphomas associated with immune deficiency and dysregulation: including polymorphic LPDs arising in immune deficiency/dysregulation: including EBV-positive mucocutaneous ulcer and inborn error of immunity-associated lymphoid proliferations and lymphomas.

Precursor T-cell Neoplasms

This short section includes T-lymphoblastic leukaemia/lymphomas not otherwise specified and early T-precursor lymphoblastic leukaemia/lymphomas, again emphasising the close genetic relationship between these two forms of blood cancer.

Mature T-cell Lymphomas

Reflecting the B-cell section, there are numerous subsections of this group of diseases, which together killed 311 people in England and Wales in 2023, although some include only one entry, such as hepatosplenic T-cell lymphoma, peripheral T-cell lymphoma not otherwise specified and both nodal and extra-nodal NK/T lymphomas. Other subsections are as follows:

- Primary cutaneous T-cell lymphomas: including those which are CD4-positive small or medium T-cell LPD, acral CD8-positive LPD, mycosis fungoides, CD30-positive T-cell LPD, lymphomatoid papulosis, anaplastic large cell lymphoma, subcutaneous panniculitis-like T-cell lymphoma, gamma/delta T-cell lymphoma, CD8-positive aggressive epidermotropic cytotoxic T-cell lymphoma, and peripheral T-cell lymphoma not otherwise specified.
- Intestinal T-cell and NK cell lymphoid proliferations and lymphomas: including indolent T-cell lymphoma of the gastrointestinal tract, indolent NK cell LPD of the gastrointestinal tract, enteropathy-associated T-cell lymphoma, monomorphic epitheliotropic intestinal and T-cell lymphoma, not otherwise specified.
- Anaplastic large cell lymphoma: including ALK-positive anaplastic large cell lymphoma, ALK-negative anaplastic large cell lymphoma and breast implant–associated anaplastic large cell lymphoma
- Nodal T-follicular helper (TFH) cell lymphoma: including nodal TFH cell lymphoma, angioimmunoblastic-type, nodal TFH cell lymphoma, follicular-type and nodal TFH cell lymphoma, not otherwise specified.

Key Point 6.3 It is clear from the above that lymphoma is a complex and diverse disease demanding considerable expertise.

Molecular Pathology

Given the dominance of NHL over Hodgkin lymphoma, the literature of the former is considerable, and, as a consequence, much is known about the genetic abnormalities in each particular variant. The National Genomic Test Directory for England adopts the layout of these variants, and there is considerable duplication, as the following list demonstrates:

- Investigation of a Burkitt lymphoma focuses on the three *IGH-MYC* translocations described above, using an NGS panel and FISH/RT-PCR. Rearrangements in *BCL2* and *BCL6* may be sought by FISH or a part of an NGS panel as part of that for *IGH-MYC*. CNVs in 11q may be detected by FISH, an Array or an NGS. These investigations also apply to testing for a high-grade lymphoma.
- Tests for a B-cell NHL include heavy and light chain gene rearrangements by complex targeted mutation testing and a panel or FISH for small variant detection and for *CNV* in *EZH2*. Other genes of interest include *BTK* (at Xq22.1, coding for Bruton's tyrosine kinase) and *PLCG2* (at 16q24.1, coding for 1-phosphatidylinositol 4,5-bisphosphate phosphodiesterase gamma-2, involved in signal transduction).
- Primary mediastinal B-cell lymphoma (found in the thoracic cavity between the sternum and vertebrae and between the lungs) may be investigated by *CD274* at 9p24.1, coding for programmed cell death 1 ligand 1, *PDCD1LG2* at 9p24.1, coding for programmed cell death 1 ligand 2 and *REL* at 2p16.1, coding an NFκB-related transcription factor. Techniques for all three include FISH and an NGS panel, the former focusing on CNV and structural variant detection.
- Mantle cell lymphoma focuses on a translocation between *IGH* and *CCND1*, that being t(11;14)(q13;q32), by FISH for a structural variant detection, and on *TP53*.
- Tests of follicular lymphoma include determination of t(14;18)(q32;q21), bringing together *IGH* and *BCL2*, and rearrangements in *BCL2* and *BCL6*, all by FISH or a multi-target NGS panel. Another panel may be used to look for small variants in *CARD11* at 7p22.2, coding a member of the caspase family, *CREBBP* at 16p13.3, coding a cAMP response element-binding protein, *EZH2 ARID1A*, at 1p36.11, coding a molecule participating in transcription, *EP300*, coding 22q13.2, coding a histone acetyltransferase *MEF2B* at 19p13.11, coding a transcription factor and *FOXO1*.

- Lymphomas of MALT may be assessed with examination of *MALT1* at 18q21.32, coding mucosa-associated lymphoid tissue lymphoma translocation protein 1, with a role in NFκB activation. Two *MALT1* translocations are relevant: t(11;18)(q21;q21) with *BIRC3* at 11q22.2, coding for an inhibitor of apoptosis, and t(14;18)(q32;q21) with *IGH*. Also relevant is t(1;14)(p22;q32) between *IGH* and *BCL10* at 1p22.3, coding another molecule involved in mitosis. A further gene of interest is tumour suppressor *FOXP1* at 3p13, coding a transcription factor. All genes in this section may be part of a multi-target NGS panel seeking structural variants.
- For ALK-positive large-cell lymphomas, those of a mature B phenotype may be tested for rearrangement of *ALK* at 2p23.2-23.1, coding for CD246 (anaplastic lymphoma kinase (hence *ALK*) a cell membrane receptor whose ligand is unknown), and translocations with *CLTC* at 17q23.1, coding for a clathrin, a component of organelle membranes, hence t(2;17)(p23;q23), and with *NPM1* at 5q35.1, coding for nucleophosmin, a molecule that binds nucleic acids, hence t(2;5)(p23;q35). *ALK* and its translocations may be detected individually by FISH or by an NGS panel. Those lymphomas of large cells of an anaplastic phenotype (i.e. are poorly differentiated with no clear mature phenotype) may be tested as above, except that investigation of *CLTC* is not required.
- For ALK-negative large-cell lymphomas, testing for *DUSP22* at 6p25.3 and coding for a phosphatase, and *TP53* by FISH is reported by the National Directory, alongside a multi-target NGS panel for these genes, and also for *IRF4* at 6p25.3, coding interferon regulatory factor 4.
- The two subsets of IgM-lymphoblastoid lymphoma/WM can be distinguished by the presence or absence of a mutation in *MYD88* (at 3p22.2 and coding for a molecule with a role in toll-like receptor and IL1 receptor activation), thought to be the hallmark driver of over 90% of these conditions. The L265 mutation triggers Bruton's tyrosine kinase, and so tumour cell growth. A second gene of interest in WM is *CXCR4* at 2q22.1, which codes for a chemokine receptor, mutations in which are present in ~40% of patients, one leading to resistance to tyrosine kinase inhibitor ibrutinib.

- T-cell NHLs are investigated with T-cell gene rearrangement by FISH and an NGS panel, encompassing *TRA* (at 14q11.2), *TRB* (at 7q34), *TCD* (at 14q11.2) and *TRG* (at 7p14), coding the alpha, beta, delta and gamma chains, respectively. A second NGS panel of *RHOA* at 3p21.31, coding a GTPase, *DNMT3A* at 2p23.3, coding for a methyltransferase, *IDH2*, and *TET2* at 4q24, coding for a dioxygenase, may also be used. Testing for hepatosplenic T-cell lymphoma includes rearrangement on i7q by FISH, and a multi-target NGS panel of *STAT3* at 17q21.2 and *STAT5B* at 17q21.2, both coding a transcription activator. These two genes may also be of value in the investigation of an NK cell/gamma-delta T-cell lymphoma.

Where a lymphoma is suspected, principal investigations focus on heavy and light-chain immunoglobulin gene rearrangements and T-cell receptor gene rearrangements by NGS panels. Some lymphomas have only one or two gene referrals, such as *MYD88* in intra-ocular lymphoma, and *MAP2K1* at 15q22.31, coding a signal-transduction mitogen-activated kinase in paediatric-type follicular lymphoma.

Self-check 6.1 What are the major forms of NHL?

6.2 Leukaemia

Although the second-most deadly blood cancer, leukaemia is certainly better known by the general public. Its name derives from the Greek for white, *leukos*, and blood, *heima*, the new name being disseminated by Virchow in the mid-1800s, the *white* reflecting the large number of white blood cells present in the circulation. The WHO reported the UK incidence of leukaemia in 2022 to be 10,755 persons, with 5,335 deaths, giving a crude mortality index of 50%. This contrasts poorly with data on all UK lymphomas of 18,317, 5,981 and so 33%, respectively. However, unlike lymphoma, from a clinical perspective, leukaemia can be classified on a timescale as being chronic (slowly developing, perhaps over years) or acute (rapidly developing, perhaps over months), with many cases of chronic disease deteriorating into acute disease.

6.2.1 The Aetiology of Leukaemia

A full understanding of the pathophysiology of leukaemia is impossible without an awareness of leukopoiesis. Numerous resources are available online that describe the development of all blood cells from stem cells in the bone marrow to mature cells in the circulation. Central to the process is the passage of the cell through different stages of differentiation that include various forms of 'blasts', collectively described as an immature (and so non-functioning) cell destined (in health) to be a mature functioning cell. Physiologically, these blast cells, which are generally larger than the mature functioning cells in the circulation, are only present in the bone marrow, so that their appearance in the circulation is pathological and likely to be leukaemic, or perhaps a rapid expansion of a white cell population driven by a factor such as an overwhelming bacterial infection. The transformation of the stem cell, through blasts, into the mature cell, is linked to certain genes, and a common feature of leukaemia is that the correct and ordered sequence of the activation of these genes is interrupted, often leading to a state of 'maturation arrest', with the build-up of these blast cells in the bone marrow and their subsequent appearance in the circulation. This important point will be revisited in sections to come.

To some extent it could be argued that modern haematologists carry the burden of a very convenient historical clinical classification of leukaemia as being acute or chronic, and as being myeloid/granulocytic or lymphocytic. These criteria, based on the rate of development of the disease and the appearance of the leading malignant cell on a blood film, is supported, in the main, by immunophenotyping. There is increasing evidence that these four groupings are becoming untenable, an example being myelomonocytic leukaemia, and so extending to the recognition of the close relationship with certain lymphomas with groupings such as B-lymphoblastic leukaemia/lymphoma. Nevertheless, the UK's ONS, adopting the WHO's ICD, lists myeloid leukaemia (which caused 2,686 deaths in England and Wales in 2023, that being 61.5% of all leukaemic deaths), lymphoid leukaemia (1,206 deaths, 27.6%), monocytic leukaemia (264 deaths, 6.0%), leukaemia of unspecified cell type (187 deaths, 4.3%) and other leukaemias of specified cell type (25 deaths, 0.6%). We shall therefore adopt this system in our examination of this disease, leaving plasma cell leukaemia developing from myeloma, and causing 48 deaths, in Section 6.3.6.

6.2.2 Myeloid Leukaemia

We must first consider the fine detail of the haemopoietic pathway. The pluripotential stem cell in the bone marrow gives rise to two other stem cells, a colony-forming unit (CFU) described as the common lymphoid precursor (hence CFU-L), and a second CFU described as the common myeloid precursor, the CFU-GEMM, the letters of this abbreviation coming from Granulocyte, Erythrocyte, Megakaryocytes and Monocyte. This is important because it places all these cells in a common family, i.e. myeloid, a feature we will return to. The CFU-GEMM then gives rise to a CFU-EMk and a CFU-GMo, the latter producing stem cells specific for eosinophils, basophils, neutrophils and monocytes (Figure 6.1).

This further emphasises the close developmental (and therefore, genetic) relationship between granulocytes and monocytes. It also explains neoplasia of mixed lineage, such as in myelomonocytic leukaemia. A further feature is the use of descriptors such as 'MPNs' and 'myeloid neoplasms', which include platelet malignancies and myelofibrosis, in addition to leukaemia. Some prefer the term 'myelogenous' as an alternative to 'myeloid', suggesting that the latter should be used only where the principal transformed cell is a myelocyte, as opposed to a myeloblast or a promyelocyte. Accordingly, great care must be taken when describing the nature of the cancer and the most common form of the malignant cell in the bone marrow or the blood.

Interrogation of the ONS's England and Wales database for 2023 finds that acute myeloblastic leukaemia (AML, also acute myelocytic leukaemia) to be the most dangerous, with 2,365 deaths, that being 88% of all myeloid leukaemia deaths. Chronic myeloid leukaemia (CML) was linked to 230 deaths, which is 9% of all myeloid deaths.

Acute Myeloid Leukaemia

The initial diagnostic pathway for AML is well-established, with (in order) full blood count/erythrocyte sedimentation rate, immunophenotyping and possible bone marrow examination, with input from biochemistry (urea and electrolytes, liver function tests, lactate dehydrogenase, etc.) and imaging for possible extra-marrow disease. In health, cells progress through stages of myeloblast, promyelocyte, myelocyte, metamyelocyte and the mature granulocyte, but as regards leukaemia, maturation arrest may occur at any of the early stages, leading to an excess number of blasts at that particular stage, such as, for example, acute promyelocytic leukaemia (linked to 36 deaths in England and Wales in 2023).

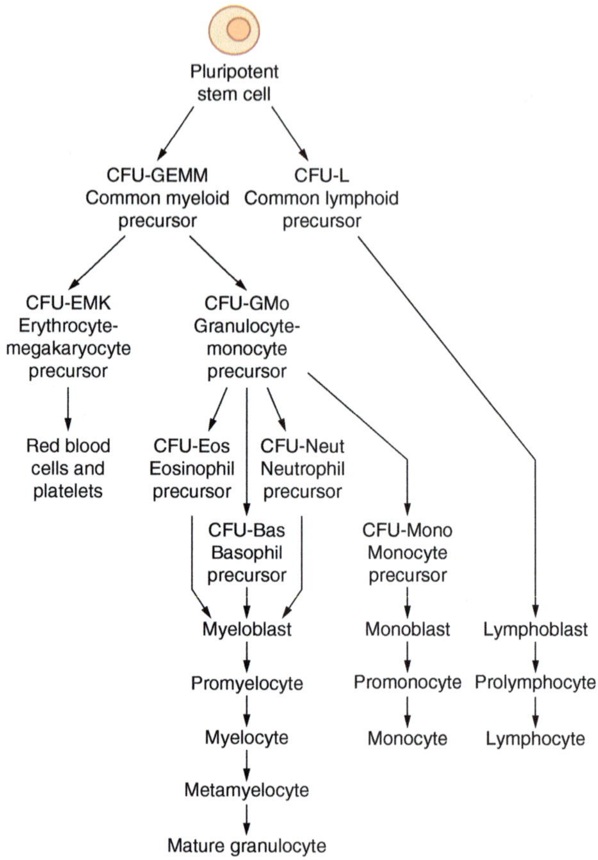

Figure 6.1 Myelopoiesis and lymphopoiesis. The four stands of haemopoiesis: from left to right – that which generates red blood cells and platelets, that which generates granulocytes, that of monocytes and that which produces lymphocytes.

However, in contrast to CML, genetic analysis of AML shows it to be, despite very similar phenotypes of the malignant cells on a blood film, a markedly diverse collection of diseases.

Self-check 6.2 Describe the major features of haemopoiesis.

Key advances in the molecular pathology of AML began in the late 1970s, with reports of abnormalities in chromosomes (such as trisomy for bands 1q25–1q32), followed by findings such as a 15;17 translocation in acute promyelocytic leukaemia, t(8;21)(q22;q22) and t(3;12)(p26;p13) translocations in AML, and *n-ras* activation in acute myeloblastic leukaemia. Reports such as that of inv(16)(p13;q22) followed in the 2000s, with the postulation that better understanding of these gene lesions would inform drug development directed towards that lesion, as it had been for CML and *BCR*::*ABL1*, which is indeed now that case (to be discussed below). In common with many diseases, a number of ncRNAs were purported to have a role in the pathogenesis of AML. A further advance in subsequent decades was the recognition that molecular pathology was a far superior tool than immunophenotyping for monitoring treatment and detecting minimal/measurable residue disease. Over the decade that followed, the increasing role of genetic abnormalities grew to such an extent that it is slowly superseding microscope-based and CD-marker based methods of classifying the disease. The most commonly mutated genes in AML are shown in Table 6.1.

Table 6.1 Selected genes mutated in AML.

Gene	Location	Product	Frequency
The internal tandem duplication variant of *FLT3*	13q12.2	CD135, a tyrosine kinase cytokine receptor	20–50%
NPM1	5q35.1	A nucleolar phosphoprotein that interacts with a ribosomal protein.	25–30%
IDH2	15q26.1	Isoenzyme 2 of isocitrate dehydrogenase	8–19%
TP53	17p13.1	p53, a tumour suppressor	8–14%
RUNX1	21q22.12	A subunit of transcription regulator core-binding factor	4–16%
The tyrosine kinase domain of *FLT3*	13q12.2	CD135, a tyrosine kinase cytokine receptor	7–10%
GATA2	3q21.3	A protein that binds the nucleotide sequence GATA	9%
IDH1	2q34	Isoenzyme 1 of isocitrate dehydrogenase	4–9%
Biallelic *CEBPA*	19q13.11	A transcription factor that binds the nucleotide sequence CCAAT	4–9%

These findings, and others, are likely to have been noted by the 2022 WHO classification of AML (Khoury et al. 2022) which focuses (a) on those forms of the disease that can be defined by genetic abnormalities and (b) on those defined by differentiation. The latter includes AML with minimal differentiation, AML with or without maturation, and leukaemia of other lineages, i.e. acute basophilic leukaemia, acute myelomonocytic leukaemia, acute monocytic leukaemia, acute erythroid leukaemia (discussed in Section 6.4.6) and acute megakaryoblastic leukaemia. The criteria for these diseases include the numbers of blasts, the extent to which they stain positively for myeloperoxidase and Sudan black stains and the presence of cell surface markers, such as CD12, CD33 and CD117, depending on the nature of the leukaemia being investigated. An additional classification system, also from 2022, is the International consensus classification of myeloid neoplasms and acute leukaemias (Arber et al. 2022), which has a similar approach.

The second aspect of the classification is genetic, of which there are numerous forms. These include gene fusions, such as between *PML* and *RARA* (present in acute promyelocytic leukaemia, and making up some 10% of all AMLs), *RUNX1* and *RUNX1T1* (present in ~4–8% of AML cases), *CBFB* and *MYH1* (at 16p13.11, coding myosin heavy chain 11), proto-oncogene *DEK* (at 6p22.3, coding a DNA-binder and *NUP214* (at 9q34.13, code nucleoporin 214), *RBM15* at 1p13.3, coding an RNA-binding protein, and *MRTFA* at 22q13.1–13.2, coding a transcription factor, and between *BCR* and *ABL1* (present in 1% of cases) and those with rearrangements in *KMT2A, MECOM* at 3q26.2, coding a transcription factor, *NUP98* and *CEBPA*. Other forms of AML include those with other defined genetic alterations and those that are MDS-related. Genes linked to these conditions include *ASXL1* at 20q11.21, coding a transcription regulator, *BCOR, EZH2, STAG2* at Xq25, coding for a subunit of cohesin, *U2AF1* at 21q22.3, *ZRSR2* at Xp22.2, *SF3B1* at 2q33.1 and *SRSF2* at 17q25.2, all four coding for a splicing factor.

The complexity of the genetic cell biology of AML is illustrated by the large number of genes with potential involvement. For example, the National Genomic Test Directory for England describes a multi-target NGS panel for the detection of small variants in *NPM1, CEBPA, RUNX1, FLT3, IDH1, IDH2, KIT, WT1, ASXL1, SRSF2, STAG2, RAD21, TP53, KRAS, NRAS, KMT2A, (MLL)-PTD, PPM1D, DDX41, PHF6* at Xq26.2, coding a transcription regulator, and *CUX1* at 7q22.1, coding a DNA-binding protein.

Table 6.2 Leading translocations and inversions in AML.

Abnormality	Genes fused
t(15;17)(q24;q21)	*PML::RARA*
t(8;21)(q22;q22)	*RUNX1::RUNX1T1*
inv(16)(p13.1q22)	*CBFB::MYH11*
t(9;11)(p21;q23)	*MLLT3::KMT2A*
t(6;9)(p22;q34)	*DEK::NUP214*
inv(3)(q21;q26)	*GATA2::MECOM*
t(1;22)(p13;q13)	*RBM15::MRTFA*
t(9;22)(q34;q11)	*BCR::ABL1*
t(3;5)(q25;q34)	*NPM1::MLF1*
t(5;11)(q35;p15.5)	*NUP98::NSD1*
t(7;12)(q36;p13)	*MNX1::ETV6*
inv(16)(p13.3q24.3)	*CBFA2T3::GLIS2*

These may be detected by an NGS panel or individually by FISH and/or RT-PCR.

There may also be cytogenetic abnormalities in numerous chromosomes, such as a complex karyotype (≥3 abnormalities), 5q deletion or loss of 5q due to unbalanced translocation, monosomy of 7, 7q deletion, or loss of 7q due to unbalanced translocation, 11q deletion, 12p deletion or loss of 12p due to unbalanced translocation, monosomy of 13 or 13q deletion, deletion or loss of 17p due to unbalanced translocation, isochromosome 17q and idic(X)(q13). Table 6.2 summarises leading translocations and inversions of relevance to AML. Unsurprisingly, our colleagues in industry offer various panels of reagents for many of these genetic abnormalities.

The importance of this genetic analysis is shown by links with prognosis:

- A good outcome with t(8;21)(q22;q22), inv(16)(q13;q22) and t(16;16)(p13;q22), a mutated *NMP1* without *FLT3*::ITD and a bZIP on-frame mutated *CEBPA*.
- An intermediate outcome with t(9;11)(p21;q23), a mutated or wild-type *NMP1* with *FLT3*-ITD, trisomy 8, trisomy 21 and most break point 11q23 lesions.
- A poor outcome with complex or monosomal karyotypes, t(6;9)(p23;q34), t(9:22)(q34.1; q11.2), t(8:16)(p11; p13), inv(3)(q21.3q26.2)

or t(3; 3)(q21.3; 26.2), t(3q26.2; v), −5 or del(5q): −7: −17/abn(17p), mutated *ASXL1, BCOR, EZH2, RUNX1, SF3B1, SRSF2, STAG2, U2AF1, ZRSR2* (at Xp22.2, coding a splicing factor) or *TP53*.

The Wilms' tumour gene (*WT1*) is linked with many cancers (as discussed in Section 5.3.4), and in AML, polymorphism rs16754 is a powerful ($p < 0.001$) predictor of outcome survival (odds ratio [95% CI]: 0.62 [0.52–0.72]), whilst mutations in *NPM1* lead to variable prognoses. Furthermore, molecular pathology is an essential tool in generating mutation-specific drug therapies, such as enasidenib (which targets mutated *IDH2*), ivosidenib (*IDH1*), midostaurin, gilteritinib, quizartinib (all tyrosine kinase inhibitors targeting *FLT3*) and venetoclax (which inhibits the protein product of *BCL2*, involved in apoptosis). Figure 6.2 shows a blood film in AML.

Chronic Myeloid Leukaemia

As with the diseases of Hodgkin and Burkitt in lymphoma, the chronic phase of this disease (CML) has a special place in the development of molecular pathology, as described in Sections 1.1.2 and 2.2.2. Briefly, the Philadelphia chromosome, where part of chromosome 22 is translocated onto part of chromosome 9 (hence t(9;22)), was reported in 1960, and

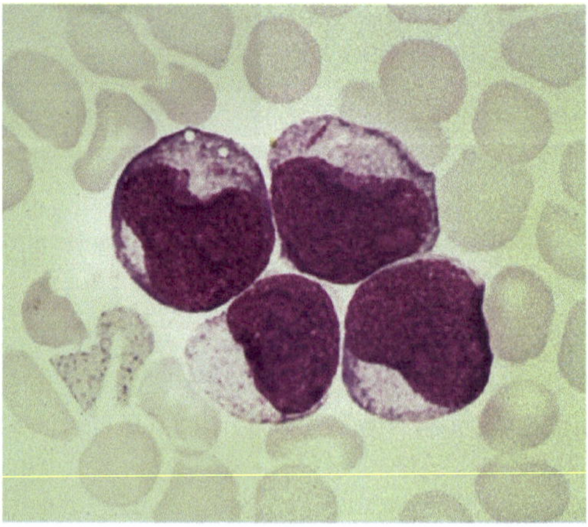

Figure 6.2 A blood film in AML. *Source:* Hoffbrand et al. (2011). Figure 13.04, p. 184, with permission from John Wiley & Sons, Ltd.

a decade later it was shown to form a new gene of the fusion of *BCR* on chromosome 9 with *ABL1* from chromosome 9, hence *BCR::ABL1*, on an adapted chromosome 22, the Philadelphia chromosome (Figure 6.3). The new gene was subsequently found to be present in chronic myeloid (or sometimes myelogenous) leukaemia (CML), a finding so robust that it became the defining marker of the disease, such that the absence of *BCR::ABL1* effectively excludes the disease. Further research identified the protein product of the oncogene, and so, in 2001, a targeted drug therapy – imatinib – for this disease. This was the first time that molecular pathology had ticked all boxes in the sequence (a) demonstration of

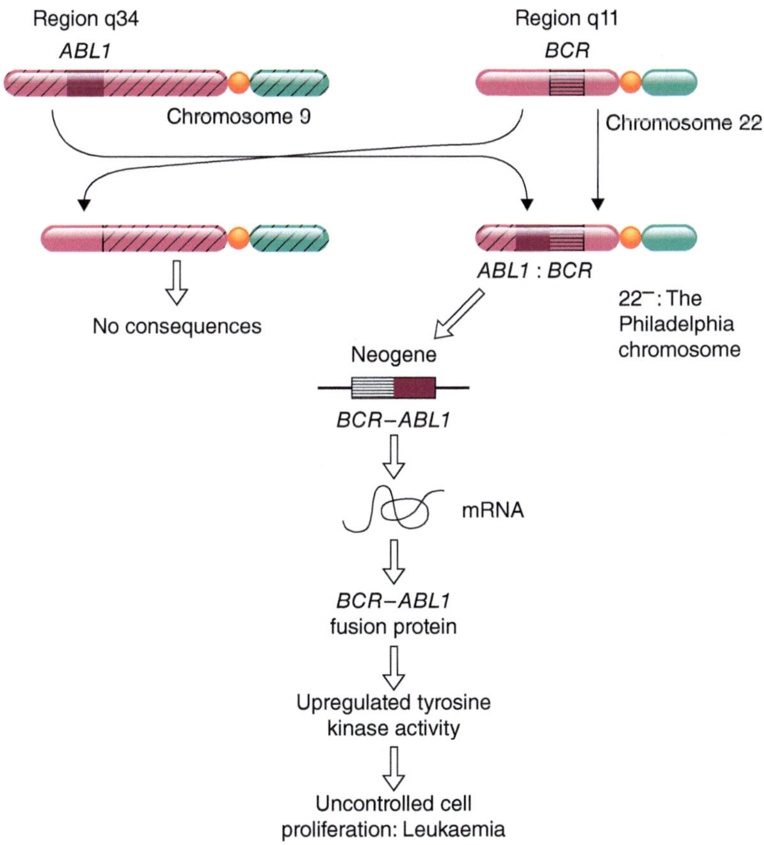

Figure 6.3 Formation of *BCR::ABL1*.

the genetic abnormality of a disease, (b) identification of its abnormal protein, (c) use of a genetic abnormality as a clinically useful biomarker and (d) development of a drug to treat the disease based on the genetic abnormality.

Despite the importance of *BCR::ABL1*, which may be detected by karyotyping, FISH, or as part of a panel, numerous other chromosomal and genetic abnormalities are important in CML. In addition, the National Genomic Test Directory for England points to a multi-target NGS panel and karyotype analysis for trisomy chromosome 8, trisomy chromosome 19, −7/7q, −5/5q, i(17q), ider(22)(q10), inv(3) and 11q23 rearrangements. FISH may also be used to assess CNV in chromosomes 8, 19, 7/7q, 5/5q, 17q and 12p. Of further interest is any structural variants in *KMT2A*, which may also be detected by FISH. The two variants of *BCR::ABL1*, p190 and p210 (the kDa molecular weights of the protein products), may be determined by PCR, whilst *SETBP1* is implicated in some cases of atypical CML. Figure 6.4 shows a blood film in CML, with the presence of various precursor myeloid cells.

> **Key Point 6.4** The elucidation of the genetic basis of CML is one of the major triumphs of molecular pathology, feeding through to diagnosis and management with tailored therapeutics.

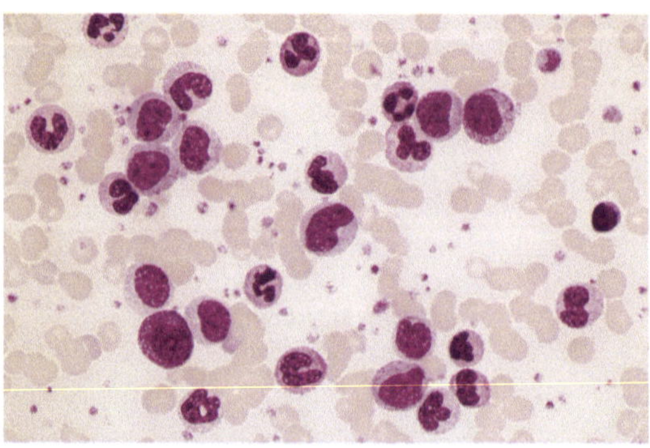

Figure 6.4 A blood film in CML. *Source:* Hoffbrand et al. (2011). Figure 14.03, p. 195, with permission from John Wiley & Sons, Ltd.

6.2.3 Lymphoid Leukaemia

Lymphopoiesis is considerably less complex than myelopoiesis (Figure 6.1), with only two pre-mature stages recognised – the lymphoblast and the prolymphocyte. However, there are at least three isotypes of the mature cell – T-cells, B-cells and NK cells, which may be distinguished by CD markers and rearrangement in genes of the T-cell receptor and the B-cell receptor (i.e. immunoglobulin light and heavy chains).

Of the 1,206 deaths due to lymphoid leukaemia in England and Wales in 2023 (29% of all leukaemias), almost three quarters (877) were linked to chronic lymphocytic leukaemia (CLL) of the B-cell type, 18% (214) to ALL, with the remainder being adult T-cell lymphoma/leukaemia (HTLV-1 associated (31 deaths), hairy-cell leukaemia (HCL, 33 deaths), and prolymphocytic leukaemia of T-cell type (28 deaths).

Acute Lymphoblastic Leukaemia

The key malignant cell in ALL is the lymphoblast, which distinguishes it from CLL, where the dominant cell resembles a mature lymphocyte on a blood film. Thus, the cell biology of an ALL places it closer to the CFU-L, whilst the malignant CLL cell is developmentally closer to the normal mature lymphocyte (Figure 6.1). Most (75–80%) ALLs are of B lymphocytes and are classified by morphology, molecular genetics and immunophenotyping. Routine blood films will show lymphoblasts, inevitably larger than lymphocytes, and others with an abnormal nucleus. Figure 6.5 shows a blood film in ALL.

Cell surface markers are almost always positive for CD17, CD22, CD24, CD79a, PAX and terminal deoxynucleotidyl transferase, whilst expressions of CD13, CD20, CD33 and CD34 can be variable. The most commonly used markers for T-cell ALL include CD1a, CD3, CD5, CD7, CD25, CD30 and CD38.

During the 1980s, the immunological classification of many blood cancers gave way to molecular genetics, with numerous translocations in ALLs, some also being reported in lymphomas. These analyses were further compounded by some ALLs exhibiting hybrid lineage markers, such as B-cell malignancies having rearrangement in T-cell receptor genes, with others expressing both myeloid and lymphoid markers. Unsurprisingly, an abnormal karyotype was detected in most ALL cases, an example being trisomy 13, whilst alterations in genes, such as *TP53,*

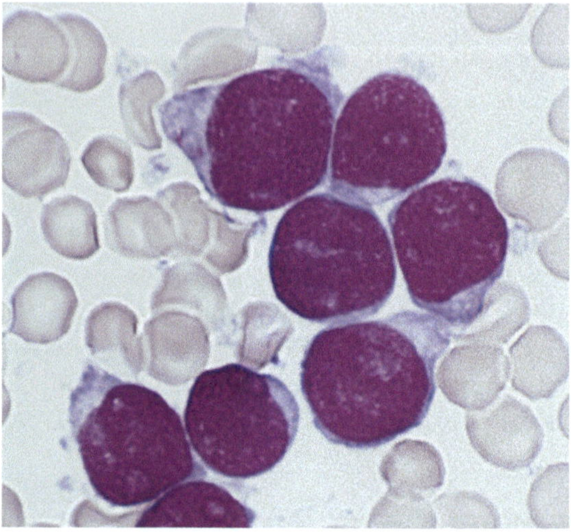

Figure 6.5 A blood film in ALL. *Source:* Hoffbrand et al. (2011). Figure 17.03A, p. 227, with permission from John Wiley & Sons, Ltd.

c-MYC and *RB1*, were published. Several reports noted chromosome rearrangement in leukaemic cell lines, such as t(14;18) and t(4;11), although the breakthrough finding was the presence of the Philadelphia chromosome. With the advent of more sophisticated NGS methods, additional gene lesions were identified and their importance evaluated. Examples include:

- t(9;22)(q34;q11), the Philadelphia chromosome, which was found in 30% of adult and 3% of paediatric ALLs. As discussed above, this lesion is present in all cases of CML.
- t(12;21)(p13;q22) brings together *RUNX1* at 21q22.12 and *ETV6* at 12p13.2 to form a neo-oncogene. Found in 25% of paediatric ALLs, it has a relatively good prognosis, with a cure rate of 90%.
- Several translocations link *KMT2A* (also known as *MLL*, for mixed-lineage leukaemia) on 11q23 with up to 80 other genes in 10% of ALLs. Partners include *AFF1* (also described as *AF4)* at 4q21 (in 36% of cases), others being *MLLT3* (at 9p21.3, also described as *AF9), ELL* (at 19p13.11, coding an elongation factor for RNA polymerase II) and *MLLT1* (also described as *ENL*, at 19p13.3 and coding a transcription regulator).

- Mutations in *NOTCH1* (at 9q34.3, coding a cell membrane receptor) are found in ~60% of T-cell ALLs and deletions of tumour suppressor *CDKN2A (*at 9p21.3) are also common.
- *PAX5* (at 9p13.2, coding a transcription factor) is a master regulator of B-cell development, with CNVs, SNPs and translocations in many B-cell malignancies.

As with AML, the National Genomic Test Directory for England points to a large number of genes of interest in ALL, in addition to those described above. Others are shown in Table 6.3 and may be investigated in multi-target NGS panels, and individually by FISH and RT-PCR, whilst karyotyping will be used to detect complex karyotypes, hyperdiploidy, high hyperdiploidy, near haploidy, low hypodiploidy, del(1)(p33p33) and iAMP21.

Table 6.3 Genes mutated in ALL.

Gene	Location	Product
IKZF1	7p12.2	A zinc-finger transcription factor
CDNK2A	9p21.3	Cyclin-dependent kinase inhibitor p16
CDNK2B	9p21.3	Cyclin-dependent kinase inhibitor p15
BTG1	12q21.33	A cell cycle regulator
EBF1	5q33.3	A transcription factor for B-cell development
CRLF2[a]	Xp22.3 and Yp11.3	Cytokine receptor-like factor 2
CSF2RA[a]	Xp22.3 and Yp11.3	Receptor for granulocyte–macrophage Colony-stimulating factor
IL3RA[a]	Xp22.3 and Yp11.3	A subunit of the interleukin 3 receptor
ETV6	12p13.2	A transcription factor
FBXW7	4q31.3	A transcription factor linked to ubiquitin
TAL1	1p33	A transcription factor
TLX1	10q24.31	A homeobox transcription factor
TLX3	5q35.1	A homeobox transcription factor
KMT2A	11q23.3	A histone methyltransferase
ABL2	1q25.2	A cytoplasmic tyrosine kinase
CSF1R	5q32	Receptor for colony-stimulating factor 1

[a]Part of the pseudoautosomal region, present on both X and Y chromosomes.

In many cases, CNV will also be sought. Multi-target NGS panels may also be used to detect variants of *NOTCH1, TP53, RB1, PTEN, KRAS, NRAS* and *HRAS*, and for complex variant detection in immunoglobulin and T-cell receptor genes. Structural variants of *JAK2, EPOR* (at 19p13.2, coding the erythropoietin receptor), *PDGFRA* and *PDGFRB* (at 5q32, coding the beta chain of the PDGF receptor) may be sought by FISH. Leading component genes in ALL are shown in Table 6.3; translocations are summarised in Table 6.4.

Self-check 6.3 Despite their lineage dissimilarity, a certain translocation can be found in both CML and in ALL. What is it, and what are the implications for treatment of these leukaemias?

Table 6.4 Leading translocations in ALL.

Abnormality	Genes fused
t(9;22)(q34;q11)	*BCR::ABL1*
t(12;21)(p13;q22)	*ETV6::RUNX1*
t(1;19)(q23;p13)	*TCF::PBX1*
t(17;19)(q22;p13)	*TCF3::HLF*
t(4;11)(q21;q23)	*KMT2A::AFF1*
t(11;19)(q23;p13.3)	*KMT2A::MLLT1*
t(6;11)(q27;q23)	*KMT2A::AFDN*
t(9;11)(p21;q23)	*KMT2A::MLLT3*
t(10;11)(p12;q23)	*KMT2A::MLLT10*
t(11;19)(q23;p13.1)	*KMT2A::ELL*
t(1;7)(p32;q11)	*TRB::TAL1*
t(11;14)(p15;q11)	*TRD::LMO1*
t(7;11)(p15;p15)	*NUP98::HOXA13*
t(10;14)(q24;q11)	*TLX1::TRD*
t(7;10)(q34;q24)	*TRB::TLX1*
t(5;14)(q35;q32)	*BCL11B::TLX3*

These may be detected by an NGS panel or individually by FISH and/or RT-PCR.

Chronic Lymphocytic Leukaemia

In contrast to ALL, the chronic variant is essentially a disease of cells that have almost completed their maturation (Figure 6.1). In some cases, the lesion is the more immature prolymphocytes (hence prolymphocytic leukaemia), which are slightly larger than standard CLL cells. This aggressive variant presents with a very high white cell count, accounting for 1–2% of all CLLs. The WHO document places CLLs in the mature B-cell neoplasm group, alongside lymphomas, and which puts monoclonal B-cell lymphocytosis and CLL/small lymphocytic lymphoma together as pre-neoplastic and neoplastic small lymphocytic proliferations. In parallel, the same document refers to T-lymphoblastic leukaemia/lymphoma, early T-precursor lymphoblastic leukaemia/lymphoma, T-prolymphocytic leukaemia and T-large cell lymphocytic leukaemia, with no referral to acute/chronic.

Research from the 1980s reported possible roles for *c-myc, k-ras* and *c-fos*, and hypomethylations in CLL, and the presence of t(2;14)(p13;q32) and t(14;19)(q32;q13), with many groups working on the rearrangement in immunoglobulin heavy and light chain genes, and in T-cell receptor genes. As immunophenotyping developed, the first choice of marker was CD5, later extended to CD23: others include weak expression of surface immunoglobulins and low or absent expression of CD79b, FMC7, CD19, CD20 and CD22. The single most important feature of a CLL blood film was defined as the presence of smudge/smear cells, these being leukaemic cells too fragile to resist the physical stress of the blood film, a feature retained to this day.

As cytogenetic and NGS tools became available, many CLLs were found to have trisomy 12, and deletions in 11q and 13q14, the latter present in over half of cases. The importance of genetic lesions is demonstrated by their link with outcome. Del13q14.3 brings a moderately good prognosis (median 17-year survival), whilst del11q23, present in 20%, carries a poor prognosis, possibly because this region hosts the DNA damage response kinase *ATM*. Del17p13 brings a particularly poor prognosis to ~7% of CLLs (median survival 7 years, mostly due to resistance to chemotherapy) and may be linked to *TP53*: even if not deleted, mutations in *TP53* are linked to CLL (and many other cancers). A commonly mutated gene is *NOTCH1* at 9q34.3, coding for a membrane receptor with epidermal growth factor-like repeats, deregulation of which may be a key carcinogenic step, and which could be a therapeutic target. A further useful analysis is immunoglobulin heavy chain variable region genes: lack of rearrangement implies an earlier stage of differentiation and a more aggressive disease. Accordingly, routine work-up of CLL includes FISH for del11q, del13q, del17p, trisomy 12, and t(11;14) in addition to *TP53* and *IgHV* mutation analyses.

These are mirrored by entries in the National Genomic Test Directory for England, with a multi-target NGS panel for the detection of small variants in *TP53, BTK, PLCG2* and *BCL2*. Another panel may be used to detect rearrangement in immunoglobulin genes. FISH is the method of choice for the detection of del(17p) (the location of tumour suppressor *TP53*) and CNVs of chromosome 11q (as *ATM* is located at 11q22.3) and 13q (as tumour suppressor *RB1* is located at 13q14.2).

A further feature of the importance of molecular pathology is the adoption by guidelines of NICE. Technology appraisal TA359 focuses on idelalisib, an inhibitor of phosphoinositide 3-kinase, alongside rituximab, in untreated CLL patients with a chromosome 17p deletion or *TP53* mutation. This recommendation has an impact on the use of venetoclax (which attaches to, and so inhibits, the apoptosis regulator Bcl-2). This gene lesion is also important in the potential use of ibrutinib, which functions by binding to tyrosine kinase product of *BTK*, thus suppressing B-cell development. Figure 6.6 shows a blood film in CLL.

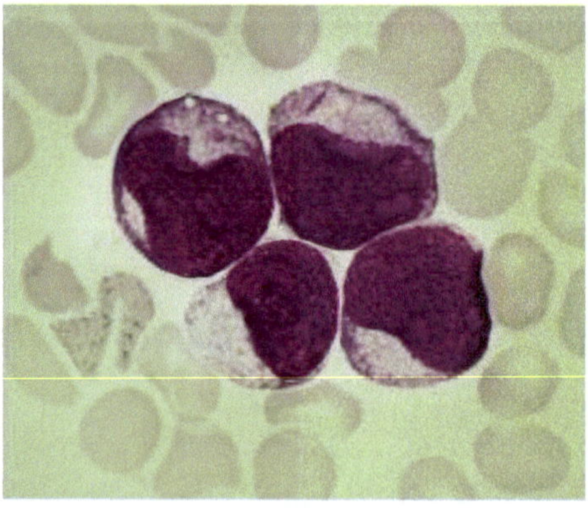

Hoffbrand et al., (2011) / John Wiley & Sons

Figure 6.6 A blood film in CLL.

Hairy Cell Leukaemia

First described as leukaemic reticuloendotheliosis in 1958, with '*pseudopods protruding from the cytoplasm, making a serrated border for the usually round, oval or polygonal cells*', it was renamed HCL in 1966, and by 1970 the expression was common. Later, the presence of immunoglobulins at the cell surface showed it to be of the B-cell lineage, with almost all cases showing rearrangement of immunoglobulin genes. Epidemiological studies reported it to account for around 2% of all lymphoid leukaemias, and in 2023, in England and Wales, it caused 33 deaths. Immunophenotyping data from the 1990s showed the presence of CD11c, CD19, CD20 and CD25, updated to the present with CD103, CD123, CD200 and PAX5, and which can be used to examine variants in HCL. However, the breakthrough report came in 2011 with the report of a mutation in *BRAF*, at 7q34, and coding for a serine/threonine kinase transcription factor.

Sanger sequencing, PCR and other methods were used to identify an SNP in exon 15 of *BRAF* leading to the V600E SNP (where glutamic acid replaces valine at amino acid position 600) leading to over-activation of the *BRAF* kinase that, in turn, promotes the Ras-Ref-MEK-ERK pathway, and so the malignancy. Several treatments (such as vemurafenib, encorafenib and dabrafenib) inhibit this pathway, and so the malignancy. Ibrutinib, a treatment for CLL described above, may also be used to treat HCL. Assessment of this SNP is the only entry in the National Genomic Test Directory for England with respect to HCL, although immunoglobulin heavy chain rearrangement may be sought for the HCL-Vh4-34 variant, and for small variants in *MAP2K1* in the non-classical variant (as in certain lymphomas, as previously described).

The 2022 WHO classification system places HCL in the 'splenic B-cell lymphomas and leukaemias' group, as a common clinical finding is splenomegaly. In addition, the same system has replaced conditions described as HCL variant (defined as a morphology between prolymphocytes and hairy cells) with the grouping 'splenic B-cell lymphoma/leukaemia with prominent nucleoli', returning to a morphology criterion. Figure 6.7 shows a blood film in HCL.

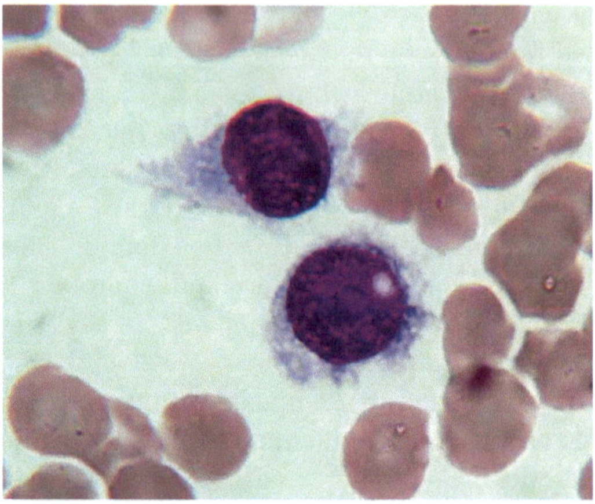

Figure 6.7 A blood film in HCL. *Source:* With permission from the Institute of Biomedical Science.

6.2.4 Other Types of Leukaemia

Monocytic Leukaemia

As the third-most frequent blood cell, transformation of precursors of the monocyte leads to the third most common form of leukaemia. Although described as a separate entity in 1928, by the 1960s, many cases, especially where acute, were described as myelomonocytic, leading to debates as to true aetiology. Nevertheless, the presence of Fc receptors, monocytic morphology and cytochemistry were used to identify acute monocytic (and some monoblastic) leukaemias. As discussed above, the close relationship between the myeloid and monocyte lineages (Figure 6.1) led to the inclusion of monocytic (subgroup M5b) and monoblastic (subgroup M5a) leukaemia in the 1976 FAB classification of AML, based on morphological criteria. The 2022 WHO guideline places acute monocytic leukaemia in the section on AML defined by differentiation, their classification closely resembling that of the FAB system, and pointing to myeloperoxidase and non-specific esterase, and immunophenotyping for CD11c, CD14, CD64 and lysozyme.

However, the advent of molecular genetics brought various gene aberrations into the picture, such as *c-fos* and *MLL*, the latter abbreviated from mixed-lineage leukaemia, later re-named as *KMT2A*, described above as being linked to ALL and AML (Tables 6.2–6.4 and nearby text). Overexpression of *IGSF4*, coding a cell adhesion molecule, has been reported as being discriminative for t(9;11)(p22;q23) patients with a monocytic/monoblastic leukaemia, whilst CNVs in 3q26.2-qter and 13q31.3 (gains) and in 2q24.2, 8p12 and 14q32 (losses) may be present. Quick and accurate diagnosis is crucial as the five-year survival rate of acute monocytic leukaemia is around 25%, compared to 89% for CLL, 73% for ALL, 72% for CML and 33% for AML. Figure 6.8 shows a blood film in monocytic leukaemia.

> **Self-check 6.4** Almost all, if not all, cases of HCL have one particular gene mutation – what is it?

Myelomonocytic Leukaemia

As described above, this variant of AML carries features of the immature forms of both granulocytes and monocytes, with the chronic form (CMML) having a high propensity to deteriorate to an acute form (AMML, which linked to 34 deaths in England and Wales in 2023), AML, and to other myeloid neoplasia. The 2022 WHO guideline refers to a myelodysplastic variant of CMML, where the white blood cell count is $<10 \times 10^9\ L^{-1}$, and a myeloproliferative variant, where the count is $>10 \times 10^9\ L^{-1}$.

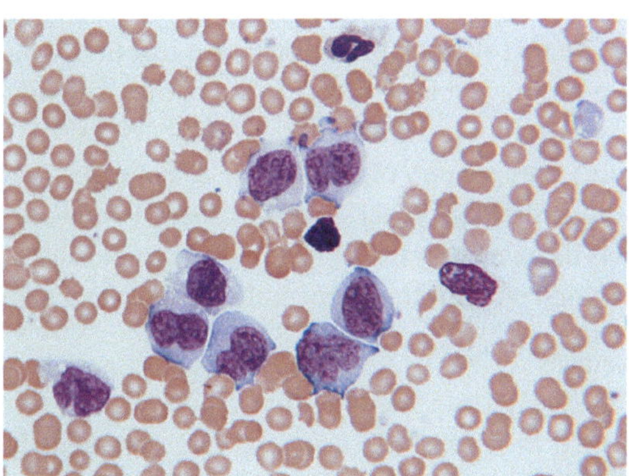

Figure 6.8 A blood film in monocytic leukaemia. *Source*: With permission from the Institute of Biomedical Science.

Although described in the 1960s, in part by the presence of lysozyme, it was not until 1976 and the FAB system that it was fully recognised (as M4), with genetic abnormalities reported in the 1980s. These included trisomy of chromosome 11, involvement of *c*-fos, inv(16)(p13q22), del(16)(22), and later, t(16;16)(p13;q22), often with an eosinophil component (and so denoted M4Eo, further supporting the complex myeloid/granulocyte aspects of the disease). More reviews also point to trisomies in chromosomes 8, 10, 19 and 21, deletions in 11q, 12p, abnormalities in chromosome 7, and complex karyotypes. FISH and NGS methods have pointed to roles for abnormalities in *SRSF2, TET2* (both present in ~55% of cases of CMML), *ASXL1* (~38%), *RUNX1* (~25%), *SETBP1* (~18%), *KRAS, FAT4* (both~16%), *EZH2* (~14%), *CBL, ARIH1* (both ~11%) and *NRAS* (~10%), with the *JAK2V617F* SNP being present in around 6% of cases.

Several of these abnormalities have clinical consequences. Mutated *ASXL1* is linked to an evolution of CMML to an acutely transformed state, and so a poor overall survival, whilst *RUNX1, NRAS* and *SETBP1* are also linked to overall survival. A gene-dosage effect is clear: in those with only a *TET2* mutation, a dual *TET2-SRSF2* or a triple *TET2-SRSF2-NRAS* mutation, median survivals were 90, 45 and 9 months, respectively, compared to an overall CMML median survival of 15–20 months. An informative case report describes a woman with CML who developed tyrosine kinase–independent resistance with mutation in *ASXL1* and *RUNX1*, with progression to CMML-like accelerated phase disease alongside an *IDH1* mutation. AMML itself is often characterised by inv(16)(p13.1;q22) and t(16;16)(p13.1;q22) that bring together *CBFB*, coding a subunit of core-binding factor, and *MYH11*, coding smooth muscle myosin, a fusion also present in AML.

An important component of CMML is a paediatric aspect (often described as juvenile MML), about which there is considerable literature, which has its own entry in the National Genomic Test Directory for England and which cites all the abnormalities described above.

NK Cell Leukaemia

Literature of the early 1980s includes discussion of a number of atypical lymphocytes such as K cells, NK cells and large granular lymphocytes, with debate regarding the precise nature of these cells and their relationship with T lymphocytes. A degree of clarity emerged with the development of CD markers, and the definition of the NK cell as those bearing CD56 (also present on certain neural cells), whilst a subpopulation also expressed CD16 (the FcγR). A report of an NK cell leukaemia reported pancytopaenia, with 96% of blasts expressing CD56.

The 2022 WHO guideline on lymphoid neoplasms recognises NK -large granular lymphocytic leukaemia, aggressive NK cell leukaemia (which has a median overall survival of seven months) and NK cell lymphoma. WES found frequent mutations in the JAK/STAT and RAS-MAPK signal transduction pathways, the most common being in *STAT3* (in ~48%), others being in *TP53* (34%), *TET2* (28%), *CREBBP* and *MLL2* (both 12%). Array CGH has found a gain of 1q23.1-q23.2 and 1q31.3-q44, with loss of 7p15.1-q22.3 and 17p13.1 in NK leukaemia compared to extra-nodal NK lymphoma. The National Genomic Test Directory for England describes a multi-target NGS panel for the detection of small variants of *STAT3* and *STST5B*.

Other Leukaemias

Reflecting their frequency in the blood, eosinophil and basophil leukaemias are very rare. The 2022 WHO document places eosinophilic leukaemia alongside CML and chronic neutrophilic leukaemia (CNL) as an MPN: a recent case report used NGS to implicate a novel mutation in *JAK1* at 1p31.1, coding for janus kinase 1. A differential diagnosis is hypereosinophilia syndrome and points to roles for *FIP1L1* (at 4q12, coding part of a polyadenylation complex), *PDGFRA* and *PDGFRB*. Basophilic leukaemia is described in Section 6.4.7.

Although technically not leukaemia, some blood cancers can be linked to sarcoma, such as a myeloid sarcoma developing from AML with links to MDS and MPNs. There may also be a sarcoma based on Langerhans cells, interdigitating dendritic cells and histiocytic cells.

A presumed late-acting gene lesion(s) in metamyelocytes is the hypothesised cause of CNL, where the blood film is dominated by phenotypically mature neutrophils, with some metamyelocytes. The key differential diagnosis is a physiological leukaemoid reaction to a severe infection, relatively easy to exclude on other grounds, e.g. levels of vitamin B_{12} and uric acid, acute presentation and fever. An analysis of 39 patients with CNL found 27 mutated genes, the most common being in:

- *ASXL1*, present in ~75% of cases and also linked to AML and MML.
- *CSF3R* at 1p34.3, coding for the receptor for granulocyte colony-stimulating factor, is present in ~62%.

- *SRSF2* at 17q25.2, coding for a splicing factor that disturbs granulocytopoiesis, is present in ~42% and linked to AML.
- *SETBP1* at 18q12.3, coding for a DNA-binding protein, present in ~40%, is also implicated in MML and atypical CML.
- *TET2* at 4q24, coding an enzyme involved in cytosine chemistry, present in 20%, also of note in T-cell NHL, MML and NK leukaemia.

6.2.5 Measurable Residual Disease

Naturally, the object of treatment is cure, or at least remission, but how is this defined? Measurable residual disease (MRD, also known as minimal residual disease) was originally defined by failure to find leukaemic cells on a blood film, with a considerable degree of false negativity. Later, it was defined by the absence of cells in multichannel fluorescence flow cytometry, probing millions of cells for relevant CD markers (e.g. CD2, CD7, CD19, CD56 and CD117 for AML) with high sensitivity, and subsequently by RT-qPCR. We now probe for specific gene aberrations (such as *RUNX1*::*RUNX1T1* and *PML*::*RARA* in AML) with even more sensitivity. The immense power of NGS and TGS is that hundreds of mutated genes can be detected, and quantified, in a single sample. However, in certain diseases, a small number of aberrant genes may be sought, such as *BRAFV600E* in HCL and t(9;22)(q34;q11.2) in CML. Table 6.5 shows those analyses the National Genomic Test Directory for

Table 6.5 Genes of value in assessing MRD in leukaemia.

Malignancy	Genes
AML	*NPM1, PML::RARA, RUNX1::RUNX1T1, CBFB::MYH11* and *BCR::ABL1*.
CML	*BCR::ABL1*.
ALL	*BCR::ABL1, ETV6::RUNX1, TCF3::PBX1, TCF3::HLF, KMT2A::AFF1, KMT2A::MLLT1, KMT2A::AFDN, KMT2A::MLLT3, IGH, IGK, IGL, TRA, TRB, TRG* and *TCRD*.
Other acute leukaemia	As AML and ALL, plus *FUS::ERG*.

Source: NHS.UK / https://www.england.nhs.uk/publication/national-genomic-test-directories/ / CC BY 4.0

England has included in testing for MRD. Notably, there is no referral to MRD in CLL or HCL, where there is considerable duplication of certain malignant genes. A major problem with assessing only blood is the question of malignant cells in the bone marrow, lymph nodes, liver and spleen, which are at present exceptionally difficult to detect.

Key Point 6.5 Despite the focus on the 'big four' leukaemias, there remains considerable clinical and scientific interest in 'minor' forms of this form of blood cancer, i.e. monocytic leukaemia, HCL, CMML, neutrophil leukaemia and others.

6.3 Myeloma and Related Diseases

This third major form of blood cancer, causing almost a quarter of all deaths, arises from a malignant B lymphocyte and its transformation into a plasma cell, focusing on the bone marrow. These diseases may also be described as plasma cell dyscrasias. The vast majority (~98%) of the 2,774 deaths reported by the ONS in England and Wales in 2023 were due to a myeloma, the remainder to a plasma cell leukaemia (PCL; 57 deaths) and a solitary plasmacytoma (24 deaths). Molecular genetics has pointed out that the same abnormality can be present in abnormal cells arising in the bone marrow (i.e. leukaemia) and in the lymph nodes (i.e. a lymphoma): the same can be said of commonality of genetic abnormalities and CD markers in myelomatous disease and lymphomas.

6.3.1 The Laboratory

A leading feature of myeloma and its related disease process is the frequent generation of abnormal antibodies (paraproteins) from the malignant plasma cell, which have a number of important clinical consequences. However, the generation of these abnormal gamma globulins is not myeloma-specific. As mentioned in Section 6.1.2, lymphocytoplasmic lymphomas such as WM can produce a paraprotein, and so may be described as a myeloma in a lymph node. Nevertheless, the extent of the production of such an abnormal protein provides an opportunity to examine the natural history of myeloma, developing from monoclonal gammopathy of undetermined significance (MGUS), through smouldering multiple myeloma (SMM) to malignant myeloma (Section 6.3.4) and, in some cases, PCL.

Established key laboratory metrics in following this progression are the extent of the paraproteinaemia, serum albumin, and plasma cells in the bone marrow, and later in the blood. Plasma cells are characterised not only by heavy SmIg staining and the standard B-cell marker, CD19, but also by the presence of CD38, CD126 (the IL6 receptor), CD56 and CD138. Much of the data that follows is derived from captured bone marrow or circulating plasma cells. In the early stages of the disease, the profile of the abnormal protein often cannot be distinguished from that in MGUS, supporting the hypothesis that MGUS is pre-myeloma.

6.3.2 Monoclonal Gammopathy of Undetermined Significance

Introduction

MGUS is present in ~3% of the over 50s, ~5% of the over 70s and ~7.5% of those over 80 years of age and is probably underdiagnosed. As the name implies, it is paraproteinaemia (predominantly of IgG) whose origin is obscure, but which makes up about half of all monoclonal gammopathies. The disease itself is often benign, with little evidence of bone marrow or peripheral white cell abnormalities, and is often found co-incidentally. However, perhaps 1–2% transform into a myeloma or lymphoma each year, a rate high enough to require outpatient surveillance. This leads to the hypothesis that MGUS is effectively pre-myeloma. In support of this is the observation that progression is linked to the size and type of the paraprotein (IgM and IgA carrying a worse prognosis than IgG) and an abnormal plasma kappa/lambda ratio (<0.26 or >1.65). Bone-related problems in those at high risk involve treatment with oral calcium and vitamin D supplements. Indeed, MGUS brings a relative risk of 2.5 for vertebral fractures.

Observational Studies

During the early 1990s, several small studies reported genetic abnormalities in myeloma but none in MGUS, although by the end of the decade evidence such as monosomy 13 was reported, as were deletions. This was followed by data on methylation of genes coding for cell cycle regulators p15(INK4b) and p16(INK4a) and somatic hypermutation of immunoglobulin heavy chain genes, whilst a genome-wide linkage family screen for MGUS (and WM) found links with chromosomes 3 and 6.

Developing laboratory methods uncovered additional abnormalities that included those of *IGH* (at 14q32) translocations t(4;14)(p16.3;q32) and t(14;16)(q32;q23), *RB1* (13q14) deletion, 1q gain, *RAS* and hyperdiploidy. Although based on a small sample size, MGUS patients were found to have 5%, 3% and 3% frequencies of t(14;20), t(14;14) or t(14;16), respectively, but far more frequent was the L265P mutation in *MYD88*, present in 13 of 24 (54%) patients, underlining the at-times close aetiology link with WM (where 93% of cases carry the mutation). Further evidence of this link was provided by the frequency of specific copy number abnormalities [+4, del(6q23.3–6q25.3), +12 and +18q11–18q23] that progressively increased from IgM MGUS (15% of cases), through smouldering WM (20%) to symptomatic WM (73%).

The National Genomic Test Directory for England refers to a multi-target NGS panel to detect small variants in *MYD88* and *CXCR4*. The role of genetics in the transition of MGUS to SMM and so myeloma is discussed in Section 6.3.4.

Predicting Disease Progression

Probing mRNA from plasma cells from 356 MGUS patients, a panel of 36 genes, including three involved in immunoglobulin processing (*IGLV1-44, IGKC,* and *IGHA1)*, but not *MYD88, CXCR4,* or other established markers such as *TP53* and *KRAS*, and protein data, was used to develop a model to predict progression to myeloma. With a follow-up of up to 10 years, the model found that those it defined as low risk had an event rate of 2%, intermediate risk of 42.5% and high risk of 82.4%.

6.3.3 Smouldering Multiple Myeloma

Introduction

This intermediate stage is characterised by a paraprotein of >30 g L^{-1} and with >10% plasma cells in a bone marrow aspirate Although asymptomatic, 10% develop malignant disease annually, with a median time to progression of 4 years. However, SMM may be overlooked if MGUS transforms rapidly, and the task of molecular pathologists is to determine any factors that can predict those at greatest risk.

Self-check 6.5 What are the key biochemical abnormalities in MGUS?

Observational Studies

A study of 123 patients with SMM reported chromosomal abnormalities in 72%, with 1q gains in 41%, 13q deletions in 42%, 17p deletions in 7% and *IGH* translocations in 42%. The most frequent abnormalities of the latter were t(4;14)(p16.3;q32) in 12%, t(14;16)(q32;q23) in 6% and t(11;14)(q13;q32) in 17%. A similar profiling study of 214 patients with SMM reported *KRAS* and *NRAS* SNPs, hyperdiploidy, a gain in chromosome 1q, deletions of 13q, 16q, 6q, 14q, 22q and 8p, and t(11;14) and t(4;14) to be the most common abnormalities. The value of inflammation was shown when an 8-genes signature of *IL8* (at 4q13.3, since renamed CXCL8), *IL10*, *IL17A* (at 6p12.2), *CCL3* (at 17q12), *CCL5* (at 17q12, both coding a chemokine), *VEGFA*, *EBI3* (at 19p13.3, coding a subunit of IL27) and *NOS2* (at 17q11.2, coding nitric oxide synthase 2) identified each condition (MGUS/SMM/symptomatic-MM) with 84% accuracy.

Predicting Disease Progression

A mutation in *KRAS* brings a shorter time to SMM progression to myeloma with a hazard ratio (95% CI) of 3.5 (1.5–8.1), whilst six genes (*IFNG* [at 12q15, coding interferon gamma], *IL2* [at 4q27, coding interleukin 2], *LTA* [at 6p21.33, coding lymphotoxin alpha], *CCL2* [at 17q12, coding a chemokine], *VEGFA* and *CCL3*) are independent predictors of patients' survival. Follow-up of 85 SMM patients found those with a *MYC* translocation or amplification to have the shortest mean time to progression of 8.4 months, compared to those free of such a lesion of 51.6 months. Other markers predicting a poor outcome were MAPK pathway mutations (14.4 vs. 60 months) and *TP53* or *ATM* alterations (15.6 vs. 50.4 months). Other predictors of a short time to progression include t(4;14), del 1p, del 8p and biallelic deletion involving *RB1*, *DIS3* (at 13q21.33, coding a nuclease), *MAX* (at 14q23.3, coding a transcript factor linked to *MYC*) and *CDKN2A*.

A number of predictor models using several factors have been reported. A scoring system (consisting of three levels) can help determine risk of progression based on certain genetic lesions, a paraprotein >20 g L^{-1}, bone marrow plasma cells >20%, and free light chain ratio <20 (kappa/lambda ratio normally 0.26–1.65). At each level, risk is greater with the presence of t(4,14), t(14,16), 1q gain and/or del13, so that those at high risk and with chromosome abnormalities have a 59% risk of 2-year progression. A more complex model used isolated bone marrow CD138+ plasma cells from 105 patients with SMM and 54,675 Affymetrix microarray system probes to define four genes – *RRM2* (at 2p25-p24), *DTL* (at 1q32, coding a ubiquitin ligase), *TMEM48* (at 1p32.3, also known as *NDC1*, coding for transmembrane nucleoporin) and *ASPM* (at 1q31, coding an assembly protein for spindle microtubules) – with the power to predict time to treatment for myeloma. With follow-up of 68 SMM patients for a maximum of 10 years, the gene score combined with M-protein and albumin levels produced low, intermediate and high risk groups that brought a 5%, 44.8% and 85.7%, respectively, risk of progression to symptomatic myeloma within 2 years.

6.3.4 Myeloma

Introduction

Globally, this disease, described by some as malignant myeloma, others as multiple myeloma, ranks 21st in incidence of all cancer, with 187,774 cases, and 17th in mortality, the cause of 121,252 cancer deaths. This gives a crude mortality index of 65%, compared to 63% for leukaemia and 43% for lymphoma. Parallel data for the UK is 6,504 cases and 3,618 deaths, and so an index of 56%, and although it has the lowest frequency of the major blood cancers in the UK, it carries the worst prognosis, with a five-year survival of about 50%. In the UK, NICE has published a guideline on the diagnosis and management of this disease (NG35).

> **Key Point 6.6** Plasma cell dyscrasias run a spectrum, starting with MGUS. The objective of clinical science, in general, and molecular pathology, in particular, is to determine which patients are at risk of progressing to the most serious end-point of PCL.

Observational Studies

The 1980s saw considerable work on immunoglobulin genes, and with reports of translocations, such as t(8;14) (also found in Burkitt lymphoma) and t(11;14), aberrations in *c-myc* (subsequently shown to be rearranged in 15% of primary myeloma tumours) and *H-ras* (as with many other blood cancers). During the 1990s, abnormalities in many other chromosomes were reported in ~25% of newly diagnosed patients, such as aneuploidy and additional translocations such as t(4;14) (p16.3;q32), which involve the fibroblast growth factor receptor 3 (*FRGR3*, also known as CD333), alongside possible roles for *IL6* with deletion or loss-of-function of *TP53* and *Rb-1*.

Crucially, evidence was building that the disease is the end result of a multistep transformation process, such as the link between monosomy 13 and the transition from MGUS to myeloma. As the present century developed, all the aberrations described in MGUS and SMM were first noted in myeloma, further supporting the general progression pathway. But additional work in molecular genetics provided a route to transformation – the t(4;14) translocation can bring together *FGFR3* next to a second gene broadly described as *WHSC1, MMSET* or *NSD2* at 4p16.3 that may result in the over-activity of the FGFR3 and so downstream activation of second messenger pathways. It also became clear that a leading lesion was a translocation of *IGH* at 14q32.33 that partners with several other genes to precipitate the transformation through different biochemical routes. The power of high-density oligonucleotide Affymetrix gene-chip microarray analysis was demonstrated by probing 118 plasma cell samples, which found 120 genes that differentiated malignant from normal plasma cells. An additional analysis reported a number of genetic lesions that are already the target of established drugs.

Predicting Disease Progression

A major clinical factor is the metastatic evolution of the disease from the bone marrow to other bone, such as the skull and ribs. This transition is linked to over-expression of a series of genes that include the receptor activator of NFκB (RANK, coded by *TNFRSF11A* at 18q21.33) and its ligand (RANKL, coded by *TNSF11*), as well as those coding osteoprotegerin and interleukins, and which results in the overactivity of osteoclasts. This pathway may also involve members of the NOTCH family, as they have roles in apoptosis and osteoclast biology. Mutations in *IL6* and its product have a place in the PI3K–AKT–mTOR second messenger pathway. Other genes and pathways, such as the B-cell-actuating pathway, BTK, those of the RUNX family, CXCR4, the WNT–Dickkopf-1 pathway, may also be important in the development of bone disease.

A study of 805 cases of myeloma, of whom 182 were untreated, 329 were in treatment and 294 at or near relapse, assessed the frequency of mutations in a number of key genes. Although there were no significant trends in the frequency of *KRAS*, *BRAF V600E*, *FGFR3*, *RB1* or *ATM/ATR*, there were clear links in others (Table 6.6), which is important data with the potential to direct future investigations.

Links with Prognosis

A study of 50 myeloma patients found low expression of *HLA-DQA1*, *VCAN* (at 5q14.2–14.3, coding versican, a proteoglycan), *BCL6* (also implicated in lymphoma development) and *BIRC3* (with a role in apoptosis), all to be linked to survival probability. An RNAseq study reported overall survival data of 787 MM patients, with progression free survival for 669, finding that high expression of *ERBB2/HER2* to be linked to poor outcome with hazard ratios (95% CIs) of 1.35 (1.1–1.66) and 1.89 (1.35–2.63), respectively, although both age and β-2-microglobulin also independently predicted a poor outcome. Perhaps unsurprisingly, a high number of circulating plasma cells, many with deletions in 17p, but with no difference in any chromosome 14 translocations, is linked to poor responses to treatment, whilst elsewhere, a higher mutational load (e.g. in *TP53*) was present in patients with recurrent relapse, and in newly diagnosed patients, mutations in *NRAS, KRAS* or *BRAF* predicted a poor response to chemotherapy.

Table 6.6 Gene frequencies in myeloma.

	Untreated	Treated	Near relapse
RAS-RAF	31.3%	35.6%	46.6%
N-RAS Q61	14.3%	16.1%	24.5%
CDNK2C	1.6%	3.0%	5.4%
TP53	6.6%	9.7%	17.0%

6.3.5 Plasma Cell Leukaemia

Introduction

A myeloma may become extramedullary, metastasising from its origin in a bone to another bone, or to the tissues or organs where it is described as a plasmacytoma. However, since blood itself is an organ, this can also be an extramedullary site, manifesting as PCL, developing in 2–4% of myelomas, and may be defined where plasma cells comprise ≥20% of the white cell count or an absolute plasma cell count $>2 \times 10^9$ cells L^{-1}. Where the leukaemia develops in an existing myeloma, it is said to be secondary (sPCL), whereas if diagnosed without a prior myeloma it is a primary leukaemia (pPCL, hence secondary/primary split ~35%/~65%). The prospects for survival are bleak, at median 8–9 months for pPCL and 2–4.5 months for a sPCL.

Observational Studies

Several abnormalities have been reported as being more frequent in PCL than in myeloma, including del-13q (~84% vs. ~40%, the most common aberration), del-17p (the location of *TP53*, around 50% of PCL cases), del-1p21, t(4;14), t(11;14) and t(14;16), whilst 33% of PCL patients have an *MYC* rearrangement. The 2022 WHO guideline on lymphoid neoplasms refers to a gain in 1q21 (with *CSK1B* over-expression of its product, regular of a cyclin kinase) as an early event in progression to PCL, whilst translocations and additional amplifications emerge later in disease progression. Differences in sPCL or pPCL, include:

- A higher rate of allelic (75% vs. 50%) and biallelic (33% vs. 11%) loss of *TP53* in sPCL.
- Hypodiploidy is more common in pPCL, hyperdiploidy in sPCL.
- A higher rate of t(11;14) in pPCL, and this translocation is the most common.
- Other *IgH* translocations (e.g. t(4;14), t(14;16), etc.) are more diverse in sPCL.

The presence of t(11;14) is important because it is a marker of *BCL2* over-expression, and so there is potential for the use of a BCL-2 inhibitor such as venetoclax.

Links with Prognosis

Given the rarity of PCL, one of the (considerable) limitations of many studies is their small sample size leading to the potential for false positives and false negatives, potentially compounded by confounders such as treatments. Nevertheless, overall, adverse outcomes are associated with del(13q14), del(17p13) and chromosome 1 abnormalities, including del(1p), gain 1q and translocations t(4;14) and t(14;16).

Self-check 6.6 What are the principal genetic abnormalities as MGUS develops through SMM to myeloma?

6.3.6 The Genetics of Disease Progression

A considerable proportion of the literature on myeloma and related disease focuses on the prediction of disease progression. Historical markers of this process include plasma proteins, but developing data examines the role of genetics, such as roles for *MYC*, *RAS*, NFκB, and cell cycle control pathways such as cyclins in the progression to malignant disease. A selection of abnormalities across the spectrum of disease development is shown in Table 6.7, some of which indicate a potentially significant pathological trend.

6.3.7 Routine Molecular Pathology

The National Genomic Test Directory for England has a section named plasma cell dyscrasias, taken by many to include the spectrum of diseases from MGUS to PCL. It describes a number of investigations:

- Multi-target NGS panels for detecting small variants in *KRAS, NRAS, BRAF, TP53, DIS3, TENT5C* (at 1p12, coding a nucleotidyl-transferase) and *IRF4*, another for detecting structural variants in fusions of *IGH* with *FGFR3, CCND3, MAF* and *MAFB*, and for *MYC* rearrangements, and a third to detect CNVs in chromosome 1p and 1q, and in chromosome 17p.
- FISH may also be used to detect CNVs in del(1p) (focusing on *CDNK2C*), gain(1q) (for *CSK1B*) and del(17p) (for *TP53*).
- FISH/simple targeted mutation testing (RT-PCR) for the *IGH* fusions described above, such as t(11;14)(q13;q32).

Table 6.7 Genetics of the MGUS–SMM–MM pathway.

	MGUS	SMM	Myeloma
Deletion in 13p	25%	35%	47%
IgH rearrangement	41%	35%	46%
t(4;14)	4%	13%	13%
	9%	–	13%
	3.1%	12.8%	10.8%
t(6;14)	1%	~1%	2%
	0%	–	1%
t(11;14)	16%	11%	14%
	12%	–	19%
t(14;16)	3%	3%	4%
	3%	–	4%
	3.1%	2.7%	3%
16q23 deletion	6%	8%	21%
t(14;20)	5%	0%	2%
	4.7%	0.7%	1.5%
	3%	–	1%
TP53 mutation	3%	~1%	10%
Hyperdiploid	42%	63%	57%
Non-hyperdiploid	58%	37%	43%

Source: Based on Checchio L, et al. Haematologica. 2009;94:1708–1713; Mikulasova A, et al. Haematologica. 2017;102:1617–1625; and Ross FM, et al. Haematologica. 2010;95:1221–1225.

6.4 Other Blood Cancer

We complete our study of blood cancer with an examination of a number of diseases that although at a relatively low frequency, nonetheless, demands our attention. A common feature is once more the transformation of cells within the bone marrow, many of which are the product of the CFU-EMk stem cell, despite the propensity of certain conditions to develop into a frank myeloid character, indicating the presence of vestiges of the GEMM stem cell (Figure 6.1, left). These are classified by the 2022 WHO guidelines as MPNs, alongside CML, CNL, CEL and juvenile MML already described above. A further aspect is the need to exclude CML for a diagnosis, as defined by *BCR::ABL1* (hence *BCR::ABL1*-negative MPNs), so the investigation of this translocation is an important part of the workup of various conditions.

This section also emphasises the common genetic basis of several of the diseases we are about to discuss, and in many cases their close relationship with each other and with other blood cancers, principally AML. Three conditions that exemplify this close relationship, often collectively described as MPNs, are myelofibrosis, polycythaemia vera (PV) and essential thrombocythaemia.

6.4.1 Myelofibrosis

This disease is characterised by the proliferation of bone marrow fibroblasts, which secrete connective tissue substrates such as collagen and reticulin, building up to cause fibrosis. Fibrocytes, originating from neoplastic monocytes and exhibiting fibroblast-like features, may also have a role, such as the release of cytokines and growth factors. The consequence of this fibrosis is bone marrow failure, an end point being reductions in the number of all blood cells, with pancytopaenia. The aetiology can be primary or secondary. The latter often forms the basis of other myeloid neoplasms of PV (most frequently) or ET (less often) (see below). Primary disease (pMF) can be further classified as pre-fibrotic or overtly fibrotic, dependent on results of a bone marrow trephine, and is an example of a clonal disorder involving a multipotent haemopoietic stem cell, with megakaryocyte hyperplasia.

Early molecular pathology pointed to deletion of 13q, with the loss of 13q14, the location of *RB1*, with mutations in *N-ras* and the potential for roles for PDGF and TGF-β. Later work pointed to del-20q, partial trisomy of chromosome 1q and trisomy of chromosome 8, with gains of chromosome 9p, the latter present in 60% of cases, and the location (at 9p24.1) of *JAK2*. A mutation in this gene, leading to JAK-V617F (valine to phenylalanine), is common in many malignancies and operates to up-regulate the STAT pathway. Molecular genetics is called upon to exclude the possibility of *BCR::ABL1*.

Further candidate genes that emerged included *MPL* at 1p34.2, coding for the thrombopoietin receptor (CD110, present in 5–10% of cases) and mutations in *CALR* (at 19p13.13), coding for calreticulin (25–30% of cases), with 10% of cases being ‘triple negative’ for the driver mutations in *JAK2, MPL* and *CALR*. Other genes, described by some as non-drivers or clonal-drivers, that were subsequently added include *K-H-N-RAS, IKZF1, TP53, DNMT3A, TET2, IDH1, IDH2, EZH2, DNMT3A, CBL, SF3B1, SRSF2* and *U2AF1*. Additional chromosome abnormalities include +8, −7/7q-, i(17q), inv(3)/3q21 and 11q-/11q23, all of which bring a poor prognosis, as do mutations in *ASXL1*, present in 50% of pMF and 30–40% of secondary MF cases. Whilst singe-gene disease is common, a second mutation is often found, with 19% having three gene mutations, 11% having four mutations and 18% having five or more. Other genetic aspects of MF are discussed in Section 6.4.5.

Survival is linked to aetiology, which is around 14 years for prefibrotic pMF, 6–7 years for overtly fibrotic pMF and 9 years for secondary MF. Around 20% progress to AML and 10% succumb to progressive bone marrow failure, whilst deterioration to a blast phase is predicted by a monosomal karyotype and alterations in *RUNX1, CEBPA*, variants of *SH2B3* (at 12q24.12, coding a signaling pathway regulator) and genes of the RAS–MAPK pathway. The principal contribution of molecular pathology to MF is that chemotherapy focuses on the *JAK*-inhibitor ruxolitinib (NICE TA386), other potential agents being fedratinib, momelotinib and pacritinib, a further option being bone marrow transplantation.

Key Point 6.7 Although the role of *JAK* is established, many of the genetic abnormalities described above are beginning to find their way into MF prognosis and treatment guidelines.

6.4.2 Polycythaemia Vera

In common with many blood cancers, PV was recognised in the Victorian era with the observation of a high red blood cell count, but this alone requires differential diagnoses that include other conditions such as erythrocytosis and erythroleukaemia (Section 6.4.6). Cytogenetics pointed to chromosomal abnormalities in the 1970s, such as del20q and trisomy 9, and the present century reinforced these findings with reports of specific gene abnormalities, such as the over-expression of p16(*INK4a*) and p14(*ARF*), and links with myelofibrosis. The importance of *JAK2* was established by the mid-2000s, demonstrating the close association of the disease with MF (as above) and ET (to follow), and is now known to be the leading genetic abnormality of this disease.

The primary laboratory feature is an increased red cell count, with raised haemoglobin, haematocrit and blood volume, implying a genetic lesion at the level of the CFU-EMk stem cell, and ~98% of cases are driven by the *JAK2 V617F*, the remainder being due to exon 12 mutations. Notably, an authoritative study failed to find pathogenic mutations in *MPL*, data that contrasts with that from MF. The link with other MPNs is the 20% risk of transformation into MF by around 15 years from a diagnosis and a 10% risk of AML within 20 years. In common with MF, treatments include *JAK* inhibitors, principally ruxolitinib, although there are others (as above). Other genetic aspects of PV are discussed in Section 6.4.5.

6.4.3 Essential Thrombocythaemia

First described less than a hundred years ago, the clonal nature of the disease was described in the early 1980s, and is diagnosed by a persistent platelet count >450 without other causes (e.g. inflammation), a subsequent finding was increased numbers of enlarged mature megakaryocytes. Cytogenetics reported abnormalities such as trisomy of chromosome 1q, del-5q and del17p, whilst other studies implicated thrombopoietin (coded for by *THPO* at 3q27.1), the major regulator of megakaryocyte function and its receptor, CD110 (down-regulated at the megakaryocyte surface), now also known to be implicated in MF (Section 6.4.1). An activating SNP in the transmembrane domain of the receptor (CD110) was present in all members of a cohort of familial ET. However, the key report in the Lancet came in 2005 regarding the presence of *JAK2* in this disease.

The genetic drivers of ET are more diverse than in PV and MF, but *JAK2* dominates (in 60% of cases), whilst mutations in *SRSF2* and *SF3B1* bring a poor prognosis. Outcome depends on age, leucocytosis, venous thrombosis and an abnormal karyotype, leading to median survival of ~18 years, with risk of progression to MF (10% risk after 15 years), AML (<5% after 20 years) and potentially an acute promyelocytic leukaemia. Cardiovascular and cerebrovascular disease causes 21% of deaths and haematological malignancy 13%. Treatment, as in PV, focuses on avoiding thrombohaemorrhagic events, possibly by venesection and aspirin alone, but high-risk patients may need active cytoreduction, such as with hydroxyurea, anagrelide or *JAK* inhibitor ruxolitinib. Extreme thrombocytosis (>2000) may be treated with 2 g hydroxyurea daily. Other genetic aspects of ET are discussed in Section 6.4.5.

The dynamic genetic status of ET and PV was examined in 49 patients and a panel of 37 genes. Of these, a mutation in *JAK2* was present in 32, *TET2* in 15, *CALR* in 13, *CBL* and *RUNX1* in 9, DDX41 in 5, *ASXL1, STAG2* and *MPL* in 4, with *CEBPA, CSF3R, DNMT3A, EZH2, GATA2, KIT, TP53, U2AF1* and *PTPN11* in a few patients. There were no mutations in *ABL1, ANKRD26, BCOR, BRAF, DDX41, ETNK1, ETV6, FLT3, GATA1, IDH1, IDH2, KIT, KRAS, NPM1, NRAS, PHF6 SETBP1, SF3B1, SRSF2, WT1*, or *ZRSR2*. After an interval of 78–141 months, there were 55 new mutations, the most common being in *CBL*, which developed in 10 patients, *DNT3A* in 8, both *DDX41* and *RUNX1* in 6 and *ASXL1* in 5 patients. A single new mutation developed in 20 patients, two mutations in 11, whilst three patients each developed three, four or five new mutations. Seven patients (14%) experienced fibrotic progression at follow-up. This data clearly demonstrates the potential, despite best-practice treatment, for additional genetic abnormalities to develop in these diseases – only 14 patients (28%) failed to develop a new mutation.

Key Point 6.8 The important cell biology aspect of *JAK2* in the MPNs is that the erythropoietin, thrombopoietin, G-CSF, stem cell factor, interleukin-3 and interleukin-5 receptors all employ this kinase for signal transduction.

6.4.4 Myelodysplasia

This condition lies outside the WHO definition of MPN but is nevertheless relevant in this section. Furthermore, it is included in the National Genomic Test Directory for England alongside MPNs and MF. Characterised by ineffective haemopoiesis due to stem cell aberrations and increased apoptosis, consequences include peripheral blood cytopaenias, often with macrocytic and other forms of anaemia. The mean age at presentation is 74 with an estimated 3-year survival of

<50%, much being due to the ~15% who develop AML. Indeed, the 2022 WHO guideline specifically refers to AML and MDS-related.

As so often, cytogenetics was the earliest method to be used, and described chromosomal lesions, such as loss of chromosomes 5 and 7, whilst it has since been shown that some can help prognosis: very good if –Y or del(11q) are present, good if del (5q), del(12p) or del (20q), intermediate if del(7q), +8, +19 or i(17q) and poor if −7, inv(3)/t(3q) and −7/del(7q). However, there are many instances of MDS following treatment with alkylating agents, such as methotrexate and bleomycin, often for other cancers. The hybridising of oncogene probes to immobilise RNA from 130 MDS cases revealed significant expression of *c-myc*, *c-ets, c-abl* and *c-fos*, with 40 patients having an abnormal karyotype.

6.4.5 Routine Molecular Pathology of MPNs

The National Genomic Test Directory for England docs not refer directly to MF, ET or PV, but has extended sections on MPNs, on MDS and on MDS/MPN, i.e. MDS/MPN, with considerable overlap. For example, each describes multi-target NGS panels to detect small variants in certain genes as follows:

- The 'core' analyses for initial diagnosis of MDS, MPN or MDS/MPN consists of *TP53*, *SF3B1*, *IDH1*, *IDH2*, *NRAS*, *KRAS*, *TET2*, *SRSF2*, *ASXL1*, *DNMT3A*, *RUNX1*, *U2AF1*, *EZH2*, *JAK2*, *SETBP1* and *CUX1*.
- If only MDS is suspected, additional analysis includes *BCOR*, *PTPN11*, *PPM1D*, *DDX41*, *PHF6*, *CUX1* and *UBA1* (at Xp11.3, coding ubiquitin-like modifying activating enzyme 1).
- If only MPN is suspected, additional analyses include *CALR*, *MPL*, *CBL*, *CSF3R*, *IKZF1*, *KIT*, *NFE2* (at 12q13.13, coding an erythroid nuclear factor), *SH2B3*, *SRSF2*, *HRAS* and *ZRSR2*.
- Should initial investigations point to MDS/MPN, analyses include the core genes, those of the MPN series, and *BCOR, PTPN11, FLT3, NF1* and *NPM1*.

Similarly, karyotyping can be used to detect a number of chromosomal abnormalities, some complex, and again, there is marked repetition. The entry for MDS includes −7/del7q, −5/del5q, i(17q)/t(17p), −13/del13q, del11q, del12p/t(12p), del9q, del17/del17p and idic(X)(q13), -Y, del20q, +8, +19, inv(3)/t(3q)/del(3q). The entries for MPN and MDS/MPN are almost identical: deletion of 4q12, −7/del7q, −5/del5q, i(17q)/t(17p), −13/del13q, del11q, del12p/t(12p), del9q, del17/del17p and idic(X)(q13), but also include t(9;22)(q34;q11) *BCR::ABL1*, del(4)(q12q12) *FIP1L1::PDGFRA*, other *PDGFRA* rearrangements, t(5;12)(q33;p13) *ETV6::PDGFRB*, other *PDGFRB* rearrangements, *FGFR1* rearrangements, t(8;9)(p22;p24) *PCM1::JAK2*, inv(3), 11q23, and rearrangements of *ABL1*, *JAK2*, *FLT3*, *RET* and *NTRK3*. Selected translocations of relevance to MPNs are presented in Table 6.8.

The use of FISH is also common between the three sections, with suggestions of determination of CNVs in chromosomes 7/7q, 5/5q, 13/13q, 11q, 12q, 17/17p, i17q, Y, 20, 8 and 19, for rearrangements in *PDGFRA, PDGFRB, FGFR1, FLT3, RET, NTRK3, JAK2, TAL1, TLX1, ABL1, KMT2A, CSF1R, IGH, CRLF2, EPOR, NUP98, BCL2, BCL6* and *KMT2A*, and for inversions such as inv(3)(q21q26), bringing together *GATA2* and *MECOM*, inv(16)(p13.1q22) *CBFB::MYH11* and inv(16)(p13.3q24.3) *CBFA2T3::GLIS2*. Table 6.9 shows the frequency of leading genetic abnormalities in these conditions.

Self-check 6.7 What is the common genetic theme in the various forms of MPNs?

Table 6.8 Selected translocations in MPNs.

Translocation	Fused genes	Translocation	Fused genes
t(1;7)(p32;q11)	*TRB::TAL1*	t(7;12)(q36;p13)	*MNX1::ETV6*
t(1;19)(q23;p13)	*TCF3::PBX1*	t(8;21)(q22;q22)	*RUNX1::RUNX1T1*
t(1;22)(p13;q13)	*RBM15::MRTFA*	t(9;11)(p21;q23)	*MLLT3::KMT2A*
t(3;5)(q25;q34)	*NPM1::MLF1*	t(10;11)(p12;q23)	*KMT2A::MLLT10*
t(4;11)(q21;q23)	*KMT2A::AFF1*	t(10;14)(q24;q11)	*TLX1::TRD*
t(5;11)(q35;p15.5)	*NUP98::NSD1*	t(11;14)(p15;q11)	*TRD::LMO1*
t(5;14)(q35;q32.2)	*BCL11B::TLX3*	t(11;19)(q23;p13.3)	*KMT2A::MLLT1*
t(6;9)(p22;q34)	*DEK::NUP214*	t(11;19)(q23;p13.1)	*KMT2A::ELL*
t(6;11)(q27;q23)	*KMT2A::AFDN*	t(12;21)(p13;q22)	*ETV6::RUNX1*
t(7;10)(q34;q24)	*TRB::TLX1*	t(15;17)(q24;q21)	*PML::RARA*

Table 6.9 Frequency of leading gene mutations in pMF, ET and PV.

Gene	PV	ET	pMF
JAK2	98%	55%	60%
MPL	0%	5–7%	7–10%
CALR	0%	25–30%	20–30%
DNMT3A	5–10%	1–5%	8–12%
IDH1	1–2%	1–2%	5–6%
IDH2	1–2%	1–2%	5–6%
ASXL1	2–7%	5–10%	15–35%
CBL	<2%	<2%	4%
SRSF2	<2%	<2%	6–14%
U2AF1	<2%	<2%	7–10%
TP53	<2%	<2%	2–5%
TET2	10–20%	1–3%	2–4%
SH2B3	2–9%	1–3%	2–4%
RUNX1	<2%	<2%	2–3%

PV, polycythaemia vera; ET, essential thrombocythaemia; pMF, primary myelofibrosis.
Source: Amended from Tashkandi H, Younes IE. Cancers. 2024;16:1679.

An objective observer may take the view that there is considerable overlap, and so confusion, regarding the four conditions in the preceding sections, as illustrated by Figure 6.9. It is equally likely that few observers will agree with this position.

6.4.6 Erythroleukaemia

Further underlining the close relationship between monocyte/granulocyte white blood cells and red blood cells (i.e. between CFU-GEMM, CFU-GMo and CFU-EMk), the 2022 WHO guideline classifies this condition as a form of AML, as acute erythroid leukaemia. An early finding was the constitutive over-expression of the receptor for erythropoietin (*EPOR*, located at 19p13.2). It was further characterised by neoplastic transformation of erythroid precursors in the bone marrow, the leading genetic abnormality being alterations in *TP53*, in a study from the Mayo Clinic, defined by FISH in 40

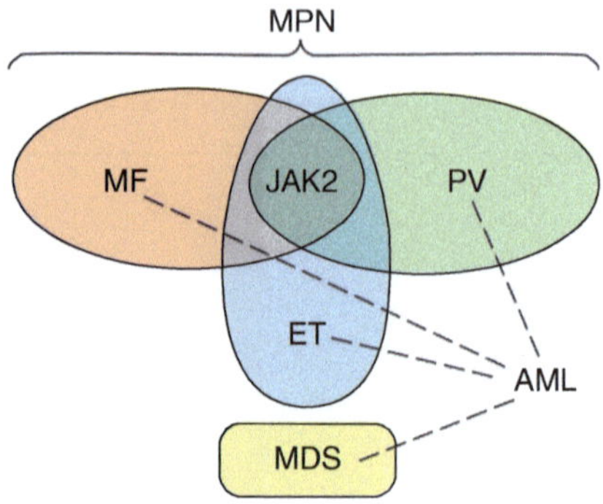

Figure 6.9 The complex relationship between the MPNs. An illustration of overlaps between different MPNs. The size of the ovoids and overlaps does not imply a numerical, clinical or other relationship. AML, acute myeloid leukaemia; ET, essential thrombocythaemia; MDS, myelodysplastic syndromes; MF, myelofibrosis; PV, polycythaemia vera. The central point of JAK (janus kinase) represents the pathological importance of this enzyme.

out of 41 cases, although it may be acquired following treatment with azathioprine or evolving from MDS. Cytogenetic changes include del(5q), −7/del(7q) and −17/17p deletions, abnormalities present in many of the diseases in previous sections. However, the key diagnostic features are morphological: the proportion of proerythroblasts (or proerythroblasts) and erythroblasts (normoblasts/nucleated red blood cells) in the bone marrow, and their appearance in the blood, whilst pancytopaenia is common.

Alongside *TP53*, likely genetic lesions include t(1;16)(p31;q24) and t(11;20)(p11;q11), and alterations in *NRAS, DNMT3A, JAK2, TET2, SF3B1, CSF3R, SPI1 (*at 11p11.2)*, KLF1* (at 19p13.13,)*, GATA1* at Xp11.23 and *GATA2* at 3q21.3, the latter four all coding for transcription factors with established roles in erythroid differentiation.

6.4.7 Basophils and Mast Cells

Basophils

As basophils are the least frequent circulating leukocyte, it is perhaps unsurprising that the 2022 Guideline of the WHO has few references to acute basophil leukaemia, classifying it as a form of AML, and neither is it referred to in the National Genomic Test Directory for England. There are numerous cases of a profound basophilia arising on the background of *JAK2*-mutated MPN, MDS/MPN and AML, which could be interpreted as a basophil leukaemia transformation, with *BCR::ABL1* being a potential marker. Other potential candidate genes include the *MYB::GATA1* fusion arising from t(X;6)(p11;q23) and a *PRKG2::PDGFRB* fusion (the former normally found at 4q21.21, coding a kinase). A series of case reports of chronic basophilic leukaemia described +8, 22q-, −7, −7/-4, t(5;12)(q31;p13), der(6), t(6:?)(q25-27;?), +mar(10) and der(17).

Mast cells

The WHO guidelines carry a section on mastocytosis, with three forms: cutaneous mastocytosis (with three subtypes), mast cell sarcoma and systemic mastocytosis. The guidelines refer to six subtypes of the latter, one of which is mast cell leukaemia, the immediate question being its difference(s) to basophil leukaemia. One answer may lie with standard blood film staining, whilst others with immunophenotype ((CD13, CD23, CD123, CD203c and CD11b), bone marrow blast counts and serum tryptase. Clarity is crucial as the median survival in one report was less than six months, thus calling for early bone marrow transplantation.

Genetically, there is focus on the D816V variant in the proto-oncogene *KIT* which is also described in the National Genomic Test Directory for England, with mention of a multi-target NGS panel or QF-PCR for this gene. The WHO document also refers to variants of *TET2, SRSF2, ASXL1, RUNX1* and *JAK2* in systemic mastocytosis, whilst other commentators focus on the same genes in mast cell leukaemia, and case studies describe del(7)(q31), and a variant of *NR2F6*, in mast cell leukaemia.

Others have used panels to detect variants in different combinations of *KIT, PHF6, RUNX1, DNMT3A, EXOC7* (at 17q25.1, coding a component of the exocyst), *NR2F6* (at 19p13.11, coding a component of a nuclear receptor), *PTPN11, CBL, SF3B1, CACNA1C* (at 12p13.33, coding a component of a calcium channel), *DNM2* (at 19p13.2, coding dynamin-2, a GTP-binding protein), *GNB1* (at 1p36.33, coding a subunit of a G protein) and *WFIKKN2* (at 17q21.33, coding a protease inhibitor). Molecular testing for *KIT* is mandatory in view of the use of the multi-kinase inhibitor midostaurin, which NICE TA728 recommends for systemic mastocytosis or mast cell leukaemia.

6.4.8 Histiocytosis

This disease is unusual in the blood cancer setting as it is certainly *of* the blood but not *in* the blood, the transformed histiocytes forming tumours in the tissues. Recognised as a terminally differentiated monocyte/macrophage, other than pancytopaenia (hence bone marrow involvement) in the 1980s, early genetic analysis pointed to a candidate gene on chromosome 5, with cases of t(2;5)(p21;q35), t(1;5)(p32;q35) and t(5;6)(q35;p12). The literature of the last century refers to tumours of the subtype of Langerhans cell histiocytosis (LCH), which also has abnormalities such as inv(13)(q21;q33), del(7)(q22) and t(7;12)(q11.2;p13), detectable in the lesion itself, in bone marrow and in peripheral blood. In the present century, *BRAFV600E* and lesions in *MAP2K1* in LCH were described, alongside the recognition that histiocytosis is a markedly heterologous group of diseases that can manifest in almost any organ.

The 2022 WHO guideline places histiocytic neoplasms alongside those of dendritic cells, referring to mutations in *BRAF*, proto-oncogene *ARAF* (at Xp11.3, coding a serine/threonine kinase), *MAP2K1*, *NRAS* and *KRAS*, with *ALK* positivity being emphasised, the most common translocation forming *KIF5B-ALK* (the former at 10p11,22 coding a kinesin). In this document, LCH is classed separately from histiocytic neoplasms, whilst in England and Wales, in 2022, a unifocal LCH was linked to 10 deaths. A malignant mast cell tumour was linked to two deaths, a sarcoma of dendritic cells to six deaths, whilst malignant histiocytosis was not linked to any deaths. However, without data on the incidence of these diseases, we are unable to speculate on the success or failure of treatment.

The National Genomic Test Directory for England's section on histiocytosis describes a multi-target NGS panel for the detection of small variants in *BRAF*, *MAP2K1*, *NRAS*, *KRAS*, *HRAS*, *ERBB3* (at 12q13.2, also known as HER3, coding for the human epidermal growth factor receptor 3), *ARAF*, *MAP3K1*, *PIK3CA* and *PIK3CD* (at 1p36.22, coding a PI3K subunit).

A second panel may be used for detecting structural variants in *EML4::ALK*, *TPM3::ALK*, *KIF5B::ALK*, *MIGA1::BRAF*, *PACSIN2::BRAF*, *RNF11::BRAF*, *CLIP3::BRAF*, *LMNA::NTRK1*, *TPR::NTRK1* and *ETV3::NCOA2*, and rearrangements of *ALK*, *BRAF* and *NTRK1*. Structural variants of many of the latter may also be detected by FISH/PCR. The importance of molecular genetics, as in many cases, is in the direction of treatment, such as vemurafenib and binimetinib for *BRAFV600E* and other roles in the MAP kinase pathway.

Self-check 6.8 Test yourself in a recall of this chapter by reviewing the learning objectives.

6.5 Summary

Together, the various forms of blood cancer are the third-most frequent cause of deaths in the UK, less than those of the lung/trachea/bronchus and colon/rectum but exceeding those of the prostate and breast. The same positions are held by a crude index of mortality to incidence, with blood cancer again in the middle. These indices are 69% (worst prognosis), 46%, 42%, 24% and 21% (best prognosis), respectively.

The most frequent blood cancer in the UK is lymphoma, dominated by the non-Hodgkin variant, there being numerous subtypes. Of these, most are of B-lymphocytes, the leading molecular pathology being *IGH* conjugated to an oncogene such as *MYC*. The minor disease in this group is Hodgkin lymphoma, whose aetiology is linked to EBV and several candidate genes.

The leading leukaemias are relatively easy to classify as acute or chronic and of lymphoid or myeloid aetiology. The molecular pathology of CML focuses entirely on *BCR::ABL1*, whereas there are many genes linked to AML. The heterologous nature of ALL is also demonstrated by the many potential genetic lesions, although *KMT2A* is often found translocated to different partners, whilst the list of candidate genes in CLL is more modest, although *TP53* is prominent. Of special note are HCL, strongly linked to *BRAF V600E*, and NK cell leukaemia, linked to variants in *JAK2*.

The third major grouping of blood cancer is a spectrum of B-cell malignancies running from MGUS to smouldering myeloma and 'classical' myeloma, with a small number transforming into a PCL. Unsurprisingly, the leading genetic lesions are based around *IGH* and a deletion of 13p. An important point regarding certain myeloid blood cancers is their inter-relationships, as exemplified by PV, ET and MSD, all of which may progress to AML. Completing the roll call of blood cancers are those of the basophil/mast cell and histiocytes.

Running through this chapter are frequent referrals to two guidelines from the WHO, each focusing on lymphoids (Alaggio et al. 2022) and myeloid cancer (Khoury et al. 2022). An additional key reference is the National Genomic Test Directory for England, which points to genes and methods applicable to the disease we have been examining.

Bibliography

Alaggio R, Amador C, Anagnostopoulos I, et al. The 5th edition of the world health organization classification of haematolymphoid tumours: Lymphoid neoplasms. Leukaemia. 2022;36:1720–1748.

Álvarez-Zúñiga CD, Garza-Veloz I, Martínez-Rendón J, et al. Circulating biomarkers associated with the diagnosis and prognosis of B-cell progenitor acute lymphoblastic leukemia. Cancers. 2023;15:4186. doi:10.3390/cancers15164186.

Arber DA, Orazi A, Hasserjian RP, et al. International consensus classification of myeloid neoplasms and acute leukaemias: Integrating morphological, clinical, and genomic data. Blood. 2022;140:1200–1228.

Duffield AS, Mullighan CG, Borowitz MJ. International Consensus Classification of acute lymphoblastic leukemia/lymphoma. Virchows Arch. 2023;482:11–26.

El Achi H, Kanagal-Shamanna R. Biomarkers in acute myeloid leukemia: Leveraging next generation sequencing data for optimal therapeutic strategies. Front Oncol. 2021;11:748250. doi:10.3389/fonc.2021.748250.

El-Tanani M, Nsairat H, Matalka II, et al. The impact of the BCR-ABL1 oncogene in the pathology and treatment of chronic myeloid leukemia. Pathol Research Pract. 2024;254:155161.

Herbst SA, Vesterlund M, Helmboldt AJ, et al. Proteogenomics refines the molecular classification of chronic lymphocytic leukemia. Nat Commun. 2022;13:6226. doi:10.1038/s41467-022-33385-8.

Khoury JD, Solary E, Abla O, et al. The 5th edition of the world health organization classification of haematolymphoid tumours: Myeloid and histiocytic/dendritic neoplasms. Leukemia. 2022;36:1703–1719.

Lupski JR, Liu P, Stankiewicz P, et al. Clinical genomics and contextualizing genome. Expert Rev Mol Diagn. 2020;20:995–1002.

Mendez-Hernandez A, Moturi K, Hanson V, et al. Hairy cell leukaemia: Where are we in 2023? Curr Oncol Reps. 2023;25:833–840.
Sun F, Cheng Y, Ying J, et al. A gene signature can predict risk of MGUS progressing to multiple myeloma. J Haematol Oncol. 2023;16:70.
Susanibar-Adaniya S, Barta SK. Update on diffuse large B cell lymphoma: A review of current data and potential applications on risk stratification and management. Am J Hematol. 2021;96:617–629.
Szymczyk A, Macheta A, Podhorecka M. Abnormal micro expression in the course of hematological malignancies. Cancer Manag Res. 2018;10: 4267–4277.
Tashkandi H, Younes IE. Advances in molecular understanding of polycythemia vera, essential thrombocythemia, and primary myelofibrosis: Towards precision medicine. Cancers. 2024;16:1679. doi:10.3390/cancers16091679.
Xu L, Wu S. New diagnostic strategy for multiple myeloma. Medicine. 2023;102:52.

Websites

Useful resource for coagulation: www.pathologyoutlines.com/coagulation.html
National Institute for Health and Care Excellence: www.nice.org.uk/
British Society for Haematology guidelines: https://b-s-h.org.uk/guidelines/
NHS England National Genomic Test Directory: https://www.england.nhs.uk/publication/national-genomic-test-directories/
The Global Cancer Observatory: htps://gco.iarc.fr/en

7

Rare, Inherited, Metabolic and Other Diseases

In this final chapter, we examine the molecular pathology of a diverse collection of conditions and diseases. Whilst not necessarily responsible directly for a great number of deaths, many do indeed bring a high degree of morbidity for the patient and so demand our attention. However, a major problem is that there are numerous examples of conditions that have a heritable aspect, many of which are also rare, and some involve metabolic pathways. Accordingly, some definitions are required:

- Inherited diseases are the most common in this chapter, and within this group, sickle cell disease (SCD) has the highest frequency, with a prevalence of 6.9% in Africa alone and a frequency of 0.037%.
- The World Health Organization (WHO) describes a rare disease as one which has a population incidence of less than 1 per 2000 live births, i.e. in 0.05%, an example being Guillain–Barre syndrome, with a global frequency of around 15 per million, i.e. 0.0015%. These conditions may also be described as orphan diseases.
- A metabolic disease is one that impacts the metabolism of the entire body or a particular organ, an example being Gilbert syndrome, which follows a loss-of-function mutation in *UGT1A1*. Many of these diseases are also described as inborn errors of metabolism.
- There are many examples of molecular pathologies that do not fit clearly in any of these three categories. Accordingly, it is necessary to place these in a fourth, 'other' category, which includes a heterologous group of conditions, with cardiovascular, respiratory, haematological, urogenital and other diseases. An example in this

Molecular Pathology: A Primer for Laboratory Scientists, First Edition. Andrew Blann.

section is *MYH7*, coding for a subunit of myosin, and a candidate gene for a form of cardiomyopathy.

A clear caveat with this almost naively simplistic four-stage classification is the likelihood of a particular condition falling into more than one category (Figure 7.1). This overlap extends to the naming of particular groups in the literature, such as by the WHO's International Classification of Diseases, 11th Revision, which includes some 5500 rare diseases and their synonyms, in addition to inherited errors of metabolism.

Fortunately, the National Genomic Test Directory for England lists what they consider to be the more important rare and inherited diseases. This important reference document brings together these diseases into well-defined clinical areas that therefore provides as formal structure for this chapter. We will examine these groups in the sequence in which they appear in the document, although most of those affecting an acutely unwell child with a likely monogenic disorder, such as Williams syndrome, have also been addressed elsewhere in this volume.

> **Key Point 7.1** There is considerable overlap between inherited, rare, metabolic and other diseases, which brings difficulties in classification.

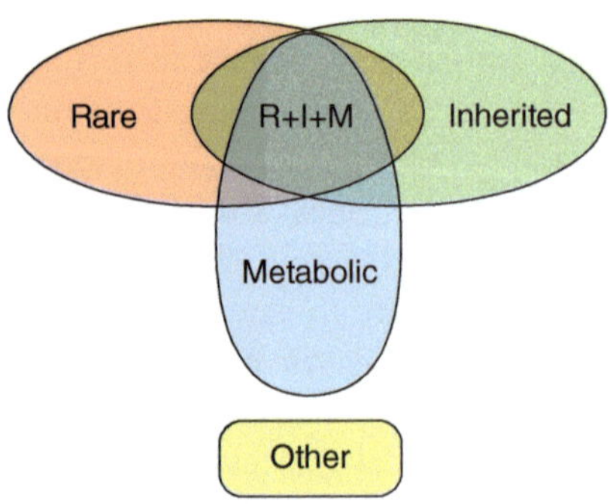

Figure 7.1 Overlaps in rare, inherited, metabolic and other diseases. Some diseases fall into one of the four categories, but most are present in two or more. The size of the three ellipses, and their overlaps, does not reflect their frequency in the population.

Learning Objectives

After studying this chapter, you should confidently be able to...

- Outline those cardiology, developmental and congenital diseases with a genetic basis
- Understand the importance of molecular pathology in endocrine, liver, immunological and intestinal diseases
- Use haemoglobinopathy as a model for certain genetic abnormalities
- Describe major lipid and metabolic diseases caused by a specific gene mutation
- Comment on the value of molecular pathology in musculoskeletal and neurological conditions
- List those renal, respiratory and dermatological diseases that have a genetic component

7.1 Cardiology

Many heart diseases are the product of long-standing aspects such as atherosclerosis, the effects of legal and illegal drugs, lifestyle choices and toxins. This chapter focuses on conditions that are genetically sporadic or heritable, many of the most severe ones being congenital, and so are essentially paediatric. In 2020, in England, the prevalence of the leading paediatric congenital anomaly was congenital heart disease, at around 67 per 10,000 births, although those that were genetic numbered 14 per 10,000 births. However, there are numerous examples of late-acting effects where genes are shown to be germline and so are present from conception.

Two major forms of genetic heart disease are discussed in the following sections, but there are other conditions, such as progressive cardiac conduction disease, that may present with structural or arrhythmic features. It is linked to mutations in several genes such as *SCN5A* (at 3p22.2), *SCN1B* (at 19q13.11), *SCN10A* (at 3p22.2), *TRPM4* (at 19q13.33) and *KCNK17* (at 6p21.2), all of which code for components of sodium, potassium or calcium cation channels, but also others such as *GJA5* at 1q21.2, coding for connexin, and *LMNA* at 1q22, coding for a member of the laminin family. We will revisit certain of these genes in later sections.

7.1.1 Structural Heart Disease

Congenital Heart Disease

With a global prevalence of 0.8–1%, congenital heart disease is the most common birth defect of the heart, although 90% of children survive to adulthood. Leading conditions include Tetralogy of Fallot, interrupted aortic arch or truncus arteriosus, atrial and ventral septal defects, endocardial cushion defects, total anomalous pulmonary venous return, tricuspid atresia and transposition complexes. Candidate genes include *MYH6*, at 14q11.2, coding for myosin heavy chain 6; *ACTC1*, at 15q14, coding for cardiac muscle actin alpha; *ELN*, at 7q11.23, coding for elastin; *NKX2-5*, at 5q35.1, *TBX1*, at 22q11.21, *TBX5*, at 12q24.21, and *GATA4*, at 8q23.1, all four coding for a transcription factor, and *CALM* genes 1–3, at 14q32.11, 2p21 and 19q13.32, all coding for a calmodulin. Several of the clinical syndromes may be present in the 22q11.2 distal deletion syndrome, many of which are also present in DiGeorge syndrome.

Bicuspid aortic valve disease and aortic coarctation are often present in Turner syndrome, whilst supravalvular aortic stenosis is linked to a deletion in 7q11.23, where *ELN* is located. Molecular genetics of the latter are required to consider the differential diagnosis of Williams syndrome.

Aneurysm

Aortic aneurysms cause 15,000–200,000 global deaths per year (causing 4857 deaths in England and Wales in 2023) and may develop over decades. A greater proportion of thoracic aortic aneurysms or dissections (357 deaths), and abdominal aortic aneurysms (2353 deaths), are due to weakness in the extracellular matrix of the arterial wall, secreted by vascular smooth muscle cells and/or fibroblasts. Relevant genes include *ACTA2* at 10q23.31, coding for actin, *MYH11* at 16p13.11, coding for myosin heavy chain 11, *MYLK* at 3q21.1, coding for myosin light chain kinase, *COL3A1* at 2q32.2, coding for type III collagen, FBN1 at 15q21.1, coding for fibrillin type 1, *VIM* at 10p13, coding for vimentin, and *TGFBR1* at 9p22.33 and *TGFBR2* at 3p24.1, together coding for the receptor for transforming growth factor-β (TGFβ). In addition, genome-wide association studies (GWAS) have provided further candidate genes, including *LRP1* at 12q13,3, coding the LDL receptor, *ULK4* at 3p22.1, coding a kinase, and *TCF7L2* at 10q25.2–25.3, coding a transcription factor.

Cardiomyopathy

In England and Wales, in 2023, this condition caused 1,857 deaths, whilst a study from the USA found dilated cardiomyopathy (DCM) to be present at a frequency of 0.0365%, with 51 candidate genes named, several of which may act in concert in 20–35% of patients. Of these 51 genes, the most important are *TTN* at 2q31.2, coding for titin, mutations being present in 20–25% of cases, *LMNA* at 1q22, coding for lamin, a component of the nuclear membrane (8% of cases), *PLN* at 6q22.31, coding for phospholamban, *RBM20* at 10q25.2, coding for an RNA-binding protein, and *SCN5A*. An arrhythmogenic variant of DCM can be linked to *DES* (at 2q35, coding for desmin), *DSP* (at 6p24.3, coding for desmoplakin), *FLNC* (at 7q32.1, coding for filamin C), *LMNA, RBM20, SCN5A* and *TMEM43* (at 3p25.1, coding for a component of the nuclear membrane).

Barth syndrome, an X-linked mitochondrial abnormality described in 200 living males worldwide, is characterised by several clinical features, of which DCM is the most common. The causal lesion is a mutation in *TAFAZZIN* at Xq28, which codes for a phospholipid-lysophospholipid transacylase, an enzyme involved in cardiolipin metabolism.

The leading aetiology of the autosomal dominant hypertrophic cardiomyopathy (HCM) variant, estimated as being present in 0.2% of young adults in the United States and <0.035% of all adults, is of altered myocardial sarcomere filaments, present in 60% of patients. This may involve any one of a dozen dysfunctional sarcomere genes, the principal ones being *MYH7* at 14q12, coding for a myosin heavy chain beta, and *MYBPC3* at 11p11.2, coding for myosin-binding protein C3, present in 70% of cases. Other candidate mutations include those in *TNNI3* (at 19q13.42, coding for troponin I3), *TNNT2* (at 1q32.1, coding for troponin T2), *TPM1* (at 15q22.2, coding for tropomyosin), *MYL2* (at 12q24.11, coding for myosin light chain 2), *MYL3* (at 3p21.31, coding for myosin light chain 3) and *ACTC1*, most of which are also linked to DCM.

In 2024, the UK Biobank reported on 200,000 people aged 40–69 who had undergone WES, of whom 1.95% were genotype-positive for DCM, 1.05% for HCM and 0.38% for arrhythmogenic right ventricular cardiomyopathy. The latter is the most common (30%) non-traumatic form of sudden death in young adults. Genes most frequently mutated were *MYH7* (25%), *TTN* (17%), *MYBPC3* (10%), *LMNA* (7%) and *DES* (7%), whilst more than one variant was present in 2.4% of cases.

In paediatric cardiomyopathy, present in perhaps 10–15 per million, the primary need is for transplantation, especially in those under 1 year, the most common forms being DCM (50%), HCM (40%), ventricular non-compaction (5%) and restrictive cardiomyopathy (<5%). Amongst 75 Chinese children with cardiomyopathy, DCM (61.3%) and HCM (17.3%) were the most common. WES and NGS found 34 pathogenic variants in 30 cases, the most common in DCM being *TNNT2, TNNI3, CDH2* (at 18q12.1, coding for cadherin 2), *TPM1, DES, GLB1* (at 3p22.3, coding for galactosidase beta 1) and *FHL2* (at 2q12.2, coding for a transcription regulator), with *MYH7, GAA* (at 17q25.3, coding for alpha-glucosidase), *MYL2, TPM1* (at 15q22.2, coding for tropomyosin 1) and *ACTC1* being common in HCM. Mutations in *ALMS1* (at 2p13.1, coding for a molecule linked to centriole function) lead to Alström syndrome, presenting at a frequency of 1–2 per million, with DCM and obesity being the common ones.

7.1.2 Arrhythmia

There are many forms of an irregular heartbeat, both of frequency and rhythm, the crucial investigations being a 12-lead electrocardiogram (ECG) and an echocardiogram, ideally with colour Doppler. The most serious arrhythmias are a leading cause of sudden cardiac death, especially in the young and those with a background of coronary artery disease.

Channelopathies

This group of diseases is based on the incorrect flow of potassium, sodium and/or calcium into and out of cardiomyocytes, often resulting in ventricular tachycardia and fibrillation, generally of the ventricles. The prevalence of long QT syndrome (LQTS) is estimated to be 0.05% in children, 0.02–0.005% in adults, but in 1.2% of patients presenting to an Accident and Emergency Unit. The majority of cases are caused by loss of function in voltage-gated potassium channels, the causative genes being *KCNQ1* at 11p15.5-15.4 (40–45% of cases), *KCNH2* at 7q36.1 (~40%), both coding for subunits of the channel, and *SCN5A* (~5%), coding for a subunit of the sodium channel. Torsade de pointes, with a ventricular tachycardia, may also be linked to a long QT interval.

Alongside *KCNQ1* and *KCNH2*, a further potassium channel gene *KCNJ2* (at 17q24.3), and calcium channel genes *CACNA1C* (at 12p13.33) and *CACNA2D1* (at 7q21.11), with *SLC4A3* (at 2q35) coding a bicarbonate/chloride exchanger, are candidate genes for the short QT syndrome. Leading genetic links with the Brugada syndrome (characterised by a raised ST section on an ECG) include *SCN5A* (as mentioned earlier). It is also a cause of sudden cardiac death. Catecholaminergic polymorphic ventricular tachycardia (VT) is a rare (<0.01% globally) condition, typically caused by mutations in *RyR2* at 1q43, or *CASQ2* at 1p13.3-11, both coding for components of the calcium channel, thereby influencing intracellular calcium levels. There may also be a place for *CALM1*, *CALM2* and *CALM3*, *TECRL* at 4q13.1, coding for a steroid reductase, and *SCN10A*, coding for a further component of the sodium channel.

Atrial Diseases

Atrial fibrillation (AF) is the leading arrythmia, causing 8,971 deaths in England and Wales in 2023 and affecting 2–4% of the global population, some 60 million people, rising to 15–20% in individuals over 80 years and in many cases leads to stroke. Although hypertension is its principal risk factor, others include smoking, obesity and high consumption of alcohol. However, a genetic component, as witnessed by family studies, led to the proposal of candidate loci at 1q21, 4q25 and 16q22, near to genes including *KCNN3* at 1q21.3, part-coding a potassium channel, *PRRX1* at 1q24.2, coding a transcription co-activator, and *PITX2* at 4q25, coding a transcription factor. Other genes, such as *FLNC* at 7q32.1, coding filamin C, *C9orf3* at 9q22.32, but also known as *AOPEP*, coding an aminopeptidase, *HCN4* at 15q24.1, part-coding a potassium channel, *SYNE2* at 14q23.2, coding a nuclear envelope protein, *SYNPO2L* at 10q22.2, coding a molecule involved in cytoskeleton binding, and *ZFHX3* at 16q22.2-22.3, coding a transcription factor, are also implicated.

A GWAS of over 35,000 persons free of cardiovascular disease in the UK Biobank project found AF to be linked to loci near to *CASQ2* at 1p13.3-11, part-coding calsequestrin, *FAF1* at 1p32.3, coding a molecule linked to the apoptosis molecule FAS, *GOSR2* at 17q21.32, coding a trafficking membrane protein, *IGF1R* at 15q26.3, coding the receptor for insulin-like growth factor, *MYO18B* at 22q12.1, coding a form of myosin, and *TTN*, *MYH6* and *MYH7*. Knowledge of the presence of mutations in these genes may provide the opportunity to target those at highest risk by addressing non-genetic risk factors.

Atrial flutter, a second form of supraventricular tachycardia, is present at a frequent one tenth of that of AF, and the two are often combined in observational studies. Nevertheless, many patients with atrial flutter have mutations in genes coding for the potassium channel as mentioned earlier, principally *KCNH2*.

Key Point 7.2 Although the aetiology of many cardiovascular diseases considers risk factors, these alone do not account for all diseases, with genetic factors playing several important parts.

7.2 Developmental Disorders

This section focuses primarily on intellectual development; those of physiology are discussed elsewhere. In many cases, unexplained impaired intellectual development in a child points to a chromosomal defect, although in many cases an abnormal phenotype may be clear, as in Williams syndrome (linked to del 7q) or Fragile X syndrome (characterised by increased CGG repeats in *FMR1*). Conditions that may appear in childhood include Angelman syndrome (caused by del 15q11-13 or mutation in *UBE3A*, coding for a ubiquitin ligases, and Prader–Willi syndrome, which results from mutations in *SNRPN*, coding for a small nuclear ribonucleoprotein, in *NDN*, also at 15q11.2, coding for necdin, and in *MAGEL2*, linked to ubiquitin ligase activity. All three genes being located at 15q11.2 (hence the importance of a del15q), and there may also be a place for small nucleolar RNAs.

Conversely, there may be phenotype dysmorphia but normal intellectual capacity, Turner syndrome (presenting at 0.05–0.02% of female births) being an example. Standard karyotyping may be sufficient, but microarray and FISH are alternatives. In certain cases, NGS sequencing may be required. In the presence of a confirmed diagnosis, both parents should be tested for autosomal recessive carrier status, as discussed in Section 7.5.

A further issue is the possibility of mosaicism, such that cells from different tissues may need to be analysed. Examples of this include Klinefelter syndrome, where some cells are XY, others are XXY, and Turner syndrome, where XX and XO cells may both be found in 30% of cases. This issue is exemplified by a case report exhibiting evidence suggestive of PTEN hamartoma syndrome: deep sequencing of blood leukocytes revealed a normal *PTEN*, but pathogenic variants of the gene were found to be present in non-tumoral stomach tissues, and in tumoral (colonic ganglioneuroma, oesophageal papilloma, fibroids of the oral mucosa) tissues.

As discussed in Chapters 1 and 2, trisomy of chromosomes 13, 18 and 21 are the most common forms of aneuploidy and can be detected by cell-free DNA in maternal blood. Leading chromosomal abnormalities linked to developmental issues are show in Table 7.1. Conditions such as Temple syndrome and Kagami–Ogata syndrome result from maternal imprinting of chromosome 14, or deletion of paternal 14q32: they present at a rate of <1 per million. Other conditions where there is multi-locus imprinting disturbance include Silver–Russel syndrome and Birk–Barel syndrome. The National Congenital Anomaly and Rare Disease Registration Service, a branch of NHS England, publishes data on the frequency of congenital anomalies, most of which have a genetic basis. Of 589,454 total births in the UK in 2020, 13,065 (2.2%) had at least one congenital anomaly, equating to one in 45 births, whilst 4775 (36.5%) had more than one anomaly, although some fail to progress to term (Table 7.1).

Self-check 7.1 What are the leading laboratory methods for determining chromosome abnormalities?

Table 7.1 Frequences of leading trisomies.

Syndrome	Abnormality	Total prevalence[a]	Live birth prevalence[a]
Down	Trisomy 21	26.5	11.5
Edwards	Trisomy 18	7.4	0.9
Patau	Trisomy 13	2.7	0.2

[a] Per 10,000.

7.3 Endocrinology

7.3.1 Diabetes

The broad aetiology of the leading causes of diabetes is established and includes chronic hyperglycaemia, gestational, autoimmunity, drug-related and genetic.

Maturity-onset Diabetes of the Young

As implied, this condition, comprising perhaps 1% (although some sources suggest up to 6.5%) of all cases of diabetes and 0.009% in the general population, resembles adult (type 2) diabetes, but presents in those with a low or normal BMI aged 25 or less, most cases being heritable. Some describe maturity-onset diabetes of the young (MODY) as monogenic diabetes. It is linked to several genes, with some defining their own subtype, as shown in Table 7.2: all (except *GCK*) are linked to beta cell destruction.

Other genes involved in MODY include *NEUROD1*, *KLF11* at 2p25.1, coding a transcription factor, *CEL* at 9p34.13, coding carboxyl ester lipase, *PAX4* at 7q32.1, coding a transcription factor, *INS*, proto-oncogene *BLK* at 8p23.1, coding a tyrosine kinase, *ABCC8* at 11p15.1, coding an ATP-binding cassette, and *KCNJ11*, also at 11p15.1, part-coding a potassium channel, some of which are proposed as being part of an additional MODY subtype, of which 14 have been described. Accurate diagnosis is

Table 7.2 Leading subtypes of MODY.

Subtype	Frequency	Gene	Location	Function
MODY1	10%	*HNF4A*	20q13.12	Hepatocyte nuclear transcription factor
MODY2	32%	*GCK*	7p13	Glucokinase: phosphorylation of glucose
MODY3	52%	*HNF1A*	12q24.31	Hepatocyte nuclear transcription factor
MODY4	<1%	*PDX1*	13q12.2	Pancreatic transcription factor
MODY5	6%	*HNF1B*	17q12	Hepatocyte nuclear transcription factor

important as many cases are misdiagnosed as type 1 or 2 diabetes, and knowledge of the correct type may direct treatment: types 3 and 4 can be treated effectively with low-dose sulfonylureas.

Neonatal Diabetes

Whilst many cases of this condition are early manifestations of MODY, it can be defined as hyperglycaemia requiring insulin treatment up to 32 weeks post-partum. Candidate genes include *ABCC8* and *KCNJ11* (as mentioned earlier, which may also be tested for in congenital hyperinsulinism, alongside *HNF1A*), whilst 6q24 methylation testing may be appropriate.

Although the leading features are a large tongue and a high birth weight, congenital hyperinsulinism is also a component of the multi-organ condition Beckwith–Wiedemann syndrome, which presents in around 0.01% live births. The leading genetic abnormalities are clustered around 11p15, where abnormal methylation is common, and where the candidate genes *CDNK1C*, coding for cyclin-dependent kinase inhibitor 1C, and *IGF2*, coding for insulin-like growth factor 2, are located.

Type 1 Diabetes

Family studies have long shown a familial tendency in both types of diabetes, although in type 1 a direct genetic link is established: the sibling risk is 6–7%, the child risk is 1–9% and the twin risk is 30–70%. The autoimmune link ties in with HLA haplotypes DRB1*0301-DQA1*0501-DQ*B10201 (DR3) and DRB1*0401-DQA1*0301-DQB1*0301 (DR4-DQ8), responsible for around half of the disease heritability. Conversely, the haplotype DRB1*1501-DQA1*0102-DQB1*0602 (DR15-DQ6) reduces the risk. Type 1 diabetes is also a leading indication of immune dysregulation, polyendocrinopathy, enteropathy, X-linked syndrome (IPEX), caused by pathogenic variants of *FOXP3*.

7.3.2 Thyroid

Congenital Hypothyroidism

Whilst the leading global cause of this condition is lack of iodine, many candidate genes have been described and can be classified as those linked to thyroid dysgenesis and those related to dyshormonogenesis. Amongst the former are *PAX8* at 2q14.1, *NKX2-1* at 14q13.3, *NKX2-5* at 5q35.1, *FOXE1* at 9q22.33, *HHEX* at 10q23.33, all coding for transcription factors, and *TSHR* at 14q24-31, coding for the TSH receptor. Genes involved in dyshormonogenesis include *TRHR* at 8q23.1, coding for the thyrotropin-releasing hormone receptor, *SLC5A5* at 19p13.11, coding for a sodium/iodine transporter, *TPO* at 2p25.3, coding for thyroid peroxidase, *DUOX2* at 15q21.1, coding for dual oxidase 2, *DUOXA2*, also at 15q21.1, coding for dual oxidase maturation factor 2, *SLC6A4* at 17q11.2, coding for a serotonin transporter, and *Tg* at 8q24.44, coding for thyroglobulin.

Hyperthyroidism

This condition has a number of different aetiologies. Twin and family studies estimate a 60–80% genetic risk of developing Graves disease, the most common form of hyperthyroidism, with women carrying a 3% lifetime risk of the disease and men 0.5% risk. The first genetic links were with HLA, such as the haplotype DRB1*0301-DQB1*0201-DQA1*0501, with *CTLA* at 2q33.2, coding a cytotoxic T-lymphocyte-associated protein 4, and with *PTPN22* at 1p13.2, coding a tyrosine phosphatase. Subsequently, GWAS and other NGS methods identified susceptibility loci at 14q31 (the location of *TSHR*), 20q11.2, and Xq21. Other findings include links with *FCRL3* at 1q23.1, coding a transmembrane receptor, whilst a study from China of over 9500 individuals reported links with intergenic regions at 4p14 and 14q32.2, with the *ABO* locus at 9q34.2, *C1QTNF6* (also known as *RAC2*) at 22q12.3-13.1, coding a TNF-related protein, *GPR174* at Xq21.1, whose product is likely to be linked to a G protein receptor, *RNASET2* at 6q27, coding for a ribonuclease, *SLAMF6 at* 1q23.2, coding a transmembrane protein, and *Tg* described earlier.

Hashimoto's thyroiditis has been linked to mutations in *LPP* at 3q27.3-28 coding a molecule involved in lipid metabolism, *SMARCA2* at 9p24.3, probably coding a transcription activator, *AATF* at 17q12, coding a molecule that may have a role in regulating apoptosis, *VAV3* at 1p13.3, coding a guanine nucleotide exchange factor, and *CTLA4* (as discussed earlier). Familial dysalbuminaemic hyperthyroxinaemia is characterised by high total serum thyroxine, despite being euthyroid. This state results from a mutant form of albumin resulting from a SNP in *ALB* at 4q13.3, which exhibits enhanced binding to thyroxine.

7.3.3 Adrenals

The most common (90–99%) form of congenital adrenal hyperplasia is characterised by a deficiency in steroid 21-hydroxylase, resulting from mutation in *CYP21A2* at 6p21.33, presenting at a frequency of <0.01% in neonates, and varying between 1 in 14,000 and 1 in 18,000.

Three types of familial hyperaldosteronism are recognised. Type I (glucocorticoid-remediable aldosteronism) is caused by an 8q24.3 cross-over between *CYP11B1*, coding steroid 11β-hydroxylase, and *CYP11B2*, coding aldosterone synthase. Type II has no known molecular mechanism. Type III is linked to numerous genes, including a gain in function in *KCNJ5* at 11q24.3, coding part of the potassium channel, *CACNA1D* at 3p21.1, coding part of a calcium channel, *ATP1A1* at 1p13.1, coding for an ATPase with a role in sodium/potassium transport, and *ATP2B3* at Xq28, coding an ATPase with a role in calcium homeostasis. Mutations in *KCNJ5, ATP1A1, ATP2B3* and *CACNA1D* have also been found in aldosterone-producing adenomas.

Primary pigmented nodular adrenocortical disease results from germline amplification of *PRKACA* at 19p13.12, coding a phosphorylating protein kinase A. The increased copy number of this gene ultimately results in hypercortisolaemia and low ACTH, with a Cushingoid phenotype. It may be part of the Carney complex, a set of heterologous clinical abnormalities.

Inherited phaeochromocytomas and paragangliomas are present at a frequency of 8 per million and may be grouped into clusters. Those of cluster 1A may be linked to variants in any of a large number of genes, such as those coding succinate dehydrogenase, *IDH1, HRAS, FH* at 1q43, coding fumarate hydratase, *MDH2* at 7q11.23, coding for malate dehydrogenase 2, *GOT2* at 16q21, coding for glutamic-oxaloacetic transaminase 2, *SLC25A11* at 17p13.2, coding a solute carrier, *DLST* at 14q24.3, coding a metabolic transferase, and *KIF1B* at 1p36.22, coding for a kinesin, cluster 1B with *VHL EGLN1* at 1q42.2, *EGLN2* at 19q13.2, both coding a metabolic hydroxylase and cluster 2 with *RET, NF1, MAX* and *TMEM127* at 2q11.2, coding a transmembrane protein.

7.3.4 Parathyroids

There are numerous diseases of this organ. Primary hypoparathyroidism may result from a mutation in *PTH* at 11p15.3, coding for parathyroid hormone (PTH), *CASR* at 3q13.33-21.1, coding a calcium sensing receptor, *GCM2* at 6p24.2, coding a transcription factor, and *GNA11* at 19p13.3, part-coding a second messenger.

Hypoparathyroidism is common in DiGeorge syndrome and in 22q11.2 deletion syndrome (present in 0.025% of live births and determined by FISH), where there is organ dysgenesis. Other related syndromes include those of Barakat (with nerve deafness and renal dysplasia, mapped to 10p14-15.1, with mutations in *GATA3*), and Eiken (with skeletal defects). Blomstrand syndrome, characterised by chondrodysplasia, is a consequence of a loss-of-function mutation in *PTHR1* at 3p21.31, coding for the PTH receptor. Mutations in *AIRE* at 21q22.3, coding the autoimmune regulator protein, may also lead to hypoparathyroidism.

Understanding disease of this organ is compounded by pseudohypoparathyroidism (where *GNAS* at 20q13.32 and *STX16* at 20q13.32, coding syntaxin 16, are involved) and pseudopseudohypoparathyroidism (linked to other lesions in *GNAS*), whilst investigation may be called for in certain cases of rickets, Albright's hereditary osteodystrophy, acrodysostosis and osteoma cutis.

Hyperparathyroidism is also described in several forms: familial, primary and secondary. The leading cause of this abnormality by far is an adenoma, another cancer being multiple endocrine neoplasia type 1 (Section 5.5 of Chapter 5, some of which can be familial), linked to *MEN1*. Other conditions include multiglandular primary hyperparathyroidism, whilst a secondary cause is familial hypocalciuric hypercalcaemia, which can be linked to *CASR*, and *AP2S1* at 19q13.32. Most cases of inherited parathyroid cancer generally involve oversecreting PTH, a common candidate gene being *CDC73* (also known as *HRPT2*) at 1q31.2, coding for parafibromin.

7.3.5 Reproduction

Chapters 1 and 2 have described the chromosomal abnormalities of the syndromes of Klinefelter (XXY), Jacobs (XYY), Patau (trisomy 13), Turner (XO) and other aneuploidies, all of which have abnormal sexual development. A key gene in the understanding of sexual development is *SRY* at Yp11.2, which codes a DNA-binding protein that activates male-specific transcription factor coding gene *SOX9* at 17q24.3.

Premature Ovarian Insufficiency

Around 10% of cases of premature ovarian insufficiency are genetic, with a strong twin concordance of 53–71%, and mothers/daughters with an odds ratio (95% CI) of 6.0 (3.4–10.7). The most common causes include X chromosome defects, notably those in *FMR1* at Xq27.3, coding for a ribonucleoprotein. WES and other methods applied to 291 members of extended families confirmed a large number of candidate genes (Table 7.3) and reported dozens more, including *USP36* at 17q25.3, coding a ubiquitin-specific peptidase, *VCP* at 9p13.3, coding for an ATPase, *WDR33* at 2q14.3, coding a molecule with a role in the polyadenylation of mRNA, *PIWIL3* at 22q11.23, coding a molecule that binds piRNAs, *NPM2* at 8p21.3, coding nucleophosmin, *LLGL1* at 17p11.2, coding a part of a cytoskeletal network, and *BOD1L1* at 4p15.33, coding a molecule that stabilises DNA repair protein RAD51.

Other genes linked to this condition include *BRCA1* and *BRCA2* (outlined in Chapter 4, Section 4.9), whilst *NR5A1* at 9q33.3, coding for a steroidogenic

Table 7.3 Genes linked to premature ovarian insufficiency.

Gene	Location	Product
HFM1	1p22.2	A helicase required for homologous recombination of chromosomes
PTPN22	1p13.2	Tyrosine phosphatase
FSHR	2p16.3	Receptor for follicle-stimulating hormone
DCAF17	2q31.1	Nuclear transmembrane protein
GDF9	5q31.1	An ovarian growth differentiation factor
HARS2	5q31.1	Histidyl-tRNA synthetase
MCM9	6q22.31	Homologous recombination repair factor
NR5A1	9q33.3	Steroidogenic transcription factor
PMM2	16p13.2	Phosphomannomutase 2
MARF1	16p13.11	Meiosis regulator and mRNA stability factor
PSMC3IP	17q21.2	Proteasome 26s subunit interacting protein
EIF4ENIF1	22q12.2	Eukaryotic translation initiation factor

transcription factor, *NOBOX* at 7q35, required for folliculogenesis, *FIGLA* at 2p13.3, and *FOXL2* at.3q22.3, both coding a transcription factor, have been identified as being linked to follicular atresia.

Gonad Development

Ovarian development requires correct functioning of genes such as *WNT4*, *RSPO1* (at 1p36.12 and 1p34.3 respectively, which both direct ovarian determination), *FOXL2* (needed to maintain granulosa cell differentiation and support folliculogenesis) and *CTNNB1*. Testis development calls for *WTI*, *NR5A1* and *GATA4* to regulate *SRY* and so *SOX9*, which also prevents the expression of genes controlling ovarian differentiation. Other relevant genes include *SOX2* at 3q26.33, coding a transcription factor linked to *SRY, SOX3* at Xq27.1, and *NR0B1* at Xp21.2, coding nuclear receptor DAX-1 that acts antagonistically to *SRY*.

Precocious Puberty

This condition presents at rates of 0.05–0.2% (depending on global geography), with a strong female dominance. Two forms are recognised: central and peripheral. The former is the leading variant (80% overall), present in 75–90% of girls and in 25–60% of boys, the principal candidate gene being *MKRN3* at 15q11.2, and others are *DLK1* and *KISS1* (coding for kisspeptin) at 14q32.2 and 1q32.1, respectively. These are associated with an overactive hypothalamic–pituitary–gonadal axis with increased gonadotropin-releasing hormone. The most common form of genetic peripheral precocious puberty is found in McCune–Albright syndrome (linked to mutations in *GNAS*), congenital adrenal hyperplasia (described earlier) and familial male-limited precocious puberty, also known as testotoxicosis, linked to *LHCGR* at 2p16.3, coding the receptor for luteinising hormone/choriogonadotropin, the pathogenic variant of which results in increased testosterone.

X-linked adrenal hypoplasia congenita, with an estimated live birth frequency of 1/12,500 (0.008%) may be linked to mutations and deletions in *NR0B1*.

7.3.6 The Pituitary/Hypothalamus

Pathology of this complex organ can, of course, lead to many diseases (as described earlier) and is dominated by cancer, with primary diagnosis based on imaging and blood tests. Pituitary organogenesis relies on genes such as *HESX1* at 3p14.3, *PROP1* at 5q35.3, *POU1F1* at 3p11.2, *LHX3* at 9q34.3, *LHX4* at 1q25.2, *PITX1* at 5q31.1, *PITX2* at 4q25 and *OTX2* at 14q22.3, many coding for transcription factors, often of the homeobox family, and so are relevant in congenital hypopituitarism, and may be autosomal dominant or recessive.

Combined pituitary hormone deficiency, present in 0.025% of live births, is linked to *POU1F, FOXA2* at 20p11.21, coding a transcription factor, *MAGEL2*, *NKX2-1*, *FGFR1* and *GH1* at 17q23.3, coding growth hormone, and microdeletions in 3q31.3. Correct functioning of the hypothalamic–pituitary–gonadal axis is vital to sex development in both sexes, with gonadotropin releasing hormone (coded for by *GNRH1* at 8p21.1) and its receptor (*GNRHR* at 4q21.2) the most crucial. Mutations in these genes are leading causes of hypogonadotropic hypogonadism and may also be present in cases of ambiguous genitalia presenting neonatally.

Self-check 7.2 What are the leading genetic causes of endocrine disease?

7.4 Ophthalmology and Audiology

The development of a retinoblastoma is perhaps one of the key features of cellular and genetic oncology, it being caused by loss of function in tumour suppressor *RB1* at 13q14.2 (Chapters 1 and 2, and Section 5.3.4 in Chapter 5). Chronic progressive external ophthalmoplegia is discussed in Section 7.12.

7.4.1 Cataracts

With an incidence of 0.012–0.136%, between 8.3 and 25% of infantile or congenital cataracts are hereditary. The leading candidate gene is *CTDP1* (also known as *CCFDN*) at 18q23, coding a phosphatase linked to a multi-factorial condition that includes microcornea, microphthalmia, micropupils and bilateral congenital cataracts. Other candidate genes include *CRYAA* at 21q22.3, *CRYAB* at 11q23.1, *CRYBA1* at 17q11.2,

CRYBB2 at 22q11.23, *CRYGC* at 2q33.3, *CRYGD* also at 2q33.3 and *CRYGS* at 3q27.3, all coding for crystallins. In the adult, several processes act independently to cause cataracts, with a number of inherited predispositions resulting from changes to the crystallin nature of the lens. One review pointed to over 50 genes associating with age-related cataract. Ophthalmological manifestations of Stickler syndrome include myopia, cataracts and retinal detachment. It is linked to variants of *COL2A1*, *COL9A1*, *COL9A2*, *COL9A3*, *COL11A1* and *COL11A2* at 12q13.11, 6q13, 1p34.2, 20q13.33, 1q21.1 and 6p21.32, respectively, all coding for different forms of collagen.

7.4.2 Albinism

This metabolic condition of the lack of melanin pigmentation may certainly be considered dermatological (Section 7.15.2), but a leading pathology is of the eye. Ocular aspects of albinism take two forms: those with effects on the eye alone (ocular albinism), and those with a generalised hypopigmentation due to impaired melanin synthesis, with characteristic ocular features (oculocutaneous albinism, OCA). The leading candidate gene for ocular albinism is *GPR143* at Xp22.2, coding for a G protein-coupled receptor, whilst there are several forms of OCA, each linked to a particular gene:

- OCA1, the most common (42% of cases) is linked to *TYR* at 11q14.3, coding for tyrosinase, the primary enzyme required for melanin production.
- OCA2 (28%), linked to *OCA2* at 15q12-13.1, coding for a melanocyte transmembrane protein.
- OCA3 (2%), linked to *TYRP1* at 9p23, coding for tyrosinase-related protein 1.
- OCA4 (MAPT) (11%), linked to *SLC45A2* at 5p13.2, coding for a melanin transporter.
- OCA6 (3%), also linked to *SLC45A2*.
- OCA7 (<1%), linked to *LRMDA* at 10q22.2-22.3, involved in melanocyte differentiation.
- OCA8 (<1%), linked to *DCT* (also known as *TYRP2*) at 13q32.1, coding for dopachrome tautomerase.

Nystagmus may present as part of albinism. An infantile form of nystagmus, estimated as present at a frequency of 0.1%, is caused by mutations in *FRMD7* at Xq26.2, whose product may be linked to vestibuloocular and optokinetic reflexes.

7.4.3 The Optic Nerve

The optic nerve (second cranial nerve) is subject to several pathological processes, such as ischaemic, inflammatory and genetic. Glaucoma, the leading condition, generally caused by high intraocular pressure, is estimated to be present in ~70 million people (0.9%) worldwide. Principal candidate genes include *CYP1B1* at 2p22.2, coding a member of the cytochrome P450 family, *LTBP2* at 14q24.3, and also known as *TIE2*, coding latent TGFβ-binding protein 2, *TEK* at 9p21.2, coding for the angiopoietin-1 receptor, *ANGPT1* at 8q23.1, coding for angiopoietin 1, *FOXC1* at 6p25.3, coding a transcription factor, and *MYOC* at 1q24.3, coding for myocilin, a component of aqueous humour.

With a prevalence of 0.002–0.008%, dominant optic atrophy is linked primarily to *OPA1* at 3q29, coding a GTPase required for mitochondrial membrane stabilisation. The presence of extraocular features such as ataxia or peripheral neuropathy suggests a diagnosis of Behr syndrome. Leber's hereditary optic neuropathy follows from mutations in certain mitochondrial genes (Section 7.11).

7.4.4 Other Ocular Diseases

Age-related macular degeneration has a prevalence of around 9% of global 45 to 85-year-olds, with some 200 million people affected, making it the fourth most common cause of blindness. Most cases are linked to genes coding for complement components (e.g. *CFH* at 1q31.3, coding complement component H), those involved in oxidant stress (*HMOX1* at 22q12.3, coding for haemoxygenase), lipid metabolism (*ABCA1* at 9q31.1, coding an ATP-binding cassette transporter), angiogenesis (*VEGFA* at 6p21.1), and *ARMS2/HTRA1* at 10q26.13, coding a serine protease.

Bardet–Biedl syndrome affects the kidneys, brain, genitals and the eye, where there is retinal rod-cone dystrophy. Over 20 genes are implicated, the leading being *BBS1* at 11q13.2, coding a molecule required for ciliogenesis. It has a frequency between 0.027 and <0.001%, depending on geography. Some two-thirds of cases of aniridia (absence of the iris) are familial, one-third sporadic, and in 90% of cases the causative gene is *PAX6* at 11p13, coding a transcription factor. It has a prevalence of around 0.002%, depending on age. Retinitis pigmentosa, with a worldwide prevalence estimated as 0.275%, can be the product of a mutation in *ROM1* at 11q12.3, coding a component of the rod photoreceptor, and *RHO* at 3q22.1, coding for rhodopsin. The X-linked form of the disease is the product of a mutation in *RHO* at Xp11.4, coding a regulator of a GTPase. There are hundreds of potential candidate genes for this condition.

There are numerous forms of corneal dystrophy. The aetiology of Fuch's corneal dystrophy is of damage to the corneal endothelium: subtypes are of early and late development. The former is linked to mutations in *COL8A2* at 1p34.3, coding for a collagen isotype, and the latter to variants of *SLC4A11* at 20p13, coding sodium bicarbonate transporter member 11, *TCF8/ZEB1* at 10p11.22, coding transcription factor 8 gene, *TCF4* at 18q21.2, coding transcription factor 4 gene, *LOXHD1* at 18q21.1, coding lipoxygenase homology domains 1, and *AGBL1* at 15q25.3, coding an ATP/GTP-binding protein. In Fleck's corneal dystrophy, the candidate gene is *PIP5K3/PIKFYVE* at 2q34, coding a phosphoinositide kinase.

Pseudoxanthoma elasticum is a multi-system disease with a prevalence of 0.002%, caused by a variant in *ABCC6* at 16p13.1, coding an ATP-binding cassette. Its ophthalmological aspects include a mottling of the retina, with angioid streaks and retinal haemorrhage. The leading ophthalmological clinical signs of the blepharophimosis, ptosis and epicanthus inversus syndrome are dysplastic eyelids, lacrimal duct anomalies, strabismus, refractive errors and amblyopia. The leading candidate gene is *FOXL2* at 3q22.3, coding a transcription factor. There is significant genetic heterogeneity in congenital fibrosis of the extraocular muscles, which has five subtypes, linked to mutants of *KIF21A* at 12q12, coding at neuronal transport protein, *PHOX2A* at 11q13.4, coding an enzyme regulator, *TUBB3* at 16q24.3, *TUBB2B* at 6p25.2, *TUBA1A* at 12q13.12, all three coding a tubulin component of microtubules (all autosomal dominant), *ECEL1* at 2q37.1, coding an endopeptidase, *COL25A1* at 4q25, coding the alpha chain of collagen type XXV, and a locus at 21q (all autosomal recessive).

7.4.5 Audiology

Around 80% of pre-lingual hearing loss is genetic, and of these, 30% are linked to other diseases/syndromes such as Alport syndrome, and 70% are non-syndromic. Of the latter, scores of mutations in autosomes have been reported, nine in sex chromosomes. The first autosomal recessive causal gene to be identified was *GJB2* at 13q12.11, coding for connexin 26, and accounts for around a fifth of cases, followed by *STRC* at 15q15.3, coding stereocilia (16%), *SLC26A4* at 7q22.3, coding for pendrin (7%), and *TECTA* at 11q23.3, coding for alpha-tectorin (5% of cases). The leading autosomal dominant gene is *KCNQ4* at 1p34.2, coding part of a potassium channel. A further gene of interest linked to hearing loss is *MT-RNR1*, part-coding a mitochondrial ribosome.

7.5 Foetal and Non-invasive Prenatal Detection

7.5.1 The Foetus and Neonate

Parental genetic testing is called for in cases of recurrent spontaneous abortion and recurrent intra-uterine deaths, with testing of the foetus, where possible. This extends to clear congenital malformations, as discussed in Sections 7.2 and 7.3.5. In England and Wales in 2023, congenital malformations, deformations and chromosomal abnormalities caused 1,516 deaths, 162 in those aged under 1 year, the leading indications being the heart, and both the Edwards and Patau syndromes. Pregnancies at risk include those with a previous child with the particular condition, and parents who are carriers.

Foetal ultrasound is a useful primary investigation for skeletal malformations, neural tube defects, hypo/hyper-organs, nuchal translucency, potential risk of severe sex-linked abnormalities and in instances of consanguinity where a monogenic disorder is considered likely. If such changes are found, or suspected, molecular genetics will be called for. Abnormal biochemical tests, such as alpha-fetoprotein, oestriol and β-hCG, may prompt genetic testing.

There are numerous instances where non-invasive prenatal testing/diagnosis is advised, such as in detecting Down, Edwards or Patau syndromes, which relies on placental (i.e. foetal) cells and so DNA being present in the maternal circulation. In most cases, GWS and microarray analysis is undertaken, whilst CGH may be used for foetal anomalies with a likely genetic cause, but if non-urgent, exon level CNV or MPLA may be required. Selected defined cases where non-invasive prenatal detection (NIPD) is likely to be appropriate are shown in Table 7.4.

Table 7.4 Selected defined cases where NIPD may be appropriate.

Condition	Gene	Location	Product
Congenital adrenal hyperplasia	*CYP21A2*	6p21.33	21-Hydroxylase
Cystic fibrosis	*CFTR*	7q31.2	Cystic fibrosis transmembrane conductance regulator
Apert and Crouzon[a] syndromes, and *FGFR2*-craniosynostosis	*FGFR2*	10q26.13	Fibroblast growth factor receptor 2
Achondroplasia and related skeletal dysplasias	*FGFR3*	4p16.3	Fibroblast growth factor receptor 3
Duchenne and Becker muscular dystrophy	*DMD*	Xp21.2-21.1	Dystrophin
Retinoblastoma	*RB1*	13q14.2	Tumour suppressor Rb1
Inherited monogenic diabetes	*GCK*	7p13	Glucokinase/hexokinase
Ectodermal dysplasia type 1	*EDA*	Xq13.1	Transmembrane protein of the TNF family

[a] Some commentators advocate additional *FEFR3* testing. *Source:* NHS.UK / www.england.nhs.uk/publication/national-test-directories / CC BY 4.0.

Absent from this list are the most common global genetic diseases – the haemoglobinopathies – described in Section 7.7, alongside other non-malignant genetic disease of blood.

7.5.2 The Parent and Adult

Central to the process of NIPD is the parent, and potentially the wider family, where candidate genes may be sought. A further aspect is screening the wider population for certain genes, examples being Tay–Sachs disease in the Ashkenazi Jewish population and haemoglobinopathy in Black individuals, where SCD has a carrier frequency of approximately 1 in 13 African Americans. A pertinent question is of the preferred level of frequency for increasingly rare conditions balanced against the related pathology, especially in high-risk populations, often geographically defined. A useful document is this respect is a resource from the American College of Medical Genetics and Genomics, which has published a list of genes relevant to this topic (Table 7.5), which for clarity shows only those with a frequency ≥0.01%. The theme of genetic screening for defined conditions arises elsewhere in this chapter and volume.

Table 7.5 Autosomal recessive genes in screening for carrier status.

Gene	Frequency	Condition
HBB	0.120	Haemoglobinopathy
XPC	0.051	Xeroderma pigmentosum
TYR	0.049	Oculocutaneous albinism type 1A and 1B
CYP21A2	0.048	Congenital adrenal hyperplasia due to 21-hydroxylase deficiency
PAH	0.046	Phenylketonuria
CFTR	0.041	Cystic fibrosis
TNXB	0.035	Ehlers–Danlos-like syndrome due to tenascin-X deficiency
HEXA	0.033	Tay–Sachs disease
GJB2	0.026	Non-syndromic hearing loss recessive 1A, dominant 3A
DHCR7	0.024	Smith–Lemli–Opitz syndrome
ATP7B	0.022	Wilson disease
ASPA	0.020	Canavan disease
ACADM	0.017	Medium-chain acyl-coenzyme A dehydrogenase deficiency
NPHS1	0.016	Finnish congenital nephrotic syndrome
PMM2	0.016	Carbohydrate-deficient glycoprotein syndrome type Ia
FKTN	0.016	Dilated cardiomyopathy, Walker–Warburg congenital muscular dystrophy
SLC26A4	0.015	Deafness autosomal recessive 4, Pendred syndrome

(Continued)

Table 7.5 (Continued)

Gene	Frequency	Condition
ERCC2	0.015	Cerebro-oculo-facio-skeletal syndrome 2, Trichothiodystrophy 1, photosensitive
DYNC2H1	0.015	Short-rib thoracic dysplasia 3 with or without polydactyly
CEP290	0.014	Joubert syndrome, Leber congenital amaurosis
GBE1	0.014	Glycogen storage disease, type IV, GBE1-related disorders
GAA	0.014	Glycogen storage disease, type II (Pompe disease)
CHRNE	0.014	Myasthenic syndrome, congenital, types 4A and 4B
G6PC	0.013	Glycogen storage disease type IA
OCA2	0.013	Oculocutaneous albinism: brown and type II
COL7A1	0.013	Recessive dystrophic epidermolysis bullosa
ABCC8	0.013	Diabetes mellitus, permanent neonatal type 3
ALDOB	0.012	Hereditary fructosuria
FANCC	0.012	Fanconi anaemia, complementation group C
GRIP1	0.012	Fraser syndrome
BCKDHB	0.012	Maple syrup urine disease
ANO10	0.011	Spinocerebellar ataxia 10
NAGA	0.011	Schindler disease, types 1 and 3
SMPD1	0.01	Niemann–Pick disease, types A and B
USH2A	0.01	Usher syndrome, type 2A
MMUT	0.01	Methylmalonic aciduria–methylmalonyl–CoA mutase deficiency
CPT2	0.01	Carnitine palmitoyltransferase II deficiency
AHI1	0.01	Joubert syndrome 3

Source: Gregg, et al., 2021 / with permission of ELSEVIER.

Key Point 7.3 NIPD is a complex matter, often sitting at the crossroads of pathology and ethics, and thus requires the skills of scientists and counsellors.

7.6 Gastrohepatology

7.6.1 Porphyria

The metabolism of haem is complex, and this section overlaps with Section 7.7.6. As the anabolic aspects of haem occur in the erythroblasts and other red cell precursors, many consider porphyria to be a haematological disease, but its catabolism is principally in the liver, therefore placing it in this group. Several forms are recognised. Acute intermittent porphyria is the product of variants in *HMBS* at 11q23.3, coding for hydroxymethylbilane synthase. Variegate porphyria, which can also manifest with an acute attack, is caused by *PPOX* at 1q23.3, coding for protoporphyrinogen oxidase.

Non-acute porphyria includes a number of clinical conditions. Porphyria cutanea tarda is the consequence of mutation in *UROD* at 1p34.1, coding for uroporphyrinogen decarboxylase. Congenital erythropoietic porphyria (Gunther disease) is caused by a variant of *UROS* at 10q25.2-26.3, coding for uroporphyrinogen III synthase. A deficiency in ferrochelatase leads to erythropoietic protoporphyria, the relevant gene being *FECH* at 18q21.31, and coproporphyria is linked to mutations in *CPOX* at 3q11.2, coding for coproporphyrinogen oxidase.

7.6.2 Cholestasis

The principal causes of cholestasis and the resulting hyperbilirubinaemia and jaundice are obstructive, and most of these are anatomical, physical and inflammatory. Once these are eliminated, molecular pathology remains. Examples of a biliary atresia form of cholestasis, causing neonatal conjugated hyperbilirubinaemia, include those linked to *ADD3* at 10q25.1-25.2, coding for an adducin, and *GPC1* at 2p37.3, coding for a proteoglycan. Another is Alagille syndrome, which has extrahepatic manifestations, and is caused by mutations in *JAG1* at 20p12.2, coding for a ligand that interacts with a NOTCH receptor, or by variants of *NOTCH2* at 1p12, coding for Notch2, a transmembrane receptor whose roles include participation in differentiation pathways.

7.6.3 Polycystic Liver Diseases

Often found alongside polycystic kidney disease, the autosomal dominant liver variant may be linked to several genes. These include *PRKCSH* at 19p13.2, coding for a subunit of glucosidase, *ALG8* at 11q14.1, coding for a glucosyltransferase, *SEC63* at 6q21, coding for a molecule involved in protein transport in the endoplasmic reticulum, and *SEC61B* at 9q22.33, also involved in endoplasmic reticulum function. Involved in both liver and kidney polycystic disease is *LRP5* at 11q13.2, coding the low-density lipoprotein receptor-related protein 5. Variants in *PRKCSH* and *SEC63* are present in 25% of cases.

7.6.4 Pancreatitis

The leading causes of hereditary pancreatitis, which generally manifests acutely, include *PRSS1* at 7q34, coding a serine protease, *CFTR*, *SPINK1* at 5q32, encoding serine protease inhibitor Kazal type 1, CASR, *CTRC* at 1p36.21, coding chymotrypsin C, *CLDN2* at Xq22.3, coding claudin-2 (involved in epithelial tight junctions), *CPA1* at 7q32.2, coding carboxypeptidase A1, *CEL* at 9q34.13, coding carboxyl ester lipase, *CTRB1* and *CTRB2*, located at 16q23.2-23.3, coding chymotrypsinogen B1 and B2, and *TRPV6* at 7q34, coding transient receptor potential cation channel subfamily V member 6.

7.6.5 Hirschsprung Disease

With an incidence of 0.02%, this condition is one of the most common congenital intestinal tract malformations. Although linked to over 20 genes, the most frequent are the proto-onco gene *RET* at 10q11.21, coding for a receptor tyrosine kinase, *EDNRB* at 13q22.3, coding endothelin receptor type B, and *PHOX2B* at 4p13, coding for a transcription factor.

7.7 Haematology

Without doubt, the greatest burden of single inherited genetic diseases are in haematology, with the haemoglobinopathies to the fore, where some 7% of the world's population are carriers, followed by mutations in genes coding for enzymes and for components of the cell membrane.

7.7.1 Haemoglobinopathies

In England and Wales, in 2023, there were 26 deaths due to sickle cell disorders and one caused by thalassaemia. One of the earliest success of molecular genetics was the complete mapping of the human haemoglobin loci on chromosomes 11 and 16. The former carries, from 5' to 3', *HBE1, HBG2, HBG1*, *HBD* and *HBB* at 11p15.4, the latter *HBZ, HBM, HBA1* and *HBA2* (and two) at 16p13.3. Errors in these genes lead to haemoglobinopathy, the leading variants being sickle cell disease and thalassaemia, the former being the first genetic disease to be fully characterised at the protein and genetic level. Disease may also follow from mutations in the locus control region or regulatory elements, located upstream from the globin gene cluster. Numerous genotypic and phenotypic forms of both diseases are known, as are hybrids between the two, described as compound haemoglobinopathy. Global data from 2010 on the frequencies of 340,000 births of major haemoglobin disorders are shown in Table 7.6.

Sickle Cell Disease

According to the Global Burden of Disease Sickle Cell Disease Collaborators, in 2021, 7.74 million people (0.098%) were living with SCD, and there were 515,000 births with a total SCD mortality burden of 376,000. This figure includes 81,100 children aged younger than 5, which ranks 12th in all-cause mortality. This burden falls principally on

Table 7.6 Global frequences of major haemoglobinopathies.

Disorder	Frequency
Homozygous (SS) sickle cell disease	63.9%
HbSC	16.1%
β-Thalassaemia major	6.8%
HbE/β-thalassaemia	5.6%
HbS/β-thalassaemia	3.3%
HbH	2.8%
Hb Bart's hydrops (α^0/α^0)	1.5%

sub-Saharan Africa, with 5.68 million cases, and some nations seeing SCD incidence at birth of 1–2%. In this sub-continent, 300,000 or more children are born with SCD each year, and, sadly, around 75% face mortality by age 5. The heterozygous state of HbAS confers a reduction of 86% in the likelihood of severe malaria with an odds ratio of 0.14 (95% confidence interval 0.12–0.16) at a significance levels of 1.6×10^{-225}. Modelling the balance of this heterozygote advantage, given these frequencies today, NGS has shown the mutation has a single origin, arising some 7300 years ago in West Africa and spreading to other locations thereafter.

The principal cause of the disease is SNP rs334 in *HBB* leading to a change in codon GAG to GTG, with a normal glutamine being replaced with valine. Other SNPs lead to HbC, where glutamine is replaced by lysine at position 6 in the amino-acid chain, to HbD, a glutamine to glutamic acid transposition at position 121, and to HbE, where glutamine is replaced by lysine at position 26. The phenotype of SCD/sickle cell anaemia (SCA) may also arise from other genetic lesions in the β-globin locus, the leading being deletions and inversions, all generating to an incompetent haemoglobin. Genes known to be involved in HBB regulation include *CCND3* at 6p21.1, coding cyclin D3, *CTSA* at 20q13.12, coding cathepsin A, *ERCC2*, ZFPM1 at 16q24.2, coding FOG1, a cofactor for transcription factor coded by *GATA1*. Others include *GATA2, KLF1* at 19p13.13, coding an erythroid transcription factor, *KLF3* at 4p14, coding a transcription factor, *PCIF1* at 20q13.12, coding an adenosine methyltransferase, *PLTP* at 20q13.12, coding a phospholipid transfer protein, *MMP9* at 20q13.12, coding matrix metalloproteinase 9, *NFE2* at 12q13.13, coding a transcription factor, *SOX6* at 11p15.2, coding a transcription factor linked to SRY, *SP1* at 12q13.13, coding a transcription factor, and *TNNC2* at 20q13.12, coding troponin C.

Several other genes interact with those of the haemoglobins that have an impact into the haemoglobinopathy phenotype, the clinical course of SCA being high variable. Perhaps the most studied is residual expression of HbF, increased levels of which reduce the severity of the disease such that an effective method for increased levels is a much-sought-after clinical tool. GWAS has shown that variants in *BCL11A* (at 2p16.1, coding a potent transcription repressor for *HBG1* and *HBG2*), in the 6q intergenic region of *HBS1L::MYB*, and in the extended β-globin locus, have a 20–50% variation effect in levels of HbF in SCA patients and in healthy

European Caucasians. Further sites of interest are rs7482144, the Xmnl site of *HBG2* and *KLF1*, whilst *SUPT5H* haploinsufficiency reduces *HBB* expression. Commercial panels for most of these genes are commercially available, one platform being that of Illumina Ampliseq. Genes with influences on clinical aspects of SCD/SCA are shown in Table 7.7

Thalassaemia

According to the Global Burden of Disease project, in 2021, 1.31 million people were living with thalassaemia, with 120,000 births, and 11,100 deaths, of whom 4200 were children. There were a total of 5.38 million people born with thalassaemia trait in 2021 (0.07% of the global population) and almost 360 million (4.6%) living with thalassemia trait.

In contrast to SCD, the leading genetic lesions in thalassaemia, established over 40 years ago, are deletions, although other lesions such as insertion/deletion and duplications are known. Amongst these are a 4.2 kb section, a 20.5 kb section, and geographically specific variants

Table 7.7 Genes influencing sickle cell disease phenotype.

Gene	Clinical link
UGT1A1 promoter repeats	Cholelithiasis
RPS24, IL1A, TBC1D1, COMT	Vaso-occlusive disease
COMT, NR3C1, TBC1D1, IL1A, KCNJ6, FAAH9	Paediatric veno-occlusive pain
APOL1 variants, *BCL11A, HMOX1, MYH9*	Renal dysfunction
VCAM1, IL4R, TNF-α, LDLR, ADRB2, ATG, GOLGBI, ENPPI, PON1, TEK	Stroke
IL-1β, MTHFR, BMP6	Osteonecrosis
NOS3, NOS1, COMMD7, HMOX1	Acute chest syndrome
KL, TGFβ/SMAD pathway genes	Priapism
IL-1β, TGFβ/BPM pathway genes (*ACVRL1, BMPR2,* and *BMP6*), *MAPK8, eNOS*	Pulmonary hypertension
KL, TEK, TGFβ/BPM pathway genes	Leg ulcers
TGFβ/BPM pathway genes, *MBL2, MPO, DARC*	Bacteraemia
APOE, TNFα, GOLGB1, PON1	Large vessel disease

such as SEA, THAI (α^0-thalassaemias) and FIL (β^0-thalassaemia), often of 30–38 Kb and which may also delete *HBG2*. However, the most common variant causing α-thalassaemia is a 3.7 kb deletion in a section of *HBA*, whilst deletions in *HBB* causing β-thalassaemia vary from relatively small (~1 kb) to extensive (over 400 kb). The latter generally lead to two types: the complete (β^0-thalassaemia) and partial (β^+-thalassaemia, perhaps 5–30% of normal) absence of globin protein. TGS such as PacBio is particularly valuable in investigating rare and unusual thalassemia genotypes, whilst Gap-PCR and PCR reverse dot hybridisation can be used for screening, and in many cases, commercial primers are available.

Compound Haemoglobinopathy

As shown in Table 7.1, there are numerous relatively common cases where two gene lesions are inherited, and of these, 76.5% of births carry HBSS and HbS/β^0, 19.6% HbSC and 3.9% HbS/β^+. A study from the USA reported on 1882 births, in which 56% had HbSS or HbS/β^0, 28% had HbSC and 10% had HbS/β^+ or another SCD type.

Intuitively, co-inheritance brings worse outcome, and in many cases this is correct, but combined α-thalassaemia and SCD is beneficial in that levels of bilirubin are not as high, implying less haemolysis (likely to be linked to reduced HbS polymerisation) and/or liver pathology. This extends to a reduced incidence of numerous clinical events such as cardiomyopathy, renal dysfunction, stroke, cholelithiasis and stroke. The presence of a SNP in *ADCY9* at 16p13.3, coding adenylate cyclase 9, alongside α-thalassaemia in paediatric SCA is linked to decreased stroke risk. However, there are undesired effects of this α-thalassaemia and SCD co-inheritance, such as a blunted response to hydroxyurea, an important treatment that increases levels of HbF.

Routine Molecular Pathology of the Hemoglobinopathies

The high degree of awareness of SCD and thalassaemia is such that routine molecular pathology is likely to be called upon at an early stage to define the precise nature of the genetic lesion. This is because other methods such as gel-based and capillary electrophoresis, HPLC and isoelectric focusing are limited by their inability to reliably define the genetic lesion, such as HbSS from HbS/β^0-thalassaemia. NGS and TGS are adept at determining uniparental isodisomy, where two copies of a gene are inherited from one parent, and none from the other, and of investigating mosaicism, both particularly important in the haemoglobinopathies. A study of 240 children from Kuwait reported a considerable discordance between HPLC and the Illumina Ampliseq NGS method, finding 184 cases of HbSS by HLPC and 159 by NGS, with 54 and 68 cases of HbS/β-thalassaemia, respectively. A comparison of certain features of molecular pathology with respect to haemoglobinopathy is shown in Table 7.8.

Apart from diagnosis and management, a further aspect of molecular pathology of the haemoglobinopathies is of preconception, antenatal and neonatal screening programs (Section 7.5.2). A utopian view is that, using these methods, the disease can be all but eliminated, whilst a pragmatic view seeks to educate and reduce their incidence. There is considerable literature on this subject on the value of different technologies, and most population screening programs for routine haematology metrics, turning to molecular pathology on those strongly suspected of harbouring a hemoglobinopathy.

Table 7.8 Molecular pathology of haemoglobinopathy.

	Conventional methods	**Next-generation sequencing**	**Third-generation sequencing**
Basic techniques	Routine FBC and morphology, electrophoresis, HPLC	Routine FBC and morphology	Routine FBC and morphology
Molecular pathology	Sanger sequencing, ARMS-PCR, Gap-PCR, MPLA.	DNA extraction, library construction, pyrosequencing, Ion Torrent, Illumina, SOLiD, etc.	DNA extraction, single molecule real-time sequencing etc. (PacBio, Oxford nanopore)
Costs	Low	Moderate	High
Turnaround time	Short	Long	Moderate
Accuracy rate	Relatively low	Moderate	High

Source: Zhan L, et al., 2023 / Frontiers Media S.A. / CC BY 4.0.

Self-check 7.3 How would you describe the key differences in the molecular pathology of SCD and thalassaemia?

7.7.2 Red Cell Enzymopathies

An issue with this section is the lack of good data from studies using molecular genetics. Many cases are defined by biochemical and other methods, such that interpretation is difficult. A common model is to screen using a non-DNA method, moving to DNA methods in those deemed positive by those other techniques. Two enzymopathies dominate this section.

Glucose-6-phosphate Dehydrogenase Deficiency

Glucose-6-phosphate dehydrogenase (G6PD, E.C. 1.1.1.49), coded for by *G6PD* at Xq28, is a key component of the cytosolic pentose phosphate pathway. It was sequenced by the Maxam/Gilbert method from cDNA clones and a 17mer probe deduced from an amino acid sequence in 1986. Its deficiency, hence G6PDD, results in failure to generate NADPH + H^+, in part required for defence from oxidants, one consequence of which is a haemolytic anaemia, often precipitated by ingestion of fava beans and a variety of pharmaceuticals.

G6PDD is the most common enzymopathy, affecting over 400 million people worldwide (5.8%), and is linked to 33,000 deaths, and so greatly exceeding the prevalence of hemoglobinopathies, but not their mortality. However, the larger proportion of this data will have been formed from enzymatic and other methods: in one analysis using only a DNA method, the prevalence estimate was 11%. DNA-defined frequencies vary markedly from 1–7% in parts of Italy, 2–16% in southern China and Tiawan, and up to 30% in parts of Africa. A study from 1988 reported various SNPs accounting for reduced enzyme activity, rising to 60 mutants by 1994, and to 230 by 2024. A study from China found 99 instances of any one of eight gene mutations in 120 children with G6PDD defined by another method, although this theoretical false positive rate of 17.5% may not be correct as other mutants may have been present. Similarly, in a study of 87 Malaysians who tested positive for G6PDD by non-DNA methods, 69 (79.3%) carried one of seven G6PD mutations, that being 10.4% of all 662 subjects in the study. These studies underline the care required of interpretation, as biochemical phenotype may not match genotype.

Pyruvate Kinase Deficiency

Pyruvate kinase (PK, EC 2.7.1.40), coded for by *PKLR* at 1q22, catalyses the transfer of a phosphate group to ADP, thereby generating ATP. Deficiency of the enzyme (hence PKD) therefore results in lack of ATP, the leading source of energy for the red cell, and thus haemolytic anaemia. Long defined by a biochemical method (as with G6PD), in 1988, *PKLR* was cloned from a liver cDNA library and sequenced by Sanger's method.

As with estimates of the frequency of G6PDD, that of PKD was first defined enzymatically, with heterozygous rates of 1.09% and 1.4% in a white population. SNPs and deletions in *PKLR* were subsequently reported, whilst a gene frequency study gave an estimate of 0.055% in a white population, although others have reported frequencies of 0.005% and <0.001%. By 2005, the number of mutations had risen to 158, with focus on the European nations and the USA. The parallel with G6PDD extends to the mutation variant – a study from Mozambique found missense mutation G829A to be present in 41.7% of PK deficient isolates, and to be present at frequencies between 1.3% and 3.4% in other sub-Saharan countries.

7.7.3 Combined Haemoglobinopathies and Enzymopathies

Since there are numerous cases of combined haemoglobinopathy, it comes as no surprise that there is also co-inheritance of other genetic diseases. A small non-DNA study from Yemen of 31 SCD patients found 7 (26%) patients to also have G6PDD, although the latter was reported as having no effect on the clinical course of the former. A small study from Brazil reported 6 of 37 men (16%) with SCD also had the G6PD 202G>A mutation, compared to 3 of 37 controls (9%), whilst equivalent data in women was 16% and 24%, respectively. In a study of 810 samples from India, 9.75% were suffering from SCA defined by non-DNA methods, but using RT-PCR for three variants, G6PDD was present in 17.2%, whilst 67% (53 cases) of those with SCA (79 cases) also tested positive for G6PDD. The combination of combined SCD and G6PDD probably leads to an increased risk of more severe anaemia and possible increased risk of cerebrovascular disease.

> **Key Point 7.4** The large number of mutants present in genes coding for red cell enzymes G6PD and PK frustrates attempts to correctly assess population frequencies and phenotypes.

7.7.4 The Red Cell Membrane

We conclude our exploration of haemolytic anaemia by looking at disease of the membrane of the erythrocytes that bring about their destruction. The geographical concentrations of several support the hypothesis that they have developed in response to malaria.

Hereditary spherocytosis is the most common hereditary haemolytic anaemia worldwide, and in Caucasians, it has a prevalence of 0.02–0.04%, with some 75% being autosomal dominant. Others include hereditary elliptocytosis (with an incidence of 0.6–3% in West Africa) and hereditary pyropoikilocytosis (also frequently diagnosed in infants of African ancestry). Leading lesions in these conditions include those in *ANK1* at 8p11.21, coding for ankyrin (50–60% of cases), in *SPTA1* at 1q23.1 or *SPTB* at 14q23.3, coding for α- or β-spectrin (together 20%), *SLC4A1* at 17q21.31, coding for band 3 (15–20%), and in *EPB42* at 15q15.2, coding for protein 4.2. Southeast Asian ovalocytosis (principally found in and nearby Malaysia) is also caused by a deletion in *SLC4A1*, which confers a resistance to cerebral malaria, with a mean allelic frequency of ≤1%.

Hereditary stomatocytosis has a dehydrated form, linked to *PIEZO1*, at 16q24.2, and coding a mechanosensitive ion channel, and an overhydrated form, linked to *RHAG* at 6p12.2, coding for Rh-associated glycoprotein (CD241), which functions as an ammonia transporter. The Gardos channelopathy is associated with *KCNN4* at 19q13.31, coding a calcium-activated potassium channel.

7.7.5 Other Forms of Anaemia

Leaving aside the anaemias described earlier, there are other forms of inherited disease of this nature. Most (75%) cases of Diamond–Blackfan anaemia, presenting in around 6 per million live births, are due to mutations in any of a dozen genes coding for ribosomal proteins, such as *RSP19* at 19q13.2 and *TSR2* at Xp11.22 (although some consider the latter to be a non-ribosomal). Other candidate genes include *GATA1*,

TEATR3 at 16q12.1, *EPO* at 7q22.1, coding for erythropoietin, and *ADA2/CECR1* at 22q11, coding for adenosine deaminase 2.

Fanconi anaemia, resulting from bone marrow failure (and so a potential thrombocytopenia), is part of a heterologous phenotype that includes short stature, microcephaly and abnormal skin pigmentation. With an incidence of <0.001%, it has a prevalence of 1–9 per million. Abnormalities in over 20 genes have been reported, many collected together into the FANC family, with variants in *FANCA* at 16q24.3, *FANCB* at Xp22.2 (also known as *FAAp95*), *FANCC* at 9q22.32, *FANCD1* at 13q13.1 (also known as *BRCA2*), *FANCD2* at 3p25.3, *FANCE* at 6p21.31, *FANCF* at 11p14.3, *FANCG* at 9p13.3 (also known as *XRCC9*) and others. The products of these genes, alongside others, form a complex that recognises and repairs interstrand cross-links. Mutations in these genes therefore lead to failure to repair damaged DNA and so loss of function. Mutations in *FANCA, FANCC* and *FANGG* account for around 85% of cases.

This section overlaps with Section 7.6.1 on porphyria, the cause of other forms of anaemia. *TMPRSS6* at 22q12.3 codes for a transmembrane serine proteinase, also known as matriptase-2, mutations lead to an iron-refractory ion-deficient anaemia. Anaemia may also arise from mutation in many of the genes whose products synthesise haem, of which there are many examples (Table 7.9). Failure of these genes to produce their product leads to conditions such as sideroblastic anaemia and erythropoietic protoporphyria. Figure 7.2 summarises major genetic disease of the red blood cell.

Table 7.9 Genes involved in haem synthesis.

Gene	Location	Product
SLC19A1	21q22.3	A folate transporter
SLC2A38	3p22.2	A glycine carrier
ALAS2	Xp11.21	Aminolaevulinic acid synthase
ABCB6	2q35	Transporter of coproporphyrinogen III
FXN[a]	9q21.11	Frataxin: stores and transports iron
FLVCR1	1q32.3	Choline and haem transporter
GLRX5	14q32.13	Glutaredoxin 5
ABCB7	Xq13.3	Regulator of iron/sulphur complexes

See also Section 7.6.1. [a] See Section 7.13.

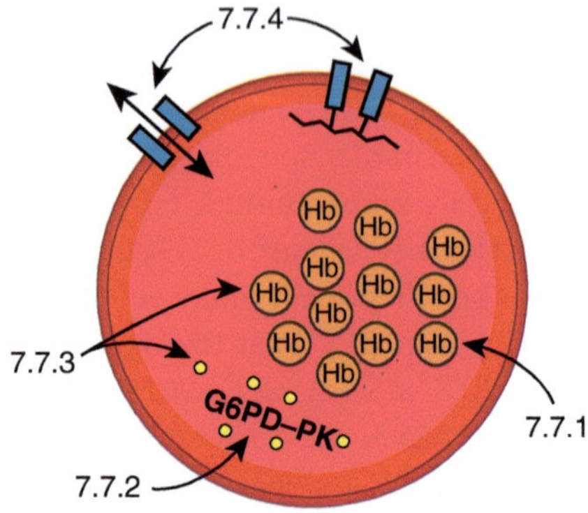

Figure 7.2 Molecular pathology of the red blood cell. At the top, section 7.7.4 refers to defects in the cytoskeleton and electrolyte channels; on the lower left, section 7.7.2 considers enzymopathies, whilst at the lower right, section 7.7.1 describes haemoglobinopathy. 7.7.3 refers to abnormalities in both haemoglobinopathies and enzymopathies.

7.7.6 Haemorrhagic Diseases

Diseases of the coagulation pathway may be classified as those causing haemorrhage (inherited disorders reported as present in 7.5 million people, 0.05% of the global population) and those that predispose to thrombosis, with sub-groups of those affecting coagulation factors and those affective platelets. However, certain diseases cause excessive bleeding (haemorrhage) because of abnormal blood vessels, an example being hereditary haemorrhagic telangiectasia. Over 80% of cases are caused by mutations in *ENG* at 9q34.11, coding for endoglin, and *ACVRL1/ALK1* at 12q13.13, coding for a receptor for TGFβ. Telangiectasia is also a feature of the clinically heterologous Bloom syndrome, caused by mutations in *BLM* at 15q26.1, coding a helicase, potentially *TOP3A* at 17p11.2, coding a topoisomerase, and *RMI1* at 9q21.32 and *RM12* at 16p13.13, both coding molecules cooperating with topoisomerase. Telangiectasias are also a feature of the neuromuscular disease ataxia telangiectasia, discussed in Section 7.13. Of further note are petechia, small haemorrhagic lesions generally resulting from capillary malformation or damage.

Haemophilia

Without doubt, the leading disease of this group is haemophilia, often described as haemophilia A, which is long recognised as having a heritable component, but it may develop spontaneously. Generally reported with a prevalence of 0.02–0.01% of live male births, the gene *F8*, at Xq28, being cloned and sequenced in 1984, is particularly large at 187 kb with 26 exons and 25 introns.

The consequence of this large gene is that it is susceptible to mutations, with over 2000 variants known, many of which cause the potentially life-threatening haemorrhage. Work from the 1980s and 1990s used restriction fragment length polymorphisms, relying on SNPs in introns 18, 19, 22 and 25 to abolish restriction endonuclease sites. More recent methods have identified other lesions, the most common (45%) of these being an intron 22 inversion, whilst an inversion in intron 1 is present in 3.5% of cases. SNPs include 1139A>G (Asp380Gly) in exon 8, 1313T>C (Ile438Thr) in exon 9, 5706T>G (Phe1902Cys) in exon 17 and 6406C>T (Arg2016Trp) in exon 19. These and other mutations are often used in diagnosis of a suspected proband, in carriers and in prenatal diagnosis. In a large study of 2401 haemophiliacs, NGS methods detected a clinically reportable lesion in 98.1%, and no potentially causative DNA variant was being found in the remainder.

An important clinical issue in haemophilia is the development of inhibitors to replacement factor VIII. Although inhibition is independent of the particular genetic lesion, there is a place for *HLA DRB1*15;01* and for *GRID2IP* at 4q22.1-22.2, coding for a glutamate receptor.

Factor IX Deficiency

Deficiency of factor IX (EC 3.4.21.22 when activated), described in 1952 as Christmas disease, or haemophilia B, is present in the UK at a frequency of 0.003% and in the USA at 0.002%. The gene, *F9*, located at Xq27.1, was cloned and sequenced from a 1.25 kb cDNA derived from liver mRNA by the Maxam–Gilbert method in 1982. It was subsequently shown to contain 8 exons spanning around 34 kb, with other analyses, such as probe hybridisation, demonstrating deletions (some of the entire gene) and SNPs that led to *F9* deficient haemorrhage. A recent review reported over 1000 unique variants, of which 73% were SNPs, 16.3% deletions, 3.3% insertions, 1.5% indels and 0.4% duplications. A study from Sweden of 77 patients with haemophilia B found 4 total deletions, 3 exon deletions, 6 partial deletions (<10 bp) and 64 SNPs. However, approximately 50% of cases are sporadic. Perhaps intuitively, a mutation linked to haemorrhage should protect from thrombosis, which is indeed the case, as the G allele in an A>G SNP in rs6048 in *F9* brings a 15–43% decreased risk of deep vein thrombosis.

Other Factor Deficiencies

Deficiencies in each of the remaining coagulation factors have been described, such as of *F7* at 13q34 and of *F13A/F13B* at 6q25.1 and 1q31.3, respectively, both factors being deficient at a frequency of around 2 per million. Deficiency in Factor V coded for by *F5* at 1q24.2 and Factor X deficiency, caused by mutations in *F10* at 13q34, each has a frequency of around 1 per million. Around 200 cases of combined Factors V and VIII deficiency have been reported, the most common mutations being in *LMAN1* at 18q21.32, and *MCFD2* at 2p21. In certain circumstances, molecular pathology is often required to determine if the deficiency is genetic or acquired.

Von Willebrand Factor

The most frequent coagulation molecule deficiency is in von Willebrand factor (vWf), coded for by *vWf* at 12p13.31 with an overall frequency of 1%. However, there is marked variation in the clinical forms of the deficiency, arising from the large size of the gene (178 kb, 52 exons) and multi-domain nature of the protein product, frequencies ranging from asymptomatic (the most common, at 65–80% of cases) to severe and often life-threating, present at 1 per million. Molecular pathology can help define the precise nature of the genetic lesion.

> **Self-check 7.4** Discuss the proposition that the majority of global genetic disease is haematological.

Platelet Disorders

Haemorrhage due to the thrombocytopenia of cancer and/or chemotherapy has been described in Chapters 4, 5 and 6, but there are several other potential causes:

- Thrombocytopenia with absent radius (the forearm bone) is often caused by mutations in *RBM8A* at 1q21.1, coding an RNA-binding protein. Decreased number of megakaryocytes may be found, and there are numerous other extraskeletal features.

- Thrombocytopenia may be present in Schwachman–Diamond syndrome, with 90% of patients having biallelic mutations in *SBDS* at 7q11, coding for a ribosome maturation factor.
- Mutations in *MYH9*, at 22q12.3, coding for a variant of myosin, are linked not only to thrombocytopenia but also to large platelets. This mutation may also be a part of the May–Hegglin anomaly and other syndromes such as that of Epstein and Fechtner.
- Congenital amegakaryoctic thrombocytopenia results from loss-of-function mutations in thrombopoietin (coded for by *THPO* at 3q27.1) or its receptor (coded for by *MPL* at 1p34.2), genes also linked to myelofibrosis and essential thrombocythaemia (Section 6.4).
- Bernard–Soulier syndrome may be caused by failure of *GP9* (at 3q21.3), *GP1BA* (at 17p13.2), *GP1BB* (at 22q11.21) or *GP5* (at 3q29) to correctly code molecules that form the GpIb-IX-V complex, the receptor for vWf. Failure of this receptor/ligand reaction results in platelets not able to form a thrombus, and so haemorrhage. It has a prevalence of less than 1 per million.
- Wiskott–Aldrich syndrome is characterised not only by thrombocytopenia and small platelets but also by immunodeficiency and eczema. The disease is caused by mutations in *WAS* at Xp11.23, whose product has multiple roles in signal transduction and the early stages of actin polymerisation in haemopoietic stem cells.

The platelet count may be normal, but haemorrhage may develop because of pathology of platelet function, two of which are the most well developed. WES and RNA profiling have showed that grey platelet syndrome is caused by mutations in *NBEAL2* at 3p21.31, coding for a protein required for alpha-granule synthesis. It is present at approximately 0.5 per million. Glanzmann thrombasthenia has an incidence of 1 per million, but in areas of high consanguinity this may increase fivefold. It is caused by mutations in *ITGA2B*, at 17q21.31, or in *ITGB3*, at 17q21.32, that together code for gpIIb/IIIa, the fibrinogen receptor.

7.7.7 Thrombotic Diseases

As with haemorrhage, we consider abnormalities in coagulation factors and in platelets. However, inherited platelet disorders resulting in thrombosis are rare, one being that both *MPL* and *THPO* gain-of-function mutations are a source of familial thrombocytosis.

The expression *thrombophilia* which translates to 'clot-loving' was developed as an umbrella term to bring together the increasing number of abnormalities in the general thrombosis arena. Although there are acquired forms of thrombophilia, as regards inheritance, it can be caused by loss of function of inhibitors and by gain of function of coagulation molecules.

Loss of Function of Coagulation Inhibitors

These include protein S deficiency, caused by a mutation in *PROS1* at 3q11.1, antithrombin deficiency, linked to SERPINC1 at 1q25.1, and protein C deficiency, caused by variants of *PROC* at 2q14.3. Both qualitative and quantitative defects in all three are known, and these defects allow the coagulation pathway to be incorrectly activated, and so promote thrombosis. Alterations in these molecules bring an increased risk of venous thrombosis (Table 7.10).

Table 7.10 Prevalence and relative risk data for the classical inherited thrombophilias.

Inherited thrombophilia	Prevalence in the general population (%)	Prevalence in the VTE population (%)	Relative risk for a first VTE (%)
Antithrombin deficiency	0.02–0.2	1–5	5–10
Protein C deficiency	0.2–0.4	2–5	5–8
Protein S deficiency	0.03–0.13	1–3	2–8
FVL heterozygous	3–15	12–20	3–8
FVL homozygous	0.004–0.02	0.1–1.5	40–80
F2 G20210A heterozygous	1–3	3–8	1.9–3.8
F2 G20210A homozygous	0.1	2.0	2.2–20.7
FVL and *F2* G20210	0.01–0.1	3.0	20–58.6

Source: Adapted From Moore (2024).

Gain in Function of Coagulation Factors

Two molecules dominate this section. Factor V Leiden (FVL) is a variant of normal factor V that is resistant to the inhibitor protein C and is caused by a mutation in *F5*, whilst mutation G20210A in *F2* at 11p11.2 results in an increased level of prothrombin. Increased levels of factors VII, IX and XI are linked to an increased risk of thrombosis, but it is unclear whether any of these increases are under genetic control. Table 7.10 shows the increased risk of venous thrombosis brought by these abnormalities.

7.7.8 Iron Metabolism

The leading disease in this group is hereditary haemochromatosis, suspected in the presence of abnormalities in blood markers such as transferrin saturation and ferritin, and characterised by increased levels of iron, deposited in various organs. Several types are described, although different commentators list these in alternative forms. The leading (>80%) causative gene is a *C282Y* SNP in *HFE* at 6p22.2, coding for a membrane receptor with roles in iron metabolism. A second mutation in *HFE* forms variant H63D. Compounds of the two variants have been described as a third, producing S65C.

Juvenile forms of the disease are caused by variants of *HJV* (or *HFE2*) at 1q21, coding for a molecule linked to the membrane receptor for bone morphogenetic protein (BMP), and by *HAMP* at 19q31, coding for the primary iron regulator molecule hepcidin. BMP6 itself, coded for by *BMP6* at 6p24, is a further component of the regulation of hepcidin.

Loss of function of *FPN* (or *SCL40A1*) at 2q32.2 leads to reduced levels of its product – ferroportin. Mutation in *TfR2* at 7q22.1 results in loss of transferrin receptor 2, whilst *SCL11A2* at 12q13.12 codes for divalent cation transporter 1, a molecule responsible for transporting iron from the intestines into the enterocyte. Although involved in copper metabolism, variants in *CP* at 3q24-25.1, coding for caeruloplasmin, also lead to iron overload.

Key Point 7.5 Despite the fact that *FVL* brings an established risk of venous thrombosis, the corollary of the very high incidence (heterozygote frequency: ~5%) in the general Caucasian population demands that there must be an evolutionary advantage, the most likely being a reduced risk of blood loss after childbirth.

7.8 Immunology

7.8.1 Severe Combined Immunodeficiency

One of the first major immunodeficiency diseases to be well characterised, severe combined immunodeficiency (SCID), is now recognised as having several forms. These are those caused by adenosine deaminase deficiency, resulting from lost-of-function variants of *ADA* at 20q13.12, a sex-linked form caused by a mutation of the gamma-chain of the IL-2 receptor, coded by *IL2RG* at Xq13.1, and by a variant of *PNP* at 14q11.2, coding for purine-nucleoside phosphorylase. Other forms are linked to *RAG1* and *RAG2*, both at 11p12, the alpha chain of *IL7R* at 5p13.2, *JAK3* at 19p13.11, and in complement components.

7.8.2 The Haemophagocytic Syndromes

These diseases, some of which are described as haemophagocytic lymphohistiocytosis, are characterised by excessive and persistent immune activation of T lymphocytes, natural killer (NK) cells, macrophages and histiocytes. The familial form can be linked to a number of genes, such as *PRF1* at 10q22.1, coding for perforin 1, *STX11* at 6q24.2, coding for syntaxin 11, *STXBP2* at 19p13.2, coding for syntaxin-binding protein 2, and *UNC13D* at 17q25.3, coding for Munc13-4. An X-linked variant follows from an abnormality in *SH2D1A* at Xq25, which codes for a signalling lymphocyte activation molecule. The same chromosomal locus is the location of *XIAP*, which codes for a protein that binds caspases and thus inhibits apoptosis.

7.8.3 Hereditary Angioedema

With an overall global frequency of around 0.001%, hereditary angioedema (HAE) has three major forms, comprising 80–85% of all cases. The most frequent are those with a quantitative (type 1) or qualitative (type 2) defect in C1-esterase inhibitor caused by a mutation in *SERPING1* at 11q12.1. Type 3 disease results from an increased production of bradykinin, itself from a FXII (coded by *F12* at 5q35.3) with abnormal sensitivity to plasma. The syndrome may also arise from hereditary alpha tryptasaemia, 95% of cases being linked to a D816V mutation in *KIT*. Other genes linked to HAE include *PLG* at 6q26, coding for plasminogen,

ANGPT1 at 8q23.1, coding for angiopoietin 1, *KNG1* at 3q27.3, coding for kininogen 1, *MYOF* at 10q23, coding for myoferlin, and *HS3ST6*, coding for heparan sulfate-glucosamine 3-sulfotransferase 6.

7.8.4 Other Immunological Diseases

Autoimmune polyendocrine syndromes result from a defect in *AIRE* at 21q22.3, coding an autoimmune regulator protein. The closely related variant, IPEX (described in 7.3.1), has an enteropathy and is linked to mutations in *FOXP3* at Xp11.23, coding a transcription regulator. It may also be described in the young as monogenic inflammatory bowel disease.

Lymphoproliferative syndromes include those with absent SAP (SLAM-associated protein) expression, coded for by *SH2D1A* at Xq25, those with a defective *FAS* at 10q23.31, coding CD95, the FAS ('death') receptor that promotes apoptosis. Other genes linked to this syndrome include the FAS ligand, coded for by *FASLG* at 1q24.3, and *CASP10* at 2q33.1, coding for caspase 10.

X-linked agammaglobulinaemia is the most common condition of B-lymphocyte maturational arrest, with an incidence of ≤0.001%, resulting from a mutation in *BTK*, also linked to B-lymphocyte disease such as certain lymphomas and myelomas. Autosomal agammaglobulinaemia is most commonly linked to mutations in immunoglobulin genes, *BLNK* at 10q24.1, coding a molecule with roles in B cell signalling, genes of the *PI3K* family, located at various sites and together coding for phosphoinositide-3-kinases, and in *TCF3* at 19p13.3, coding for a transcription factor. Immunodeficiency may also be present in Wiskott–Aldrich syndrome (Section 7.7.4).

7.9 Lipid and Metabolic Diseases

7.9.1 Lipids

Familial Hypercholesterolaemia

As befits one of the leading genetic diseases predisposing to cardiovascular disease, familial hypercholesterolaemia (FH), with a heterozygous frequency estimated at 0.4%, is well known to the scientific community and, to a lesser extent, to the general population. The leading cause is mutations in *LDLR* at 19p13.2, coding the receptor for low-density lipoprotein cholesterol (LDL). Mutations in *APOB* at 2p24.1, coding for apolipoprotein B (and important component of many lipid particles) account for around 5% of FH cases, whilst an abnormality in *PCSK9* at 1p32.3, coding for PCSK9, which helps degrade the LDLR, is present in <1% of cases.

Other Lipid Diseases

Other mutations linked to lipid disease include those *LPL* at 8p21.3, coding for lipoprotein lipase, deficiency in which leads to familial chylomicronaemia syndrome, and *LDLRAP1* at 1p36.11, coding for an adaptor protein for the cytoplasmic domain of the LDLR. There may also be a role for *LMF1* at 16p13.3, coding for lipase maturation factor 1, and for *GPIHBP1* at 8q24.3, coding for glycosylphosphatidylinositol-anchored high-density lipoprotein-binding protein 1. A family of 12 proteins that are related to the LDL receptor have a number of roles in various pathologies. One of these, *LRP8*, at 1p32.3 (close to *PCSK9*), codes for the apolipoprotein E receptor 2, whose ligand is ApoE, itself coded for by *APOE* at 19q13.32, and which may have a role in Alzheimer's disease.

Hypercholesterolaemia may also be found in sitosterolaemia, linked to *ABCG5* and *ABCG8*, both at 2p21, coding for subunits of an ATP-binding cassette, mutations in which result in hyperabsorption of dietary sterols, such as sistosterol. Smith–Lemli–Opitz syndrome results from a mutation in *DHCR7* at 11q13.4, coding for 7-dehydrocholesterol reductase, leading to high levels of its substrate 7-dehydrocholesterol and so low levels of its product, serum cholesterol.

Key Point 7.6 Screening for FH is a common practice and often prompted by measurement of serum cholesterol, itself frequently in response to a family report or cardiovascular event. In the UK, documents CG71, QS41, IND203, IND204 and others from the National Institute for Health and Care Excellence are dedicated to this condition.

7.9.2 Metabolic Diseases

Often described as inborn errors of metabolism (IEM), this group includes a very heterologous family of abnormalities, many of which fall within a particular subtype, and many of which involve enzymes. Globally, they are estimated as being present in 509 per million live births

(0.051%, 28.9% of all IEMs). The clinical aspects of these conditions may arise from an excess of the substrate and/or lack of the product. Classification can be by biochemical nature, clinical features or cell biology, and some fall into more than one category.

Disease of Amino Acids

Globally, this is the largest group of IEMs, with a frequency of 0.015%; phenylketonuria, presenting at a rate of ~0.007%, is the most common. It results from loss of function of phenylalanine hydroxylase, coded by *PAH* at 12q23.2. Reputedly, the second most common is glycine encephalopathy, with an incidence of 0.002–0.008%, resulting mostly (70–75%) from a variant of *GLDC* at 9p24.1, coding for glycine hydrogenase, leading to hyperglycinaemia.

Leucine is catabolised by iso-valeryl-CoA dehydrogenase, coded by *IVD* at 15q15.1, lesions which lead to isovaleric acidaemia. Four enzymes, *GLDC*, *BCKDHB* at 6q14.1, *DBT* at 1p21.2, and *DLD* at 7q31.1, code for the subunits of the branched-chain alpha-keto acid dehydrogenase complex, which catabolises leucine, isoleucine and valine. Increased levels of these amino acids lead to the autosomal recessive maple syrup urine disease, present in around 6 per million.

Glutaric acidaemia/aciduria is caused by a lesion in *GCDH* at 19p13.13, coding for glutarly-CoA dehydrogenase. Ornithine is an intermediate in the urea cycle and is decarboxylated to citrulline in the mitochondrion by ornithine transcarbamylase, coded by *OTC* at Xp11.4. The syndrome presents as a frequency of around 0.004%. Defects in *CBS* at 21q22.3, coding for cystathionine-β-synthase, or in *MTR* at 1q43, coding for methionine synthase, lead to increased levels of serum and urine homocysteine. A closely linked enzyme, methylenetetrahydrofolate reductase, coded for by *MTHFR*, is associated with venous thrombosis.

Lysosomal Disorders

Examples of diseases of this organelle, all of which are autosomal recessive and which are the second largest group of IEMs (0.0013% of live births, 26% of all IEMs), include the following:

- Gaucher disease, the most common lysosome storage disease (~20 per million), is caused by lack of glucosylceramidase (also known as beta-glucosidase). Some 400 mutations in *GBA1* at 1q22 have been described, making exact molecular diagnosis difficult.

- Lysosomal acid lipase deficiency is caused by mutations in *LIPA* at 10q23, resulting in lipid deposits.
- Niemann–Pick disease has a number of forms. Mutations in *SMPD1* at 11p15.54, coding sphingomyelin phosphodiesterase, are present in types A and B, whilst type C is linked to variants of *NPC1* at 18q11.2, and *NPC2* at 14q24.3, the products of both genes regulating lipid transport.
- Pompe disease is caused by deficiency of acid alpha-glucosidase, coded by *GAA* at 17q25.3, and leads to a build-up of glycogen. This condition can also be classified as a glucose storage disease, as described in the following section.
- Tay–Sachs disease (~5 per million) is caused by a mutation in *HEXA* at 15q23, coding for hexosaminidase A that digests gangliosides. Major defects include brain and nervous system pathology.
- The most common forms of neuronal ceroid lipofuscinosis result from mutant variants of *CLN3* at 16p12.1, coding a lysosomal transmembrane molecule (battenin), *PPT1* at 1p34.2, coding for palmitoyl-protein thioesterase, and *TPP1* at 11p15.4, coding for tripeptidyl peptidase. Their biochemical characteristic is deposition of lipofuscin in various organs, e.g. the eye and nervous system.
- Cystinosis, the accumulation of cystine in the lysosome, results from a mutant of *CTNS* at 17p13.2, coding for a cystine transporter.
- Krabbe disease is caused by a mutant form of *GALC*, located at 14q31.3, and normally coding for galactosylceramidase, resulting in excessive amounts of glycosphingolipids. The latter also results from mutations in *PSAP* at 10q22.1.
- There are two leading forms of mucolipidosis. Type 2 is the result of abnormal *GNPTAB* at 12q23.3 and type 3 of abnormal *GNPTG* at 16p13.3. Both code for a subunit of N-acetylglucosamine-1-phosphate transferase, which produces a mannose-6-phosphate lysosomal recognition marker.
- Sandhoff disease results from a mutation in *HEXB* at 5q13.3, coding for a subunit of hexosaminidase, which causes a build-up of a ganglioside (a compound of a sialic acid and a glycosphingolipid).

There are several forms of mucopolysaccharidosis, with individual frequencies between 1 and 20 per million, totalling around 40 per million, which are summarised in Table 7.11.

Table 7.11 Leading mucopolysaccharidoses.

Type	Syndrome	Gene	Location
1	Hurler–Scheie	*IDUA*	4p16.3
2	Hunter	*ISD*	Xq28
3	Sanfilippo (types A–D)	*SGSH, NAGLU, HGSNAT, GNS*[a]	17q25.3, 17q21.2, 8p11.21, 12q14.3[a]
4	Morquio (A, B)	*GALNS, GLB1*[a]	16q24.3, 3p22.3[a]
6	Maroteaux–Lamy	*ARSB*	5q14.1

[a] Respectively.

Carbohydrate Disorders

There are several diseases pertaining to galactose. Deficiency of galactose-1-phosphate uridyl transferase, with a UK frequency of around 0.002%, results from a mutation in *GALT* at 9p13.3, thus increased serum galactose. A mutation in *GALK1* at 17q24, coding for galactokinase, leads to high levels of galactose in the blood and urine.

Deficiency of uridine diphosphate galactose-4-epimerase leads to any one of a series of clinical features that include impaired growth and infantile jaundice. The causative gene, *GALE*, is located at 1p36.11. Lack of aldolase B, coded by *ALDOB* at 9q31.1, results in accumulation of fructose-1-phosphate. Fabry disease results from deficient activity of α-galactosidase, itself due to a variant of *GLA* at Xq22.1. Some describe it as a lysosomal storage disease.

There are numerous forms of glucose/glycogen storage and metabolic disease, with frequencies varying from around 50 per million (type 0) to 1 in a million (type 7; Figure 7.3), where the place of hexokinase, generating glucose-6-phosphate, is shown.

In Lewis disease, type 0, a mutation in *GYS1* at 19q13.33 and/or *GYS2* at 12p21.1, fails to correctly produce glycogen synthase. In type 1 (von Gierke disease), deficiency in glucose-6-phosphatase (with three isoforms) or glucose-6-phosphate translocase (coded for by *SLC37A* at 11q23.3) leads to accumulation of glycogen, whilst type 2 (Pompe disease) is described earlier. Type 3 (Cori disease) is linked to *AGL*, at 1p21.2, a complex gene coding for a dual enzyme with amylo-alpha-1, 6-glucosidase and 4-alpha-glucanotransferase activity, responsible for

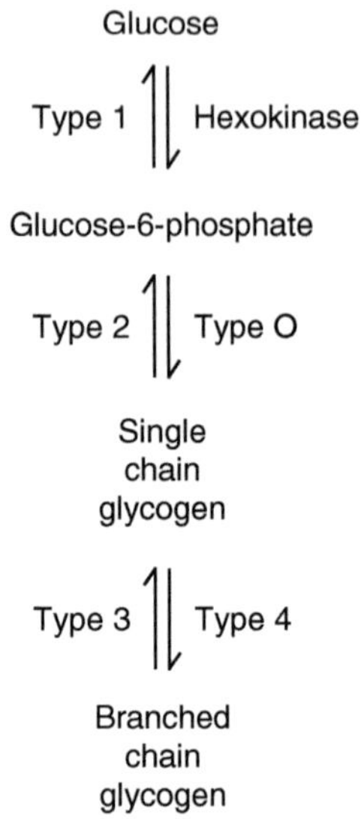

Figure 7.3 Glucose/glycogen metabolism. Types refer to the group of glycogen storage disease described in the text.

debranching glycogen. In contrast, Anderson disease, type 4 GSD, resulting from a mutation in *GBE1* at 3p12.2, there is failure to build branches onto a growing glycogen molecule.

The leading manifestation of type 5, McArdle disease, is of skeletal muscle pathology, being caused by lack of myophosphorylase, of which there several isoforms coded by various genes, the most common being *PYGM* at 11q13.1. The enzyme breaks glycogen down to glucose-1-phosphate. Type 6, Hers' disease, is caused by a lack of glycogen phosphorylase, coded for by *PYGL* at 14q22.1, and so insufficient glucose-1-phosphate. Loss of function of *PFKM* at 12q13.11, coding for muscle phosphofructokinase, resulting in myopathy, haemolysis and other pathologies, is the key feature of Tarui disease.

Deficiency in glucose-6-phosphate dehydrogenase has been described in Section 7.7.2, causing a haemolytic anaemia.

Other Metabolic Diseases

Medium-chain acyl-CoA dehydrogenase deficiency is caused by a defect in *ACADM* at 1p13.1, leading to a build-up of its substrates of fatty acids.

Wilson disease is characterised by a build-up of copper in various organs and is the consequence of a lesion in *ATP7B* at 13q14.3, which codes for a copper transporter. Copper is also at the centre of Menke disease, caused by an abnormal variant of *ATP7A* at Xq21.1, but results in copper deficiency.

Gilbert syndrome is linked to a mutant form of *UGT1A1* at 2q37.1, which codes for uridine diphosphate glucuronosyltransferase. Low levels of this enzyme result in high levels of serum bilirubin, which may cause a transient jaundice.

Dietary trimethylamine is normally catabolised by flavin-containing monooxygenase-3, coded for by *FMO3* at 1q24.3; defects in the gene lead to excess trimethylamine, which appears in the urine, giving its characteristic smell of fish. Rickets or osteomalacia can result from variants of *PHEX* at Xp22.11, coding for an enzyme that regulates phosphates, leading to hypophosphatemia, and so bone pathology. One of the leading metabolic disease of purines and pyrimidines, Lesch–Nyhan syndrome, with a prevalence of 2–3 per million, is caused by a lesion in *HPRT1* at Xq26.2, coding for hypoxanthine–guanine phosphoribosyl transferase.

The frequencies of the leading metabolic disease are shown in Table 7.12

7.10 Mitochondrial Diseases

This section includes those products of nuclear genes (such as pyruvate dehydrogenase) that act within the mitochondrion, the organelle itself having 37 genes. As sperm mitochondria does not enter the oocyte, all these genes are maternal. There are several lesions in the components of the electron transport chain, all of which are composed of subunits. In complex 1, these focus on NADH: ubiquinone oxidoreductase (coded for by several genes of the *NDUFA* family), complex 2 on succinate: ubiquinone oxidoreductase (coded for by *SDHA, SDHB, C560, DHSD)*, complex 3 on ubiquinol: cytochrome-c oxidoreductase (also coded for by numerous genes), complex 4 on cytochrome-c oxidase (many genes of the *Cox* and *Coa* family) and complex 5 on F_OF_1-ATP-synthase (eight genes of the *ATP* gene family).

Mitochondrial encephalopathy, lactic acidosis, and stroke-like episodes (the MELAS syndrome) can be caused by mutations in *MT-ND1* or

Table 7.12 Inborn errors of metabolism.

Disorder	Percentage of all errors (%)
Phenylketonuria	20.6
Fabry disease	8.8
Chronic progressive external ophthalmoplegia	4.3
All other mitochondrial diseases	4.1
Homocystinuria	3.9
X-linked adrenoleukodystrophy	3.8
Pompe disease	3.6
Galactosaemia	2.7
Mitochondrial encephalomyopathy with lactic acidosis and stroke-like episodes	2.6
Trimethylaminuria	2.4
McArdle disease	2.3
Ornithine transcarbamylase deficiency	2.2
Hypophosphataemic rickets	1.4
Maple syrup urine disease	1.1

Source: Modified from Sirrs et al. (2016).

MT-ND5, coding for subunits in NADH dehydrogenase, or in *MT-TM*, *MT-TL1* or *MT-TV*, coding for transfer RNAs. Errors in other genes coding for transfer RNAs, *MT*-TE and *MT-TK*, lead to maternally inherited diabetes and deafness (the MIDD syndrome).

Chronic progressive external ophthalmoplegia (CPEO, linked to Kearns–Sayre syndrome and Pearson syndrome) may be caused by malfunctioning ADP/ATP translocase, coded by *SLC25A4* at 4q35.1, by abnormal DNA polymerase subunits gamma 1 and 2, coded by *POLG1* at 15q26.1 and *POLG2* at 17q23.3, and by loss of a mitochondrial helicase activity, coded by *TWNK* at 10q24.31. Other mitochondrial genetic abnormalities are as follows:

- Pyruvate dehydrogenase (PDH) deficiency (with a frequency of ~0.002% of live births) results from mutations in *PDHA1* at Xp22.12 (most commonly), in *PDHA*2 at 4q22.3, or in *PDHB* at 3p14.3.

- The MERRF syndrome (myoclonic epilepsy with ragged red fibres) results from altered transfer RNA genes *MT-TK, MT-TL1, MT-TH, MT-TS1, MT-TS2* or *MT-TF*.
- The NARP syndrome (neuropathy, ataxia, retinitis pigmentosa) is caused by mutations in *MT-ATP6*, coding for ATP synthase F_0 subunit 6.
- Leigh syndrome is a consequence of mutations in *MT-ND2*, *MT-ND3*, *MT-ND5*, *MT-ND6* and *MT-CO1*, coding subunits of NAD dehydrogenase and a cytochrome oxidase.
- The mitochondrial neurogastrointestinal encephalopathy (MNGIE) syndrome is caused by a variant of *TYMP* at 22q13.33, coding thymidine phosphorylase.
- Thiamine metabolism dysfunction syndrome 2 is linked to variants in *SLC19A3* at 2q36.3, coding a thiamine transporter.
- Maternally inherited cardiomyopathy results from the 4300A>G SNP in *MT-TI*, coding for a transfer RNA linked to isoleucine.
- Leber's hereditary optic neuropathy results from mutations in genes coding for components of mitochondrial NADH dehydrogenase, such as *MT-ND1* and *MT-ND4*.

7.11 Musculoskeletal Diseases

7.11.1 Diseases Linked to the Fibroblast Growth Factor Receptor

Several conditions follow from abnormalities in the fibroblast growth factor receptor (FGFR). Mutations in *FGFR1* at 8p11.23 cause craniosynostosis, a major part of Pfeiffer syndrome, characterised in this case by an abnormal skull. However, this particular skeletal abnormality (and others) may also be caused by variants of *EFNB1* at Xq13.1, *EFR* at 19q13.2, *TWIST1* at 7p21.1, and *SMAD6* at 15q22.31, in addition to other isotypes of the FGFR.

The second receptor, *FGFR2* at 10q26.13, is also linked to craniosynostosis and other abnormalities, and is part of Apert, Crouzon, Jackson–Weiss and Pfeiffer syndromes, with abnormalities such as syndactyly and abnormalities of the jaw. Variants of *FGFR3* at 4p16.3 are often the cause of Muenke syndrome, achondroplasia, hypochondroplasia and thanatophoric dysplasia, where there is gain in function.

Self-check 7.5 Study Table 7.5 and Table 7.12. Which of the inborn errors are candidates for a genetic population screening exercise, and why?

7.11.2 Muscular Dystrophy

This condition presents with a progressive weakness of muscles and has some 50 variants, with a global prevalence of around 36 per million. The most common forms are those of Becker muscular dystrophy (BMD) and Duchenne muscular dystrophy (DMD) that together account for over half of all cases. The leading aetiology is mutations in dystrophin, coded for by *DMD* at Xp21.2-21.1, composed of 79 exons and 2.5 million base pairs, and so the largest gene in our genome. The global prevalence of DMD is in the region of 46 per million, with an incidence of 0.03% of male births, that of BMD being 16 per million with an incidence ten times smaller, and the latter having a less-severe clinical phenotype.

Other forms of muscular dystrophy include distal muscular dystrophy, linked to a mutation in *DYSF* at 2p13.2, coding for dysferlin, a lesion also common in the limb–girdle subtype. The myotonic variant is linked to altered *DMPK* at 19q13.32, coding for a protein kinase, or to *CNBP* at 3q21.3, coding an RNA-binding protein, whilst the oculopharyngeal variant is caused by a mutated *PABPN1* at 14q11.2, coding for a poly(A)-binding protein. Facioscapulohumeral muscular dystrophy is caused by an aberrant *DUX4* at 4q35.2, coding a transcription activator.

7.11.3 Other Genetic Musculoskeletal Diseases

Osteopetrosis (where bones are excessively dense) can be linked to *CLCN7* at 16p13.3, the most common, coding for a subunit of a chloride channel, *TCIRG1* at 11q13, coding a regulator of T lymphocytes, or to *IKBKG* at Xq28, coding for a regulator of NFκB. The reverse, osteogenesis imperfecta, is where the bones are too weak and badly formed. Being an umbrella term for many varieties, many genes can be involved, such as those of the *COL* family coding for different forms of collagen, or for cartilage formation, such as *CRTAP* at 3p22.3.

Cleidocranial dysplasia is likely to be caused by abnormal *RUNX2* at 6p21.1, coding for a transcription factor, whilst amelogenesis imperfecta, manifesting as a dental pathology, may be linked to a number of genes, and thus a multi-target NGS panel would be required. Exostoses are bone spurs, and many such abnormalities may be linked to *EXT1* at 8q24.11, *EXT2* at 11p11.2, both coding for a glycosyltransferase, or *EXT3* at 19p. Van der Woude syndrome, characterised by clefts in the lip and palate, in mostly caused by mutations in *IRF6* at 1q32.3, which codes for interferon regulatory factor 6.

Ehlers–Danlos syndrome has at least 13 subtypes, each with their own particular genetic aberration, with frequencies ranging from 0.04 to 0.001% of live births. Accordingly, investigation calls for a multi-target NGS panel, the most common form being the hypermobile form, with a frequency of >80%. In many cases, there are changes in the extracellular matrix, such as of proteoglycans and collagens. Similarly, Marfan syndrome, with an estimated prevalence of 0.025%, 75% being inherited, takes many forms, with cardiovascular, musculoskeletal and ocular aspects, although the leading candidate genes are *FBN1* (in >80% of cases) and those of the TGFβ family. An important differential diagnosis is Loeys–Dietz syndrome, which also has lesions in genes for TGFβ and its receptor.

There are numerous genes that contribute to height, making it polygenic, and thus are heritable. These include *SHOX*, at Xp22.33 and Yp11.32, and if mutated or only one copy is present, leads to a short stature, whilst multiple copies lead to greater height than expected.

Key Point 7.7 There are many genetic aspects of inflammatory connective tissues diseases, including those of HLA and cytokines, as demonstrated in rheumatoid arthritis (such as *HLA-DRB1*, bringing a threefold increased risk) and systemic lupus erythematosus (such as *HLA*-DQB1, bringing a twofold increased risk).

7.12 Neurological Diseases

These conditions may crudely be classified as those primarily of nerves and those principally of the central nervous system, notably the brain. However, there are instances where both suffer the same pathology, and where the same gene(s) influence both parts.

7.12.1 Diseases of the Nerves

Ataxia

Of the varying forms of ataxia, characterised primarily by progressive loss of controls of muscles controlling gait, ataxia telangiectasia is present at the highest frequency, being linked to *ATM*, also linked to breast, prostate and pancreatic cancer. It has a global prevalence between 5 and 25 per million. Friedreich's ataxia is linked to mutations in *FXN*, whilst hereditary ataxia may be linked to variants of *RFC1* at 4p14, coding for a DNA polymerase accessory protein. Ataxia is also a key component of hereditary spastic paraplegia, of which there are dozens of subtypes liked to their own particular gene and presenting at different ages (infant/child/teenage/adult). Some are X-linked (such as *LICAM* at Xq28, coding a transmembrane neuronal adhesion molecule), some are autosomal dominant (*ATL1* at 14q22.1, coding a GTPase), and some are autosomal recessive (*CYP7B1* at 8q12.3, coding a cytochrome P450 hydroxylase). This large number of candidate genes inevitably calls for an extensive multi-target NGS panel or panels.

Spinal Muscle Atrophy

With an incidence of 0.01–0.017%, this disease has four subtypes. Type 1 is the most common (50% of cases), the diagnostic tests being mutations in *SMN1* and *SMN2*, both at 5q13.2, coding for survival motor neurone protein, and most patients express two copies of *SNM2*. Other types involve the same two genes, but present with different pathological features. An X-linked variant involves *UBA1* at Xp11.3, which codes for ubiquitin-like modifier activating enzyme.

Other Neurological Diseases

There are several forms of adult-onset neurodegenerative disorders. The autosomal dominant Huntington disease is caused by an increased number of CAG repeats in *HTT* at 4p16.3, coding for huntingtin, the mutation leading to abnormal inserts of glutamine molecules. Hereditary neuropathy is linked to *PMP22* at 17p12, coding for peripheral myelin product 22, a transmembrane protein on Schwann cells where it supports myelination, whilst multiple system atrophy is characterised by abnormal *SNCA* at 4q22.1, coding for α-synuclein and leading to oligodendroglial cytoplasmic aggregates of this protein.

Dystonia (excessive muscle contractions, abnormal repetitive movements and postures) has over 20 variants. The most common candidate genes include *ANO3* at 11p14.3-14.2, possibly coding part of a calcium channel, *GNAL* at 18p11.21, coding a subunit of guanine nucleotide-binding protein, *THAP1* at 8p11.21, coding a molecule with a potential role in apoptosis, *TOR1A* at 9q34, coding an ATPase, and *VPS16* at 1p36.22-21, coding a vacuolar protein. Other potential candidates include *KMT2B, PANK2, TSPOAP1, ATP123* and others.

Charcot–Marie–Tooth disease, a hereditary sensory and motor neuropathy, has a prevalence of 0.04–0.01%, the first genetic lesion to be reported in this condition being *PMP22*. Other causal genes include *GJB1* at Xq13.1, coding for connexin 32/gap junction protein β-1, *MPZ* at 1q23.3, coding for a protein component of the myelin sheath, and *MFN2* at 1p36.22, coding for mitofusin-2, a membrane component.

7.12.2 The Central Nervous System

Leukodystrophy

This group of heterologous and rare conditions of the white matter of the brain and spinal cord presents in the range from 1 to 100 per million. Leading conditions include X-linked adrenoleukodystrophy, generally the product of a mutation in *ABCD1* at Xq28, coding for an ATP-binding cassette that results in deposition of very long-chain fatty acids. A neonatal variant of this disease is caused by a mutation in *PEX1* at 7q21.2 or in *PEX6* at 6p21.1, both coding for molecules with links to peroxisomes, leading to Zellweger syndrome.

Late juvenile metachromatic leukodystrophy is linked to a variant of *ARSA* at 22q13.33, coding for arylsulfatase A, and Canavan disease is caused by a lack of aspartoacylase that results from a defect in *ASPA* at 17p13.2. The biochemical aetiology of Krabbe disease is described in Section 7.10.2, regarding lysosomal disorders.

Epilepsy

Whilst epileptic seizures have been reported for millennia, currently affecting some 40–60 million (0.5–0.6%) people globally, the genetic aetiology of many cases remains unclear. Its prevalence in the UK in 2016 was 220,000, that being 0.33% of the population. One of the first candidate genes to be reported was those of the gamma-aminobutyric acid

(GABA) metabolic pathway; others included sodium channel genes *SCNA2* at 2q24.3 and *SCN8A* at 12q13.3 (both gain of function), and *SCN1A* and *SCN2A*, both at 2q24.3 (both loss of function), and potassium channel genes *KCNQ2* at 20q13.3 and *KCNQ3* at 8q24. There may also be role for *GRIN2A* at 16p13.2, coding part of a glutamate receptor.

X-linked epilepsies are heterogenous, and some overlap with intellectual disability syndromes; candidate genes include *PCDH19* at Xq22.1, coding protocadherin 19, *CDKL5* at Xp22.13, coding cyclin-dependent kinase 5, *MECP2* at Xq28, coding a methyl-CpG binding protein, *ARX* at Xp21.3, coding a transcription factor, *DCX* at Xq23, coding doublecortin, and *FLNA* at Xq28, coding filamin A. Progressive myoclonus epilepsy is linked to variants of *CSTB* at 21q22.3, coding for cystatin, *EPM2A* at 6q24.3, coding for laforin, a glycogen phosphatase, and *EPM2B* at 6p22.3, coding for malin, an E3-ubiquitin ligase, *NEU1* at 6p21.33, coding for neuraminidase 1, and *GOSR2* at 17q21.32, coding for Golgi SNAP receptor complex member 2. Section 7.11 contains details of a subtype, the MERRF syndrome.

Other Central Nervous System Diseases

The pathology of CADASIL (cerebral autosomal dominant arteriopathy with subcortical infarcts and leukoencephalopathy) arises from an abnormality in *NOTCH3* at 19p13.12, coding for a transmembrane receptor that may have a role in signalling pathways. Tuberous sclerosis is characterised by growth in the brain and other organs, leading to neurological symptoms such as seizures. The condition, with an incidence of 0.017–0.007%, is caused by mutated forms of *TSC1* and *TSC2*, which are genes also important in low-grade gliomas.

Hereditary neuropathy or pain disorder may be caused by a variant of *RFC1* at 4p14, coding for a component of a DNA polymerase. With a prevalence of 9 per million, the congenital myaesthenic syndrome, with dysfunctional neuromuscular junctions, has 35 causal genes. The most common are *CHRNE* at 17p13.2, coding a subunit of the AChR ε subunit, *COLQ* at 3p25.1, coding a collagen-like section of acetylcholinesterase, *RAPSN* at 11p11.2, coding a protein associated with a synapse receptor, *DOK7* at 4p16.3, coding a molecule required for synapse formation, and *GFPT1* at 2p13.3, coding glutamine–fructose-6-phosphate transaminase.

The leading candidate genes for holoprosencephaly, a structural abnormality of the brain, affects 0.4% of conceptions and between 0.0125 and 0.006% of live births. It may be present in several other syndromes, such as that of Stromme, linked to *CENPF* at 1q41, coding a centromere protein, but may also be non-syndromic, caused by any one of ten genes, such as *SHH* at 7q36.3, coding a signaling molecule. Up to 40% of cases of hydrocephalus, the accumulation of cerebrospinal fluid in the brain, have a genetic basis, with over 100 genes mutated in syndromic cases, 10% of which are X-linked. Leading candidate genes include *L1CAM* at Xq28, coding a cell adhesion molecule, *AP1S2* at Xp22.2, coding part of a complex present on Golgi complex vesicles, *MPDZ* at 9p23, coding a signaling protein, and *CCDC88C* at 14q32.11-32.12, coding a regulator of a signalling protein.

The prevalence of microcephaly in Europe is estimated as 0.015% of live births; the severe form can present in several ways. It is often found in aneuploidy, in certain large chromosomal deletions, with dozens of single-gene defects such as in *XRCC4* at 5q14.2, coding for a DNA repair molecule, in *LIG4* at 13q33.3, coding for a DNA ligase, and in *MCPH1* at 8p23.1, coding for microcephalin, and in some syndromes:

- Seckel syndrome is caused by biallelic variants in *ATR*, *RBBP8*, *CENPJ*, *CEP152*, *CEP63*, *NIN*, *DNA2*, *TRAIP* and *NSMCE2*.
- Osteodysplastic primordial dwarfism type 2 is caused by biallelic variants in *PCNT*.
- Meier–Gorlin syndrome is linked to biallelic variants in *ORC1*, *ORC4*, *ORC6*, *CDT1*, *CDC6*, *GMNN*, *CDC45L* and *MCM5*.

7.12.3 Myopathy

Cardiomyopathy is discussed in Section 7.1.1.

The congenital myopathies are a group of diverse conditions characterised by progressive muscle weakness, with an all-age prevalence of 15 per million, but 27 per million in children. There are links to over 30 genes, some in particular sub-groups. For example, variants of *RYR1* at 19q13.2, coding for the ryanodine receptor, and *SEPN1*, also known as *SELENON*, at 1p36.11, coding for a selenoprotein, cause central core disease. *ACTA1* and *TPM2* cause nemaline myopathy, whilst *MTM1* at Xq28, coding myotubularin 1, and *DNM2* at 9p13.3, coding tropomyosin 2, are implicated in centromere myopathies. Both recessive and dominant variants in *CACNA1S* at 1q32.1, coding part of a calcium channel, can cause dihydropyridine receptor congenital myopathy. An important subtype of the myofibrillar myopathies is hereditary myopathy with early respiratory failure, caused by variants in *TTN* and is also important in cardiovascular disease (Section 7.1.1).

Distal myopathies are those affecting the hands and feet, with many genes (*ACTN2*, *CAV3*, *CRYAB*, *DNAJB6*, *DNM2*, *FLNC*, *HNRNPA1*, *HSPB8*, *KHLH9*, *LDB3*, *MATR3*, *MB*, *MYOT*, *PLIN4*, *TIA1*, *VCP*, *NOTCH2NLC*, *LRP12* and *GIPS1*) linked to an autosomal dominant form, and four (*ADSSL*, *ANO5*, *DYSF*, *GNE*) causing an autosomal recessive form.

Malignant hyperthermia results from excessive hypermetabolism, generally precipitated by certain medications. It is linked, with variable penetration, to variants of *RYR1* at 19q13.2, coding the ryanodine receptor, which part-regulates calcium flux, *CACNA1S* (as mentioned earlier) and *STAC3* at 12q13.3, coding a protein rich in cysteine (and which is also linked to a myopathy).

Arthrogryposis is characterised by abnormal contraction of joints and is linked to over 30 genes but may be part of other conditions such as muscular dystrophy. Leading candidate genes include *FKBP65*, *PLOD2*, *PLOD3*, *VPS33B* and *VIPAS39*, and collagen-coding genes include *COLL11A2*, *COL6A1*, *COL6A2* and *COL6A3*. Distal forms are linked to *TNN12*, *TNNT3*, *TPM2*, *MYH3* and *MYH8*.

Self-check 7.6 Which of the genetic neurological diseases do you consider to be the most important?

7.13 Renal Diseases

A study suggests that around 20% of chronic kidney disease (CKD) that presents before the age of 25 is linked to mutations in any one of 200 genes. In turn, clinically significant non-syndromic congenital anomalies of the kidney and urinary tract (CAKUT) consist of perhaps half of all cases of early-onset kidney disease in children. Renal cancer is discussed in Section 4.8.4.

7.13.1 The Leading Causes

One of the most well-known genetic nephrology disease is autosomal dominant polycystic kidney disease (ADPKD), reputed to be the basis of 44% of monogenic cases of CKD. Often classified as a ciliopathy, it has an estimated prevalence of 0.1–0.25%. It is caused primarily by mutations in *PKD1* at 16p13.3 (85% of cases), coding polycystin 1, a likely G protein-coupled receptor, and *PKD2* at 4q22.1 (10–15%), coding for polycystin 2, a calcium-permeable non-selective cation channel. A small number of families express abnormalities in *DNAJB11* at 3q27.3, coding a heat shock protein, or *GANAB* at 11q12.3, coding a subunit of glucosidase.

As described in Section 7.8, Alport syndrome can bring not only hearing loss but also a renal aspect with childhood-onset haematuria, later-onset proteinuria and glomerulonephritis, and may be the cause of 23% of genetic CKD. Mutations in collagen-coding genes *COL4A3* and *COL4A4*, both at 2q36.3, or *COL4A5* at Xq22.3 result in a thin layer of collagen in the basement membrane, causing the clinical features, in common with thin basement membrane disease. Hereditary systemic amyloidosis has numerous manifestations, such as cardiomyopathy, and in the kidney it causes proteinuria and nephrotic syndrome, said to be linked to 4.1% of monogenic CKD. It may result from the accumulation of proteins such as liver-generated transthyretin, coded for by *TTR* at 18q21.1.

7.13.2 Diseases Linked to Electrolyte Disturbances

A study from the USA reported an astonishing rate of nephrocalcinosis or nephrolithiasis of 1 in 11 adults, although the paediatric rate was more modest at 0.054%, with genetics being responsible for between 11.5 and 32% of cases. Leading candidate genes include *CASR* at 20q13.2, coding a member of the cytochrome P450 family, and *CYP24A1* at 3q13.33-21.1, coding for the calcium sensing receptor. Renal phosphate wasting with hypercalciuria is linked to *SLC34A1* at 5q35.3, *SLC34A3* at 9q34.3 (both coding a sodium/phosphate co-transporter) and *SLC9A3R1* at 17q25.1, coding a sodium hydrogen exchanger.

Nephrocalcinosis may also be present in Bartter syndrome, characterised by failure of electrolyte homeostasis in the kidney, caused by variants of *SLC12A1* at 15q21.1, *KCNJ1* at 11q24.3, *CLCNKB* at 1p36.13, *BSND* at 1p32.3, *CLCNKA* at 1p36.13 and *MAGED2* at Xp11.21. Many of the same disturbances are present in Gitelman syndrome, linked to *SLC12A3* at 16q13, coding a sodium-chloride cotransporter. Renal tubular acidosis may result from abnormalities in *SLC4A1* at 17q21.31, coding the red blood cell band 3 anion exchanger, *ATP6V1B1* at 2p13.3, coding a subunit of an ATPase, *ADCK2* at 7q34, coding a kinase, and *WDR72* at 15q21.3, coding a molecule that may have a role in calcium transport.

7.13.3 Other Causes

The pathology of focal segmental glomerulosclerosis focusses primarily on damage to the podocyte and resulting sclerosis with loss of barrier function. Aetiology can be linked to the three collagen genes (*COL4A 3-5*) described earlier, but also to *LMX1B* at 9q33.3, coding a transcription factor, *LAMB2* at 3p21.31, coding a subunit of laminin, *APOL1* at 22q12.3, coding apolipoprotein L1, and *NPHS1* at 19q13.12, coding for the transmembrane protein nephrin. Pathogenic variants in podocyte expressed genes such as *NPHS2* at 1q25.2, coding for podocin, and *TRPC6* at 11q22.1 coding a cation channel, affect podocyte structure or integrity. The closely related diffuse mesangial sclerosis may be linked to mutations in *WT1* at 11p13, coding the Wilms' tumour protein, and to *LAMB2*, mentioned above.

Several renal diseases are linked to aberrant complement activity, such as the nephropathy linked to the loss in inhibitors that are the root of pathology such as C3-related membranoproliferative glomerulonephritis. Such inhibitors include Factor H, coded by *CFH* at 1q31.3, Factor I, coded by *CFI* at 4q25, and by *CD46* at 1q32.2, coding a further regulator of complement. Around 90% of cases of congenital nephrogenic diabetes insipidus are caused by loss-of-function mutations in *AVPR2* at Xq28, coding vasopressin receptor 2, which impacts into aquaporin-2 activity.

7.14 Respiratory Diseases

7.14.1 Cystic Fibrosis

Once described as the most common life-threatening inherited disease of children, advances in medical care have brought the estimated median age of survival to 50, the leading pathology being pulmonary, followed in many cases by pancreatic pathology. It has an incidence of 0.03–0.05% of live births (dependent on geography) and a carrier frequency estimated at 1 in 30, and accordingly, carrier status is often determined. The causative gene, *CFTR* at 7q31.2, codes for the CF conductance regulator, the leading mutation (ΔF508, present in ~85% of cases) coding a malformed protein that fails to correctly regulate the movement of sodium and chloride ions. Other commonly reported mutant genes code for S466X, G85E and N1303K, and although many of the 2000+ other mutations cause the same clinical effects, in a minority, a minor degree of homeostasis remains. Indeed, in one study, 15% of adults in a registry were diagnosed after the age of 16.

7.14.2 Other Respiratory Diseases

Central congenital hypoventilation is linked to *PHOX2B* at 4p13, coding a transcription factor required for neuronal development: its principal clinical feature is transient nocturnal respiratory arrest. Genes linked to pulmonary arterial hypertension may be grouped as those of bone morphometric receptor/TGFβ (such as *BMPR2* at 2q33.1-33.2, coding bone morphogenetic protein receptor type 2, the most common lesion present in 60% of cases, and *ACVRL1* at 12q13.13, coding a serine/threonine kinase receptor forTGFβ, those of a channelopathy (*ATP13A3* at 3p29, coding an ATPase, and *KCNK3* at 2p23.3, coding part of a potassium channel), those coding a transcription factor (such as *SOX17* at 8q11.23, and *KLF2* at 8q11.23), and those as yet ungrouped (such as *NOTCH3* at 19p13.12 and *GGCX* at 2p11.2).

The consequences of primary ciliary dyskinesis, which has a prevalence of ≤0.01%, include bronchiectasis and pneumonia. It may be caused by a large number of genes, including *GGCX, FOXJ1, CCNO, GAS8, CCDC103, CFAP221, STK36, RPGR, DNAH9, GAS2L2, NEK10, SPEF2, HYDIN, TTC12, RSPH4A* and *LRRC56*. The leading cause of familial pneumothorax is mutation in *FLCN* at 17p11.2, coding for folliculin, which inhibits a form of lactate dehydrogenase (and which also has a role in multiple benign skin tumours), whilst there may also be place for *FBN1* at 15q21.1, which codes for fibrillin.

Abnormal loss-of-function variants of *SERPINA1* at 14q32.13 are the leading cause of alpha-1-antitrypsin deficiency, which has a prevalence between 0.01 and 0.05%, although the coded product, a general-purpose antiprotease, will also inhibit other enzymes such as elastin. Other genes, such as *NOS3* at 7q36.1, coding for nitric oxide synthase, and *CHRNA3* at

15q25.1, coding for a subunit of an acetylcholine receptor, may modulate the clinical aspects of the condition.

The leading causes of surfactant deficiency, responsible for infant respiratory distress syndrome, include loss-of-function mutations in *ABCA3* at 16p13.3, coding for an ATP-binding cassette. However, there are also places for surfactant protein B, coded by *SP-B*, also known as *SFTPB*, at 2p11.2, surfactant protein C, coded by *SP-C*, also known as *SFTPC*, at 8p21.3, and for *NKX2.1* at 14q13.3, coding a transcription factor. Alveolar capillary dysplasia (with misalignment of pulmonary veins) is linked to *FOXF1* at 16q24.1, coding a transcription factor, whilst pulmonary alveolar microlithiasis is caused by variants of *SLC34A2* at 4p15.2, coding for the sodium-dependent phosphate protein 2B.

Key Point 7.8 CF is often used as an example of how a simple single gene mutation can cause widespread disease, of how an initial biochemical diagnosis can be confirmed, and that determination of carrier status (perhaps 1/30) is an important genetic tool in public health.

7.15 Dermatological Diseases

The ocular and metabolic aspects of albinism are discussed in Section 7.4.2.

7.15.1 Xeroderma Pigmentosum

Although very rare, with an estimated incidence of between 1 and 4 per million (but higher in India), this debilitating skin disease brings a risk of skin cancer, with a clear pathophysiological process. It has a number of subtypes, most of which are linked to mutations in genes for repairing damage to DNA (e.g. excision repair) as may be caused by ultra-violet light, or other processes in DNA biology. These genes include *XPA* at 9q22.22, *XPB*, also known as *ERCC3*, at 2q14.3 (present in 30% of cases), *XPC* at 3p25.1 (27%), *XPD*, also known as *ERCC2*, at 19q13.32 (15%), *XPE*, also known as *DDB2*, at 11p11.2, *XPF*, also known as *ERCC4*, at 16p13.12, and *XPG* at 13q33.1.

Some of these genes, and others, are also the cause of trichothiodystrophy (focusing on the hair, and where *MPLKIP* at 7p14.1 may be important) whilst Cockayne syndrome is also linked to mutations in genes coding for excising repair proteins, e.g. *ERCC8* at 5q12.1 and *ERCC6* at 10q11.23. Investigation of all these genes is likely where there is cutaneous photosensitivity with a likely genetic cause.

7.15.2 Other Dermatological Diseases

Segmental overgrowth disorders can involve many organs and tissues, such as the brain and vascular stem, and including the skin. A leading potential cause is a gain in function of candidate gene *PIK3CA* at 3q26.32, coding a subunit of a kinase. By contrast, ectodermal dysplasia, also manifesting in several forms, has a large number of candidate genes, including *EDA* at Xq13.1 and *EDAR* at 2q13, coding for ectodysplasin A and its receptor, three genes (*PKP1, PKP2* and *PKP3* at 1q32.1, 12p11.21 and 11p15.5, respectively) coding for plakophilin, alongside *PORCN* at Xp11.23, coding for an acyltransferase.

Congenital epidermolysis bullosa, presenting in the range of 2–20 per million, is most often linked to *COL7A1* at 3p21.31, coding a chain of collagen type VII, and *KRT14* at 17q21.2, coding keratin 14. A variant, with congenital skin fragility, may be linked to *COL17A1*, at 10q25.1, coding a chain of collagen XVII, *ITGB4* at 17q25.1, coding an integrin subunit, and *LAMA3* at 18q11.2, *LAMB3* at 1q32.2 or *LAMC2* at 1q25.3, all three coding a laminin subunit. These seven genes are likely to form a multi-target NGS panel.

The various forms of ichthyosis are linked to particular genes. The vulgaris type (with a frequency of perhaps 0.33%) is linked to *FLG* at 1q21.3, coding for filaggrin, whilst the X-linked form (with a frequency of 0.04%) involves *STS* at Xp22.31, coding for a sulfatase. The lamella form (~10 per million) in linked to several genes: *TGM1* at 14q12, coding a keratinocyte transaminase, *NIPAL4* at 5q33.3, coding a magnesium transporter, *ALOXE3* at 17p13.1, coding a lipoxygenase, and *ABCA12* at 2q35, coding an ATP-binding cassette.

Bullous ichthyosis may be caused by *KRT1* and *KRT2*, both at 12q13.13, and *KRT10* at 17q21.2, all coding a form of keratin, whilst palmoplantar keratodermas may also be linked to *KRTI*, and to *KRT9* also at 17q21.2, coding for keratin 9. Erythrokeratoderma may be caused by a mutation in *GJB3* at 1p34.3, coding for gap-junction β-3 protein, whilst autosomal recessive primary hypertrophic osteoarthropathy is the product of variants of *HPGD* at 4q34.1, coding for a prostaglandin dehydrogenase, or *SLCO2A1* at 3q22.1-22.2, coding for a prostaglandin transporter.

The leading candidate gene for incontinentia pigmenti, occurring in 20–25 per million, is *IKBKG* at Xq28, coding for an inhibitor of NFκB, whilst epidermodysplasia veruciformis has been described as an inborn error of immunity, with the later development of squamous cell carcinomas. The aetiology involves the human papilloma virus and mutations in *EVER1*, also known as *TMC6*, and *EVER2*, also known as *TMC8*, both at 17q25.3 and coding a different transmembrane channel-like protein, and *CIB1* at 15q26.1, coding a calcium and integrin-binding protein. There may also be roles for *GATA2*, *CXCR4* and *DOCK8*, at 9p24.3, coding a protein involved in intracellular signaling networks.

Self-check 7.7 Test yourself in a recall of this chapter by reviewing the learning objectives.

7.16 Summary

This chapter, by its very nature, has sought to collect together a heterologous group of conditions where molecular pathology plays a part in diagnosis and often in management. However, there is deliberate omission of rare cancers, such as Hodgkin's lymphoma, which presents at a rate of around 0.003% of the UK population annually, as these are described elsewhere. Nevertheless, as there are some 6000–7000 rare diseases, reputed to be present in 5% of the global population, they collectively constitute a major group and span almost the complete spectrum of human disease, to which the many diverse sub-sections we have discussed will attest.

Bibliography

Applegarth DA, Toone JR, Lowry RB. Incidence of inborn errors of metabolism in British Columbia, 1969–1996. Pediatrics. 2000;105:e10.

Brambullo T, Colonna MR, Vindigni V, et al. Xeroderma pigmentosum: A genetic condition skin cancer correlated—A systematic review.

BioMed Res Internat. 2022;2022(1):8549532. https://doi.org/10.1155/2022/8549532.

Gavriilaki M, Chatzikyriakou E, Moschou M, et al. Therapeutic biomarkers in Friedreich's ataxia: A systematic review and meta-analysis. Cerebellum. 2024;23:1184–1203.

Gregg AR, Aarabi M, Klugman S, et al. Screening for autosomal recessive and X-linked conditions during pregnancy and preconception: A practice resource of the American College of Medical Genetics and Genomics. Genet Med. 2021;23:1793–1806.

Grixti L, Lane LC, Pearce SH. The genetics of graves' disease. Revs Endocrin Metab Disorders. 2024;25:203–214.

Moore GW. Thrombophilia screening: Not so straightforward. Semin Thromb Haemost. 2024. doi:10.1055/s-0044-1786807.

Puapatanakui P, Miner JH. Alport syndrome and Alport kidney diseases – elucidating the disease spectrum. Curr Opin Nephrol Hypertens. 2024;33:283–290.

Sawhney JPS, Madan K. Familial hypercholesterolaemia. Indian Heart J. 2024;76:S108–S112.

Scotet V, L'Hostic C, Ferec C. The changing epidemiology of cystic fibrosis: Incidence, survival and impact of the CFTR gene discovery. Genes. 2020;11:589. doi:10.3390/.genes11060589.

Sirrs S, Hollak C, Merkel M, et al. The frequencies of different inborn errors of metabolism in adult metabolic centres: Report from the SSIEM Adult Metabolic Physicians Group. J Inherit Metab Dise Reps. 2016;27:85–91.

Swystun LL, James PD. Genetic diagnosis in haemophilia and von Willebrand disease. Blood Rev. 2017;31:47–56.

Tsilifis C, Slatter MA, Gemmery AR. Too much of a good thing: A review of primary immune regulatory disorders. Front Immunol. 2023;14:1279201. doi:10.3389/fimmu.2023.1279201.

Zhan L, Gui C, Wei W, et al. Third generation sequencing transforms the way of the screening and diagnosis of thalassemia: A mini-review. Front Pediatr. 2023;6(11):1199609.

Zhanga Y, Adamob M, Zoua C, et al. Management of hypertrophic cardiomyopathy. J Cardiovasc Med. 2024;25:399–419.

Websites

Alzheimer's disease: https://www.alzheimers.org.uk/about-dementia/types-dementia/alzheimers-disease

Foundation for Prader-Willi Research: www.fpwr.org

Down's syndrome: www.downs-syndrome.org.uk

Sickle cell disease: www.sicklecellsociety.org

Thalassaemia: www.ukts.org

Umbrella site for numerous genetic diseases: https://geneticalliance.org.uk/our-members/

Audiology: https://hereditaryhearingloss.org/

Rare diseases in the UK: https://digital.nhs.uk/ndrs/data/data-outputs/rare-disease-and-condition-prevalence

National Institute for Health and Care Excellence: www.nice.org.uk

Genes Index

b

d

e

g

j

k

l

m

n

t

TTN 261, 262, 264

Text Index

Molecular Pathology: A Primer for Laboratory Scientists, First Edition. Andrew Blann.

n

V

W

X

y

z